T0140167

Lecture Notes in Computer Science 12634

More information about this subseries at http://www.springer.com/series/7409

Qiaohong Zu · Yong Tang ·
Vladimir Mladenović (Eds.)

Human Centered Computing

6th International Conference, HCC 2020
Virtual Event, December 14–15, 2020
Revised Selected Papers

 Springer

Editors
Qiaohong Zu
Wuhan University of Technology
Wuhan, Hubei, China

Yong Tang
South China Normal University
Guangzhou, China

Vladimir Mladenović
University of Kragujevac
Kragujevac, Serbia

ISSN 0302-9743   ISSN 1611-3349   (electronic)
Lecture Notes in Computer Science
ISBN 978-3-030-70625-8   ISBN 978-3-030-70626-5   (eBook)
https://doi.org/10.1007/978-3-030-70626-5

LNCS Sublibrary: SL3 – Information Systems and Applications, incl. Internet/Web, and HCI

This Springer imprint is published by the registered company Springer Nature Switzerland AG
The registered company address is: Gewerbestrasse 11, 6330 Cham, Switzerland

Preface

This volume presents the proceedings of the 14th Human-Centered Computing (HCC) conference, which was held online from 14–15 December 2020. Throughout the extraordinary year of 2020, we witnessed many extraordinary and far-reaching events unfolding in front of us. In the midst of the unforeseen turmoil, human-centred computing became a term that attracted an unprecedented level of attention—while we were restricted physically, technologies played an important role in closing the gaps between people, drew closer societies separated by geographic boundaries and political disagreements and demonstrated the power of unity through virtual communities.

In 2020, as members of the HCC family, we experienced how human communities adapted to and thrived in the new "norm"; we proudly participated in the technology provisioning that enables hundreds of thousands of people to receive better health and civic services; we accelerated the adaption of technologies that bridge ever-growing digital divides which had become more prominent during the COVID-19 pandemic; and we advocated the shift of paradigms to promote resilience and technology reach. We are 11 months into the pandemic and quite likely we have only just seen the tip of the iceberg of many socio-technological changes that this unfortunate event will trigger—it will also give raise to, hopefully, new HCC discussion topics and research directions.

At the end of 2020, we successfully reunited albeit virtually to look back on this extraordinary year and exchanged what we had achieved on various fronts of HCC research directions. HCC2020 brought together a wide range of state-of-the-art research from different disciplines. In total, HCC2020 received 113 full paper submissions. All submissions went through a very strict reviewing process (at least two peer reviews and one meta-review by senior programme committee members). In the end, the programme committee accepted 28 long papers and 20 short papers to be included in the HCC2020 post-conference proceedings. The quality of HCC2020 was ensured by the arduous efforts of the programme committee members and invited reviewers, to whom we owe our highest gratitude and appreciation.

Since its inception, the HCC conference faced the gravest challenges in 2020 and the community emerged with increased strength. Here, we owe our deepest gratitude to the conference authors who happily adapted to the new conferencing paradigm. We are also grateful to all members of the Technical Programme Committee and conference Organisation Committee, without whom we would not have enjoyed another year of fruitful discussion. The most significant task fell upon their experienced shoulders. This year, we had to invite extra reviewers, who did a great job under enormous pressures— we thank them for their contributions and persistence. Last but not least, our special thanks go to Springer's editorial staff for their hard work in publishing these post-conference proceedings in the LNCS series.

January 2021

Qiaohong Zu
Yong Tang
Vladimir Mladenović

Organization

Conference Chairs

Yong Tang   South China Normal University, China
Vladimir Mladenović   University of Kragujevac, Serbia

Programme Committee Chair

Bin Hu   Lanzhou University, China

Organisation Committee

Qiaohong Zu   Wuhan University of Technology, China
Philip Moore   Lanzhou University, China

Publication Committee

Bo Hu   University of Southampton, UK
Jizheng Wan   Coventry University, UK

Partners

University of Kragujevac, Serbia
South China Normal University, China
Wuhan University of Technology, China
University of Maribor, Slovenia

Reviewers

James Anderson	Luis Carriço	David Dupplow
Jose Albornoz	Jingjing Cao	Haihong E
Natasha Aleccina	Qinghua Cao	Talbi El-Ghazali
Angeliki Antonio	Guohua Chen	James Enright
Juan Carlos Augusto	Tianzhou Chen	Henrik Eriksson
Yunfei Bai	Yiqiang Chen	Chengzhou Fu
Roberto Barcino	Lizhen Cui	Yan Fu
Paolo Bellavista	Aba-Sah Dadzie	Shu Gao
Adam Belloum	Marco De Sá	José G. Rodríguez García
Marija Blagojević	Matjaž Debevc	Mauro Gaspari
Natasa Bojkovic	Luhong Diao	Bin Gong
Zoran Bojkovic	Monica Divitini	Horacio González-Vélez

Chaozhen Guo
José María Gutiérrez
Chaobo He
Fazhi He
Hong He
Andreas Holzinger
Bin Hu
Cheng Hu
Changqin Huang
Zongpu Jia
Mei Jie
Hai Jin
Lucy Knight
Hiromichi Kobayashi
Ines Kožuh
Roman Laborde
Hanjiang Lai
Thomas Lancaster
Victor Landassuri-Moreno
Liantao Lan
Bo Lang
Agelos Lazaris
Chunying Li
Jianguo Li
Shaozi Li
Wenfeng Li
Xiaowei Li
Zongmin Li
Xiaofei Liao

Hong Liu
Lianru Liu
Lizhen Liu
Yongjin Liu
Alejandro Llaves
Yanling Lu
Hui Ma
Haoyu Ma
Yasir Gilani
Mohamed Menaa
Marek Meyer
Danijela Milošević
Dragorad Milovanović
Vladimir Mladenović
Maurice Mulvenna
Mario Muñoz
Aisha Naseer
Tobias Nelkner
Ivan Nikolaev
Sabri Pllana
Xiaoqi Qin
Klaus Rechert
Uwe Riss
Andreas Schraber
Stefan Schulte
Susan Scott
Beat Signer
Matthew Simpson
Mei Song

Xianfang Sun
Yuqing Sun
Wenan Tan
Menglun Tao
Richard Taylor
Shaohua Teng
Boris Villazón-Terrazas
Coral Walker
Maria Vargas-Vera
Jizheng Wan
Qianping Wang
Yun Wang
Yifei Wei
Ting Wu
Zhengyang Wu
Toshihiro Yamauchi
Bo Yang
Yanfang Yang
Linda Yang
Zhimin Yang
Xianchuan Yu
Guanghui Yue
Yong Zhang
Gansen Zhao
Shikun Zhou
Shuhua Zhu
Tingshao Zhu
Gang Zou

Contents

Dynamic Pick and Place Trajectory of Delta Parallel Manipulator

Qiaohong Zu[1], Qinyi Liu[1(✉)], and Jiangming Wu[2]

[1] School of Logistics Engineering, Wuhan University of Technology, Wuhan 430061, China
674792020@qq.com
[2] Central South University, Changsha 410083, China

Abstract. Based on the high-speed and stable demand of the dynamic pick-and-place of a Delta parallel manipulator in industrial production, a 5-3-5° multi-segment polynomial is used to design the speed laws of the joints of the manipulator, combined with the constraint of motion, the shortest period, and vibration control to construct a multi-target and multi-constrain. The nonlinear motion trajectory planning model is obtained, and the optimal solution of trajectory planning is obtained by using the optimized gravitational search algorithm. The results show that the speed and acceleration of the motion trajectory reach extreme values, and the motion trajectory is continuously smooth, meeting the expected planning requirements. The effectiveness of the motion trajectory model in shortening the operation cycle and the vibration of the control mechanism is verified, and the effectiveness of the optimized gravity search algorithm in the convergence control and global optimization is verified.

Keywords: Parallel manipulator · Trajectory planning · Multi-stage polynomial · Gravitational search algorithm

1 Preface

The development of the robot industry has played an important role in promoting the third industrial revolution, and the proposal of "industry 4.0" has even prompted many countries to carry out research on emerging technologies in the field of intelligent robot manufacturing [1]. Delta parallel manipulator is one of the modern production equipment which widely used in the industrial production. It has compact and simple structure, flexible and fast operation. It can be seen in food, medicine, logistics, life science, chip manufacturing and other industries.

At present, the study of the Delta manipulator mainly concentrated in the end of the actuator motion trajectory model design and processing algorithm. It has smooth dynamic trajectory which can improve the working efficiency and stability of the Delta manipulator, further stimulate their working performance in the industrial manufacturing, it has important practical significance and value of engineering application on the promotion of advanced manufacturing production. Costantiesu D and Croft E A [2] have studied the relationship between the velocity variation law of the manipulator and the

© Springer Nature Switzerland AG 2021
Q. Zu et al. (Eds.): HCC 2020, LNCS 12634, pp. 1–11, 2021.
https://doi.org/10.1007/978-3-030-70626-5_1

vibration of the mechanism, and found that when the velocity function is continuous for at least two times, the acceleration is continuous to ensure the smoothness of the motion trajectory, and the vibration of the mechanism can be suppressed by controlling the change range of acceleration. In the literature [3, 4], polynomial was used to plan the motion law of the actuator for the impact, vibration and other problems that tend to occur when the trajectory of the Delta manipulator is in right-angle transition. The simulation results show that the polynomial speed design can smooth the transition part in a reasonable period of motion. In literature [5], 4-3-4, 3-5-3 and 5-5-5 piecewise polynomials are respectively selected to interpolate the trajectories in joint space, and the continuity, maximum and whether there is a limit of multi-order derivatives of different polynomials are analyzed. The results show that 4-3-4 piecewise polynomials have certain advantages in motion period and acceleration control.

In this paper, the dynamic picking and dropping trajectory planning problem of the manipulator will be transformed into a nonlinear mathematical model according to the optimization needs of Delta manipulator in operating time and mechanical vibration. Based on the characteristics of Gravitational Search Algorithm (GSA), which is easy to realize and it has strong global Search ability, a Fuzzy Parameter gravitation search Algorithm (FPGSA) designed by Fuzzy control parameters is proposed to balance the convergence speed and global Search ability of the Algorithm. The nonlinear mathematical model is solved by gravity search algorithm before and after optimization, and the optimal solution of trajectory planning problem is obtained respectively.

2   Kinematic Analysis of the Delta Manipulator

2.1   Analysis of Configuration

The positive and negative solutions of Delta manipulator position are to obtain the specific mapping relationship between joint Angle and actuator displacement. The specific position of the actuator is the positive solution of the mechanism. On the contrary, joint Angle is obtained based on the position of the end-effector, which is the inverse solution of the mechanism [3]. Figure 1 is a simplified schematic diagram of the structure of the common Delta manipulator, which consists of a moving platform with end-effector, a static platform with fixed robot position, three active rods and the corresponding three groups of follower rods.

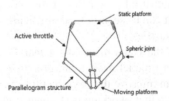

Fig. 1. Structure diagram of Delta parallel manipulator

2.2 Workspace Analysis

The workspace of the Delta manipulator actuator is the intersection of the activity space of each moving branch chain, which is an important indicator of robot mechanism design [7]. In the workspace, a reasonable motion region is divided for dynamic pick up and drop, so as to provide a basic environment for actuator trajectory planning. Analysis formula (1), if there is a solution to this equation, then:

$$(2a)^2 - 4(c - b)(b + c) \geq 0 \tag{1}$$

$$c^2 - \left(a^2 + b^2\right) \leq 0 \tag{2}$$

Relevant parameters in Eq. (3) are substituted into Eq. (5) to obtain:

$$K_i(x,y,z) = [(x_p\cos\alpha_i + y_p\sin\alpha_i - \triangle r)^2 + (x_p\sin\alpha_i - y_p\cos\alpha_i)^2 + z_p^2 + 1^2 - m^2]^2$$
$$- 4 1^2 \left[(\triangle r - x_p\cos\alpha_i - y_p\sin\alpha_i)^2 + z_p^2 \right] \leq 0 \tag{3}$$

When inequality (3) is 0, it represents the boundary of the workspace. Because of the uniform Angle distribution of the three moving branch chains of the Delta manipulator, the rotating coordinate system $T_i(x_i, y_i, z_i)$ is established. The position of point P in the coordinate system T_i is:

$$\begin{cases} x_p^i = x\cos\alpha_i + y\sin\alpha_i \\ y_p^i = -x\sin\alpha_i + y\cos\alpha_i \\ z_p^i = z \end{cases} \tag{4}$$

The $\alpha_i = \frac{2i-2}{3}\pi$ ($i = 1,2,3$), in the coordinate system T_i, space boundary $K_i(x,y,z) = 0$, after finishing can be simplified as:

$$\left[1^2 - \sqrt{\left(x_p^i - \triangle r\right)^2 + z_p^{i2}} \right]^2 + y_p^{i2} = m^2 \tag{5}$$

Equation (4) is the intersection of three standard toris equations. A calculation example is selected to draw the workspace of the Delta manipulator in MATLAB. The structure of the Delta manipulator and the parameters of joint rotation Angle are shown in Table 1.

The parameters of the Delta manipulator in Table 1 are imported into the space solution program to obtain the 3d space scope of the working area, as shown in Fig. 2.

Table 1. Basic parameters of the Delta manipulator

Parameter	Figure
Radius of static platform (mm)	290
Length of driving rod (mm)	260
Length of follower (mm)	850
Radius of Moving platform (mm)	50
Range of joint angles (°)	2.9–112.6

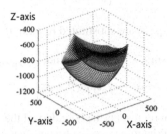

Fig. 2. Schematic diagram of Delta manipulator operating area

3   Gravity Search Algorithm Based on Fuzzy Control

3.1   GSA

GSA is a swarm intelligence optimization search algorithm proposed by Rashedi et al. [10]. Based on the law of universal gravitation and Newton's second law of motion, which USES particle motion to represent the solution process. It is widely used in communication systems, digital image processing and energy analysis [11]. The position of particles in the gravitational search algorithm corresponds to the solution of the optimization problem. Under the action of gravity, particles with a large mass attract other particles, which gradually gather from the disordered state to the point with a large mass, and finally gather to the optimal solution region [12]. Select an optimization model of d-dimensional space and update the position, velocity and acceleration of particle I in the d-dimensional by using Eqs. (6)–(8):

$$v_i^d(t + 1) = rand_i \times v_i^d(t) + a_i^d(t) \tag{6}$$

$$x_i^d(t+1)= x_i^d(t)+v_i^d(t+1) \tag{7}$$

$$a_i^d(t)= F_i^d(t)/M_{pi}(t) \tag{8}$$

$F_i^d(t)$ is the magnitude of the force exerted on particle i in d dimension, $M_{pi}(t)$ is the inertial mass of particle i.

$$F_i^d(t) = \sum_{j \in kbest, j \neq i}^{N} rand_j F_{ij}^d(t) \tag{9}$$

$$F_{ij}^d(t) = G(t)\frac{M_{pi}(t) \times M_{ai}(t)}{R_{ij}(t) + \varepsilon}(x_j^d(t) - x_i^d(t)) \tag{10}$$

$F_{ij}^d(t)$ is the gravitational attraction of particle j on particle i. $G(t) = G_0 e^{-\alpha t/T}(\alpha = 20G_0 = 100), R_{ij}(t)$, is the Euclidean distance between individual i and individual j, is a small value constant. The inertial mass of particle i is solved as follows:

$$m_i(t) = \frac{fit_i(t) - worst(t)}{best(t) - worst(t)} \tag{11}$$

$$M_i(t) = m_i(t)/\sum_{i=1}^{N} m_i(t) \tag{12}$$

$fiti(t)$ is the fitness function of particle i at time t, and $best(t)$ and $worst(t)$ are the optimal fitness value and the worst fitness value. The fitness function can be set as:

$$best(t) = \min_{i=1, 2, ..., N} fit_j(t) \tag{13}$$

$$worst(t) = \max_{i = 1, 2, ..., N} fit_j(t) \tag{14}$$

Information between GSA particles is transmitted by the law of universal gravitation, which has strong global search ability. But the local search ability is poor, the convergence is not ideal, through the optimization of the algorithm flow design, to make it more suitable for the problem.

3.2 FPGSA

The gravitational constant G can be regarded as the search step of the algorithm, and its change has an impact on the convergence speed of the algorithm. The selection of G value is related to the location of the optimal solution and the range of the solution set, which cannot be obtained based on the qualitative index, otherwise the algorithm cannot effectively deal with the fuzzy problem. Due to the uncertainty of the value of parameter G, fuzzy control is carried out. Fuzzy control is a nonlinear control strategy commonly used in automatic control systems, which mainly includes input fuzzy, deductive reasoning and fuzzy decision [13]. The corresponding specific steps are as follows:

(1) The input is determined by the number of iterations it. Last iteration parameter $\alpha(it-1)$, richness R_N and development level A_N, the output is the value of this iteration α. Richness R_N represents the position distribution of the population within the solution range. A higher population richness can prevent the algorithm from falling into the local optimal solution in the iterative process, which can be obtained according to Eq. (15):

$$R_N = (\sum_{i=1}^{N} \sqrt{\sum_{j=1}^{D} (X_{ij}^{it} - \overline{X}_j^{it})^2})/(N \times R_L) \tag{15}$$

Where N is the number of individuals in a population, D is the dimension of the particle, the $\overline{X}$ said individual distribution center, the R_L is the distance between two most remote particle in a population. The value range of R_N is $(0,1)$. The larger the value of R_N is, the more abundant the population is. Design development level A_N by fitness function:

$$A_N = (f_a(it-1) - f_a(it))/f_a(it) \tag{16}$$

$f_a(it\text{-}1), f_a(it)$ is the average value of the $it\text{-}1$, it iteration adaptation value. In the minimum value problem, when the value of A_N is positive, the population development is high and the algorithm tends to the optimal solution gradually. When the value of A_N is negative, the direction of the algorithm deviates from the optimal solution.

(2) Deductive reasoning, the subset characteristics of fuzzy control are not obvious. In order to accurately describe the characteristics of group changes, membership degrees of up, medium and low are set.

Table 2 shows the corresponding relationship between the value range and membership of R_N, A_N, it and the value range of matrix $(t\text{-}1)$.

Table 2. Values and grades of various parameters

Level	R_N	A_N	it	$\alpha(t\text{-}1)$
Low	$(0, 0.5)$	$(-\infty, 0)$	$(0, 15)$	$(10, 15)$
Medium	–	–	$[15, 25]$	$[15, 20]$
Up	$(0.5, 1)$	$[0, +\infty)$	$(25, 30)$	$(20, 25)$

(3) Fuzzy decision, parametric fuzzy processing rules (Table 3).

Table 3. Values and grades of various parameters

Rules		a	b	c	d
Condition	R_N	Low	Low	Up	Up
	A_N	Up	Low	Up	Low
	it	Low	Medium	Up	Up
Result	$\alpha(t)$	Low	Medium	Medium	Up

The interpretation of fuzzy control rules is as follows:

Rule a: In the early stage of the algorithm, if the population development level is high but the richness is low, and the local optimal solution may appear, then the value of the parameter decreases to reduce the convergence speed and improve the global search ability of the algorithm.

Rule b: If the population richness is low and the development level is low in the middle of the algorithm, then the value of parameter order decreases to increase the population richness and ensure the global search ability of the algorithm;

Rule c: If the population richness is high and the development trend is good at the later stage of the algorithm, then the value of parameter order should be further increased to promote the algorithm solution and improve the efficiency of the algorithm.

Rule d: If the population richness is high but the development level is low and the algorithm convergence rate is low at the later stage of the algorithm, then the value of parameter order is increased to speed up the algorithm convergence rate.

4 Trajectory Optimization Simulation

4.1 Parameter Settings

Set Delta manipulator dynamic up path for 25 mm, 305 mm long class door font path, combined with Fig. 4 describes the Delta manipulator workspace, set actuator center point P on the pick up point coordinates of P_A, placed the coordinates of the point of P_D, on the motion path to select two points P_1, P_2, the unit is mm, end get cartesian space location discrete sequence. The coordinates are pick points $(140, -90, -850)$ intermediate points $(100, -60, -825)$ intermediate points $(-64, 63, -825)$ placement points $(-104, 93, -850)$. The size parameter and the position of the actuator were substituted into the position inverse solution model to solve the rotation Angle of the driving joint corresponding to each trajectory point, as shown in Table 4.

Table 4. Joint space Angle discrete virtual sequence

Diversion (degree)	θ_1	θ_2	θ_3
Picking point	0.3109	0.9017	0.5979
Intermediate point	0.2808	0.7125	0.4976
Intermediate point	0.6234	0.2984	0.5372
Placement point	0.8049	0.3087	0.6547

4.2 Analysis of Simulation Results

In this paper, MatlabR2012a was used for simulation. GSA and FPGSA were used to solve the motion trajectory planning model respectively, and the single operation duration t (half movement cycle) and acceleration range were obtained. The single operation time was processed to obtain the optimal and average single operation time and standard beat time (CPM) in the optimal solution. The movement trajectory results are shown in Table 5.

Table 5. Trajectory planning results

Algorithm	tmin (S)	CPMmax	tavg (S)	CPMavg	Jmin/105 (rad/s3)
GSA	0.1795	167	0.1853	162	0.2411
FGSA	0.1726	174	0.1786	168	0.2845

The minimum time of a single operation for GAS and FGSA was 0.1795 s and 0.1726 s, respectively, and the corresponding standard time beats were 167CPM and 174CPM, respectively. The mean standard time of FGSA was 3.7% higher than GSA. Under the 5-3-5 multi-stage polynomial velocity rule design, the Angle, angular velocity and angular acceleration curves of the motion trajectory under the optimal solution obtained based on FGSA are shown in Figs. 3, 4, 5 and 6.

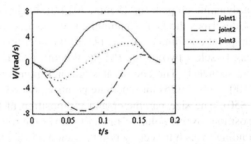

Fig. 3. Joint Angle variation curve based on FGSA

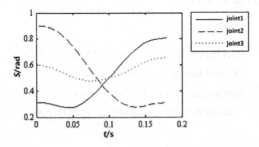

Fig. 4. The variation curve of joint velocity based on FGSA

The kinematic trajectories obtained based on FGSA are smooth and continuous, with no mutation point in velocity transformation and the range of acceleration variation is effectively controlled. Through the change of the acceleration curve of the three joints, the change amplitude of the acceleration is obtained. The smaller the amplitude of

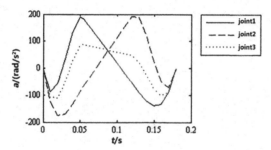

Fig. 5. Curve of joint acceleration based on FGSA

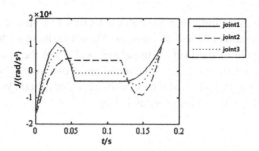

Fig. 6. Curve of joint acceleration based on FGSA

acceleration change, the more obvious the effect of the optimized motion trajectory on the vibration suppression of the mechanism, the better the stability of the motion planning, which is more conducive to the picking and lifting operation of the Delta manipulator. The objectives of operational planning include motion cycle length control and mechanism vibration suppression, adjust the weight coefficient of planning objectives, and analyze the influence between motion cycle reduction and vibration suppression effect. The changes of $k1$ and $k2$ values are shown in Table 6.

Table 6. The values of $k1$, $k2$

Group	1	2	3	4	5
k_1	0.85	0.8	0.75	0.7	0.65
k_2	0.15	0.2	0.25	0.3	0.35

The weight coefficients of each group were substituted into the objective function of the algorithm to obtain the change curves of single operation time and acceleration under different weight coefficient value groups, as shown in Fig. 7.

From Fig. 7, it can be found that there is a dual relationship between motion period reduction and vibration suppression. When the weight coefficient of the motion period is higher, the amplitude of acceleration change is larger, indicating that the vibration suppression effect is normal. When the weight coefficient of vibration suppression is

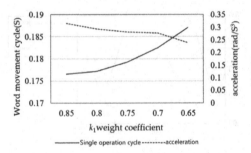

Fig. 7. Curve of single operation time and acceleration change

high, the single operation time increases, the velocity of motion trajectory planning decreases and the operation efficiency decreases. Therefore, there are high requirements for the movement cycle of the Delta manipulator, and when the requirements for the pick and drop smoothness are low, $k1$ value in the target planning is large. The stability of the picking and lifting operation of the Delta manipulator is high. When the operation time is loose, $k2$ value in the target program is large.

5 Conclusion

The design which based on the 5-3-5 multi-stage polynomial in the Delta manipulator joint angular velocity and angular acceleration can achieve the performance constraints of the manipulator extreme value, and at the same time, the displacement, velocity curve is continuous smooth, it ensure the stability of the manipulator in the picking up operation. The motion cycle of FGSA is shorter and the vibration suppression effect is better than that of GSA, which proves the validity of FGSA. By analyzing the relationship between motion cycle reduction and vibration suppression effect in object programming, the motion trajectory model can meet the demand of dynamic pick up and drop trajectory of the Delta manipulator under various working conditions.

References

1. Dong, G.: The position and trend of chinese industrial robots in the global value Chain. China Sci. Technol. Forum **2016**(3), 49–54+118 (2016)
2. Constantinescu, D., Croft, E.A.: Smooth and time-optimal trajectory planning for industrial manipulators along specified paths. J. Field Robot. **17**(5), 233–249 (2015)
3. Na, W., Dongqing, W., Zhiyong, Z.: Kinematics analysis and trajectory planning of a 3 DOF Delta parallel manipulator. J. Qingdao Univ. (Eng. Technol. Edition) **32**(1), 63–68 (2017)
4. Su, T., Zhang, H., Wang, Y., et al.: Trajectory planning method of Delta robot based on PH curve. Robot **40**(1), 46–55 (2018)
5. Yu, F.: Research on Motion Control System of Two-DOF Translational Parallel Manipulator. China Jiliang University (2016)
6. Wu, X., Qi, M., Ma, C., et al.: Application of Integrated Drive Control Technology on Delta Robot. Mechanical Science and Technology (2018)

7. Wengang, Q., Lei, W., Tianbi, Q.: Singularity analysis and space study of a three-DOF Delta robot. Mach. Tools Hydraulics **46**(9), 61–64 (2018)
8. Jiangping, M., Jiawei, Z., Zhengyu, Q., et al.: Trajectory planning method of three-degree-of-freedom Delta parallel manipulator. J. Mech. Eng. **52**(19), 9–17 (2016)
9. Li, S.,J., Feng, Y., et al.: Review of Research on Articulated Industrial Robot Trajectory Planning. Computer Engineering and Applications (2018)
10. Rashedi, E., Nezamabadi-Pour, H., Saryazdi, S.: GSA: a gravitational search algorithm. Inf. Sci. **179**(13), 2232–2248 (2009)
11. Siddiquea, N., Adelib, H.: Applications of gravitational search algorithm in engineering. J. Civil Eng. Manage. **22**(8), 981–990 (2016)
12. Qingyu, L., Xiaoming, Z.: A gravitational search algorithm with adaptive hybrid mutation. J. Chongqing Normal Univ. (Natural Science Edition) **03**, 91–96 (2017)
13. Shi, S., Wang, X., Cao, C., et al.: Adaptive fuzzy classifier with regular number determination. J. Xidian University (Natural Science Edition), 2017 (2) (2017)

Design of Aquatic Product Traceability Coding Scheme Based on EPC Label

Qiaohong Zu[1], Ping Zhou[1(✉)], and Xiaomin Zhang[2]

[1] School of Logistics Engineering, Wuhan University of Technology, Wuhan 430061, China
1113089750@qq.com
[2] Guangzhou East Railway Automobile Logistics Co. LTD, Guangzhou 510800, China

Abstract. Aiming at the existing quality and safety issues in the aquatic product industry, with famous and excellent fresh aquatic products as the research object, the traceability coding of its supply chain is discussed. This paper analyzes the main links and elements information involved in the circulation of famous and excellent aquatic products, and uses EPC coding technology and RFID anti-counterfeiting encryption technology to design the traceability code of famous and excellent aquatic products, and builds the traceability system of aquatic products. Users can track and query famous and excellent aquatic products in breeding base, logistics center, sales center and other links, realizing the information transparency of aquatic products in the supply process, which has theoretical and practical significance for improving the quality and safety of aquatic products.

Keywords: Famous and excellent aquatic products · EPC code ·
Anti-counterfeiting encryption · Traceability system

1 Introduction

As an important part of food safety, aquatic product quality safety has been paid more and more attention by people. The establishment of aquatic product traceability system can inquire all the information in the process of aquatic product circulation and improve the quality and safety of aquatic products. At present, researchers at home and abroad have done a lot of research on EPC coding in the supply chain and product safety traceability of different types. Taking livestock products as the specific research object and EPC coding as the core technology, Xiaoxian Huang designed the traceability system of livestock products based on Internet of Things technology, and realized the query and update of basic information such as breeding, slaughtering and immunization [1]. Taking agricultural capital as the specific research object, MAO Lin et al. designed the traceability of agricultural capital with EPC coding and applied anti-counterfeiting encryption technology to realize the traceability and query of agricultural capital products [2]; Taking aquatic products as the research object, Dina Xia uses EPC codes to identify the traceability codes of each process, and traces the aquatic products from breeding, processing, distribution to sales [3]. Dongyan Wu et al. took cold chain food as the object and selected the EPC-96 coding structure to encode and identify the food, and built a cold

© Springer Nature Switzerland AG 2021
Q. Zu et al. (Eds.): HCC 2020, LNCS 12634, pp. 12–18, 2021.
https://doi.org/10.1007/978-3-030-70626-5_2

chain traceability system based on RFID/EPC technology, so as to realize the traceability of the whole process of food from production, logistics to sales [4]. Because our country aquatic product industry is more complex than other industries, the case that combines aquatic product industry design is not very many. But EPC coding technology has the advantages of large enough capacity and unique distribution, which is very suitable for the single product management of famous and excellent aquatic products. Therefore, the key links affecting the quality and safety of aquatic products can be analyzed, and the security traceability of famous and excellent aquatic products supply chain can be realized by using relevant technologies, so as to meet the requirements of intelligent information management of enterprises, consumers' inquiry of aquatic product safety information and government supervision.

2 Factor Analysis of Aquatic Product Traceability Link

This paper traces is famous and excellent aquatic products fresh in the supply chain information, so the information in the processing of such aquatic products is not considered. Base on this, the main participants in the aquatic product supply chain are: raw material suppliers (fry, medicines and feed), farmers/breeding bases, logistics distribution enterprises, terminal vendors (supermarkets, specialty stores, etc.), and the government and quality inspection departments.

(1) Raw material suppliers: mainly provide basic materials needed for aquatic product cultivation, such as feed, medicine and fry, etc., the quality of which directly affects the quality of aquatic products.

(2) Breeding enterprise: The main information recorded in this link includes the selection of seedlings at the beginning of aquaculture, the use of drugs and feed during the breeding process, the detection of various indicators of aquatic products, the fishing pool of aquatic products, and grade inspection.

(3) Logistics center: The choice of distribution center is not only Considering the internal conditions of the center itself, but also combining the key elements of distribution to achieve overall selection.

(4) Sales center: sellers record their own information, receive aquatic products, and submit aquatic product circulation information and status. We can use large and medium-sized supermarkets as the end of the traceability system.

(5) Government agencies: Government agencies mainly conduct quarantine, inspection and supervision of aquatic products. For aquatic products that have problems, they will be held accountable by relevant companies in accordance with the law to protect the interests of consumers.

Responsible subject and content of responsibility are shown in Table 1.

In any link, aquatic products shall be inspected and quarantined to ensure product quality, and the responsible body can be determined by scanning EPC code in the label, so as to realize traceability of the whole supply chain.

Table 1. Traces back the responsible subject and content.

The responsibility subject	Responsibility for the content
Raw material supplier	To ensure the source and quality of feed, medicine and seeding
Breeding base	The aquaculture quality of aquatic products from the pond to the fishing pond is guaranteed
Logistics enterprises	To guarantee the quality of aquatic products in the process of logistics
Sales center	Guarantee the quality of aquatic products from shelf time to consumers
Government agencies	Conduct inspection and supervision of aquatic products

3 Design of Traceability Coding for Fresh, Famous and Excellent Aquatic Products

3.1 Choice of Encoding Type

In this paper the responsibility subjects of traceability in mostly adopt the GID-96-bit encoding structure, mainly because EPC scheme of the generic identifier is independent of any specification or identification scheme as well as EPC-global standard tag data, and the capacity of the 96-bit encoding is large enough to fully meet the requirements of aquatic product traceability.

3.2 Coding Design Scheme

The general identifier GID-96-bit coding structure is shown in Table 2:

The specific coding rules are as follows:

Manager code: The manager records the code of farm, logistics center and sales department. Seven digits mark the three links of aquaculture, distribution and sales in the aquatic product circulation process to achieve traceability.

Object classification code: Object classification code is identified with six digits. Object classification code is mainly used to identify aquatic product types, feed, drug types and freight units in aquatic product traceability.

Serial number: The serial number consists of nine digits. The first six digits of the serial number are dates in the form of YYMMDD, and the last three digits are randomly generated serial Numbers, which are used to uniquely identify the product.

In this paper, design of breeding, feed, medicine, aquatic products using EPC tags sheet is tasted, transportation vehicles, general syntax for GID - 96: urn: epc: id: gid: Manager Number. Object Class. Serial Number, are converted to decimal format design.

3.3 Coding Example

According to the above EPC field code design and batch code allocation scheme, for example: Suppose a batch of first-level red crabs were caught on May 20, 2014 at the

Table 2. GID-96 bit coding structure

Encoding structure	GID-96			
URI Template	urn: epc: tag: gid-96:M.C.S			
Total bit	96			
Logical partitioning	EPC header	Manager code	Object classification code	Serial number
Logical partition number	8	28	24	36
Code division	EPC header	Manager code	Object classification code	Serial number
URI portion		M	C	S
Coding segmentation number	8	28	24	36
Bit position	b95 b94 ...b88	b87 b86 ... b60	b59 b58 ... b36	b35 b34 ... b0
Coding method	00110101	Integer	Integer	Integer

Daming Aquatic Farm, and they were transported to a supermarket via a fishery product distribution company. The corporate code of Daming Aquatic Products is 11; the code of a logistics company is 21; the code of a supermarket is 31; the code of crabs is 01; the code of red crabs is 02; the first-class code is 03; then EPC URL The structure of is urn:epc:id:gid:112131.010203.140520001. For the convenience of reading in the system, the decimal EPC URL of GID-96 is converted to hexadecimal, and the lack or blank digits are replaced with zeros in the conversion process.

3.4 Design of Aquatic Product Safety and Anti-counterfeiting Scheme

EPC Label Encryption Anti-counterfeiting. At present, the commonly used encryption algorithms in RFID systems include DES (symmetric cryptosystem), RSA (asymmetric cryptosystem), etc. Considering the cost and speed of the tag, it is finally decided to use the symmetric public key DES algorithm to encrypt the information in the electronic tag, the encrypted information is stored in the label [5].

DES is a block encryption algorithm that uses 64-bit packets. Specifically, a 64-bit group of plaintext is input from one end of the algorithm and 64-bit ciphertext is output from the other end. Encryption and decryption use the same algorithm (except for different key arrangements). The length of the key is 56 bits, which can be any 56 bits, and can be changed at any time. All confidentiality depends on the key [6].

Code Encryption Anti-counterfeit Work Principle. Coding principle of encryption security is to farms, logistics units and sales units, product batches, attributes, production dates and so on a series of information coding through data encryption algorithms process, the plaintext into ciphertext to generate the corresponding encoding, and then put the ciphertext information written into the tag chip, the label and the data transmission between reading and writing in cipher way transmission, ensure the security of data transmission. In this way, even if criminals get tag corresponding information, it will not be exposed in the identity of any available information, in order to achieve the purpose of preventing information disclosure and information tampering [7]. At the same time, in order to store the commodity information and the encrypted code, a corresponding database needs to be established. Meanwhile, the global unique identification code TID in the label must also be stored in the database. The encrypted digital regulatory code stored in the tag is connected to the tag reader, and the EPC code in the tag serves as the index to realize the correlation and query of information in each link and the traceability of goods. The principle of DES encryption algorithm is shown in the Fig. 1.

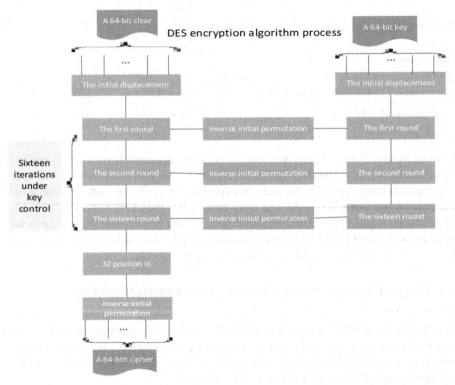

Fig. 1. DES encryption algorithm schematic

4   Application of EPC Coding in the Overall System Architecture

4.1   The Overall Architecture Design of the System

Applying EPC coding technology to the traceability system of famous and excellent aquatic products to realize the traceability management of the quality of aquatic products, find the source of problematic aquatic products, and determine the problematic links, can effectively guarantee the quality of famous and excellent aquatic products. According to the division of functions and the principle of ease of implementation, the hierarchical structure of the aquatic product traceability system based on the EPC label is shown in the Fig. 2.

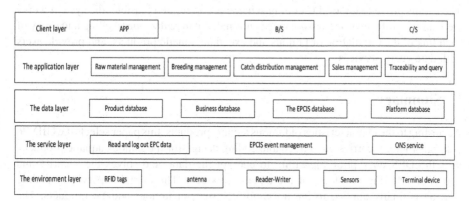

Fig. 2. The hierarchy diagram of the famous aquatic product traceability system architecture

4.2 Description of the Overall Architecture of the System

The environment layer is composed of RFID tags, readers, antennas, sensors and terminal equipment. It is mainly to collect farm information, distribution link information, sales information. In the process of information collection, the safe transmission of data is realized by means of digital signature. The content on the RFID tag is encrypted to form a uniquely identified electronic tag.

The service layer is mainly composed of EPC middleware, EPCIS and ONS server, which is mainly responsible for the data transmission and processing of aquatic product information. The EPC middleware is mainly responsible for processing the data information read by the RFID system and sending the information to the ONS server, ONS server looks up the aquatic product code based on the pre-registered EPC code and returns it to the EPIS address [8].

The data layer is mainly to transfer the collected information to the database for storage. The currently widely used transmission methods include various private networks, the Internet, wired and wireless communication networks, network management systems, and cloud computing platforms [9]. Data transmission stores data information in a database to ensure data security.

The application layer is mainly to realize the application of enterprises in all links and the management of traceability information, including raw material supply management, breeding management, fishing distribution management, sales management, and traceability query modules.

The client layer provides operations on the mobile terminal and PC, and can add, delete, modify, and check information.

Using the idea of software engineering, this system first uses Power Designer 12.5 modeling software to design the database, uses ASP.NET as the development platform, uses Microsoft SQL Server 2008 as the database for storing information, and uses the C# programming language to develop and implement C/S The reader function module in the mode and the B/S mode are based on the EPC coded aquatic product supply chain traceability platform. The system has been prototyped and tested in the laboratory. The results prove that the EPC coding scheme is correct and feasible, the system works well, and the information transmission is normal. It can realize the tracking and query of each link of the circulation of famous and excellent aquatic products, and provide users with data query services.

5 Summary

According to the characteristics of famous aquatic products, this paper selects the GID-96 coding structure, carries out a traceable coding design for the famous aquatic products, uses the DES encryption algorithm to encrypt the data, and solves the security and anti-counterfeiting problems of the aquatic products. The digital supervisory coding scheme is finally applied to the traceability system of famous aquatic products. The establishment of the aquatic product supply chain traceability platform based on EPC coding theory has been realized, and the traceability and inquiry of famous, special and excellent aquatic products have been completed in breeding bases, logistics centers and sales centers, which has theoretical and practical significance for improving the quality and safety of aquatic products.

References

1. Huang, X.: Design and implementation of animal product traceability system based on Internet of Things technology. Tianjin University (2018)
2. Mao, L., Cheng, W., Cheng, T., Jin, M., Zhong, Z., Cai, L.: Application research of agricultural materials traceability system based on product electronic code (EPC) coding. Zhejiang Agric. Sci. 59(07), 1312–1318 (2018)
3. Xia, D.: Development of aquatic product supply chain traceability platform on RFID and EPC Internet of Things. Electron. Technol. Softw. Eng. 24, 75 (2015)
4. Dongyan, W., Li, D., Yangjiang, W., Zhou, X.: Food cold chain traceability system based on RFID-EPC technology. Food Ind 38(04), 238–240 (2017)
5. Liu, Q., Hao, Y.: A RFID encryption and decryption module based on DES algorithm. Sci. Technol. Inf. 17(27), 13–15 (2019)
6. Geng, X.: Research on file encryption based on DES algorithm. Inf. Comput. (Theoretical Edition) 32(03), 44–46 (2020)
7. Sheng, K., Ma, J.: Research on traceability coding technology of traditional Chinese medicine based on EPC Internet of Things. J. Wuyi Univ. 38(09), 62–66 (2019)
8. Ling, J., Liu, D., Zhu, Y.: Tobacco anti-counterfeiting and quality traceability system based on Internet of Things. Food Ind. 12, 247–250 (2014)
9. Yang, J.Y., Zhang, C.: Research on the traceability of fresh product supply safety based on RFID technology. Food Ind. 39(05), 165–168 (2018)

DEVS-Based Modeling and Simulation
of Wireless Sensor Network

Songyuan Gu, Chen Wu, and Yuanyuan Qin[✉]

China Academy of Electronics and Information Technology, Beijing, China
gusongyuan614@163.com, wuchen5251@163.com, qinyuan98@163.com

Abstract. A modeling and simulation method for wireless sensor network (WSN) using discrete event system specification (DEVS) is proposed in this paper considering that the existing simulators cannot fully satisfy the requirements of WSN simulation. The method is for multi-layer and multi-aspect modeling which contains the component layer in sensor nodes, the sensor node layer, the wireless sensor net-work layer and the external environment where a WSN is exposed, and modules in lower layer are integrated into superior models through coupling relation and model reuse. Minimum Hop Count (MHC) protocol is chosen as the routing protocol of the WSN in the simulation experiment. It is finally demonstrated through the performance analysis of WSN using MHC that the proposed DEVS-based modeling and simulation method is feasible for WSN simulation.

Keywords: Wireless sensor network · Modeling and simulation · Discrete event system specification · Minimum-hop-count routing

Wireless sensor network (WSN) is a wireless network system composed of a number of spatial distributed sensor nodes with wireless transmission capabilities through self-organization. The fundamental purpose of WSN establishing is to comprehensively perceive and monitor target area, and to obtain spatial and temporal distributed data in target area for monitoring and analyzing the environment of target area. WSN is widely used in several domains e.g. military, industry and environment monitoring; especially it can execute data acquisition and monitoring by entering the hard to-reach areas. WSN is also the concept foundation and the underlying network of Internet of Things [1, 2].

For optimizing the design of sensor nodes and network and achieving stable and robust WSN deployment for complex environment, it is necessary to perform simulation test for performance parameters and deployment environment of a WSN. Modeling and simulation is an essential method to observe WSN behaviors under certain conditions which can analyze topology and communication protocol or forecast network performance. There are several simulators capable of WSN simulation e.g. NS2, OPNET, OMNeT++, SensorSim, TOSSIM, etc. but these simulators could partly satisfy simulation demands for WSNs and it is difficult to find a universal, customizable and modular simulator applied in simulation of behaviors and performance of a WSN and in modeling of environmental scenario a WSN exposed in [3]. In the field of WSN simulation, we need a universal simulation method which can describe internal structure of sensor nodes completely, be module-reusable and customize the external environment of a WSN. Discrete event system specification (DEVS) is able to model system

© Springer Nature Switzerland AG 2021
Q. Zu et al. (Eds.): HCC 2020, LNCS 12634, pp. 19–30, 2021.
https://doi.org/10.1007/978-3-030-70626-5_3

in multi-level, multi-granularity and provide simulators to simulate models. Therefore, this specification is employed in this paper to model modular sensor nodes, WSNs and external environments of WSNs and to perform simulation experiments for verifying the feasibility of WSN simulation using DEVS.

1 Discrete Event System Specification

Discrete event system specification is a modular, hierarchical and formal modeling and simulation mechanism based on general system theory which is founded by Bernard Zeigler [6–8]. The specification is able to characterize systems in which finite variations arise in finite intervals of time, e.g. discrete event systems, discrete time systems and continuous time systems [5]. Modeling and simulation stages are clearly divided in DEVS and a simulation experimental framework is provided in which models can be executed. In terms of modeling, DEVS divides models into two levels: atomic models (AM) and coupling models (CM); atomic model defines internal transitions under time domain variations and external input conditions as well as input and output behaviors; coupling model defines hierarchical structures between atomic models or between atomic models and lower level coupling models. General atomic model is defined as a 7-tuple:

$$AM = <X,Y,S,\delta_{ext},\delta_{int},\lambda,ta> \tag{1}$$

Where X is an input event set which includes input event values and input ports; Y is an output event set which includes output event values and output ports; S is a system state set; δ_{int} is an internal transition function in which a system state would transit from s to $\delta_{int}(s)$ when there are no external input events for a certain time; δ_{ext} is an external transition function in which a system has remained in a state for time interval e when external input event $x \in X$ arrives, then the system state would transit to $\delta_{ext}(s, e, x)$ at once and e would recover to 0; λ is an output function which outputs event $\lambda(s)$ before state transition; ta is a time advance function.

General coupling model could be defined as

$$CM = <X,Y,D,\{M_d|d \in D\},EIC,EOC,IC, select> \tag{2}$$

Where D is a coupling member name set; $\{M_d\}$ is a member model set; EIC represents external input coupling relations which connect input ports of a coupling model to input ports of internal member models; EOC represents an external output coupling relations which connect output ports of internal member models to output ports of a coupling model; IC is an internal coupling relation set which connect input ports of some internal members to output ports of other members; $select$ is a selection function which is employed to select a state transition as the one of a coupling model from the concurrent state transitions of internal members.

2 DEVS-Based WSN Modeling

Sensor nodes are capable of computing, communicating and perceiving [9]. The architecture of typical sensor nodes is shown as Fig. 1. A typical sensor node is generally composed of a sensor component, a processor component, a wireless communication component and a power component; some sensor nodes also comprise positioning components and mobile components. The sensor component is responsible for sampling target data, determining whether sampled value exceeds a threshold and transferring sampled data to processor component; the processor component is responsible for managing activities of entire node such as the processing for sampled data as well as the processing and the routing for messages to be forwarded; the wireless communication component is responsible for wireless channel monitoring, messages receiving and forwarding to a channel.

Distributed WSN nodes communicate with each other via multi-hop and thus constitute a WSN system. Each node transfers sampled data to a sink node directionally. The sink node, which has greater processing capability and more adequate power, is a gateway node connecting the WSN to Internet or a satellite communication network. Subsections below present the establishment of the models of component-layer, node-layer and network-layer based on DEVS in detail.

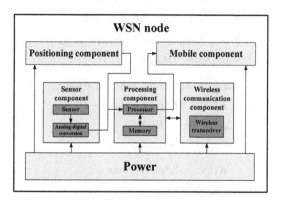

Fig. 1. Architecture of a wireless sensor node

2.1 Sensor Component Modeling

The sensor component atomic model describes perceiving and interaction behaviors of a sensor node with external environment. The structure of sensor component atomic model AM_{Sensor} is shown as Fig. 2. As the unique component interacting with external environment in a sensor node, the sensor component model reserves the port Env_info for external environment data input, which could connect to external environment model through an external input coupling relation and interact with external environment or not according to the simulation scenario. For establishing an integrated simulation scenario, diversified external environment models could be established with freedom to fulfill diversified simulation demands. An external temperature model $AM_{Temperature}$ is defined in this paper to provide external input for the WSN. Input port $Power\_info$ and output port $Power\_cons$ are utilized to interact with the power component: the power component transfers current state of itself to the sensor component through $Power\_info$; the sensor component transfers its working condition to the power component through *Power_cons* for calculating the power consumption. Output port *Sample_info* is utilized to transfer the sampled data to the processor component.

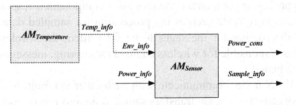

Fig. 2. Architecture of AM_{Sensor}

The formal description of atomic model AM_{Sensor} is as follows

$$AM_{Sensor} = <X, Y, S, \delta_{ext}, \delta_{int}, \lambda, ta>$$

$$X = \{\text{"Env\_info"}, \text{"Power\_info"}\}$$

$$Y = \{\text{"Sample\_info"}, \text{"Power\_cons"}\}$$

$$S = \{RUN, DEAD\}$$

$$\delta_{ext} : RUN \times \text{"Env\_info"} \rightarrow RUN$$

$$RUN \times \text{"Energy\_info"} \rightarrow DEAD$$

$$\delta_{int} : RUN \rightarrow RUN$$

$$\lambda : RUN \rightarrow \text{"Sample\_info"}$$

$$RUN \rightarrow \text{"Power\_cons"}$$

$$ta : RUN \rightarrow t_{Sample}$$

$$DEAD \rightarrow \infty$$

2.2 Processor Component Modeling

The processor component is responsible for managing each of the components in a sensor node, processing information received by the wireless communication component or sampled by the sensor component and routing for messages to be forwarded. Processor component atomic model $AM_{Processor}$ is shown as Fig. 3. Input port *Mess_info* and output port *Proc_info* are responsible for interacting with wireless communication component: the wireless communication component delivers received messages to the processor component through *Mess_info* to process them, and if continue forwarding is needed, the processor component would calculate the route; the processor component delivers messages to be forwarded to the wireless communication component through *Proc_info*. Input port *Power_info* and output port *Power_info* are utilized to transfer current state of power component and calculate power consumption as well. Input port *Sample_info* is responsible for receiving sampled value from the sensor component. Calculation and storage functions are simulated in atomic model $AM_{Processor}$.

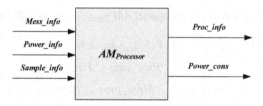

Fig. 3. Architecture of $AM_{Processor}$

The formal description of atomic model $AM_{Processor}$ is as follows

$$AM_{Processor} = <X, Y, S, \delta_{ext}, \delta_{int}, \lambda, ta>$$
$$X = \{\text{"Mess\_info"}, \text{"Sample\_info"}, \text{"Power\_info"}\}$$
$$Y = \{\text{"Proc\_info"}, \text{"Power\_cons"}\}$$
$$S = \{RUN, DEAD\}$$
$$\delta_{ext} : RUN \times \text{"Mess\_info"} \to RUN$$
$$RUN \times \text{"Sample\_info"} \to RUN$$
$$RUN \times \text{"Power\_info"} \to DEAD$$
$$\delta_{int} : RUN \to RUN$$
$$\lambda : RUN \to \text{"Proc\_info"}$$
$$RUN \to \text{"Power\_cons"}$$
$$ta : RUN \to t_{Proc}$$
$$DEAD \to \infty$$

2.3 Wireless Communication Component Modeling

The wireless communication component model represents communication behaviors of a sensor node, namely the message interactions with nodes within communication radius of current node. Wireless communication component atomic model $AM_{Transceiver}$ is shown as Fig. 4. Input port *Mess_rec* and output port *Mess_forw* communicate with other nodes through receiving and forwarding messages; input port *Proc_info* is utilized to acquire data processed from the processor component; meanwhile, output port *Mess_proc* is utilized to transfer messages to be processed to the processor component.

Fig. 4. Architecture of $AM_{Transceiver}$

The formal description of atomic model $AM_{Transceiver}$ is as follows

$$AM_{Transceiver} = <X, Y, S, \delta_{ext}, \delta_{int}, \lambda, ta>$$

$$X = \{\text{"Mess\_rec"}, \text{"Proc\_info"}, \text{"Power\_info"}\}$$

$$Y = \{\text{"Mess\_forw"}, \text{"Mess\_proc"}, \text{"Power\_cons"}\}$$

$$S = \{RECEIVE, IDLE, FORWARD, DEAD\}$$

$$\delta_{ext} : RECEIVE \times \text{"Mess\_rec"} \rightarrow RECEIVE$$

$$FORWARD \times \text{"Proc\_info"} \rightarrow FORWARD$$

$$RECEIVE \times \text{"Power\_info"} \rightarrow DEAD$$

$$IDLE \times \text{"Power\_info"} \rightarrow DEAD$$

$$FORWARD \times \text{"Power\_info"} \rightarrow DEAD$$

$$\delta_{int} : RECEIVE \rightarrow IDLE$$

$$IDLE \rightarrow FORWARD$$

$$FORWARD \rightarrow RECEIVE$$

$$\lambda : RECEIVE \rightarrow \text{"Mess\_proc"}$$

$$RECEIVE \rightarrow \text{"Power\_cons"}$$

$$FORWARD \rightarrow \text{"Mess\_forw"}$$

$$FORWARD \rightarrow \text{"Power\_cons"}$$

$$ta : RECEIVE \rightarrow t_{Rec}$$

$$IDLE \rightarrow t_{Idle}$$

$$FORWARD \rightarrow t_{Forw}$$

$$DEAD \rightarrow \infty$$

2.4 Power Component Modeling

The deployment environment of a WSN is usually very harsh where power supply for each sensor node could not be continuous, and therefore power equipment with fixed power capacity is a feasible solution. Accordingly, power consumption is a critical factor about whether a WSN could work continuously and effectively, and design and working condition of each component in sensor nodes need elaborative consideration about power consumption [10]. The power model is required in WSN modeling and simulation to simulate and analyze power consumptions of sensor nodes. Figure 5 shows the structure of the power component atomic model, where the wireless communication, processor and sensor components transfer their working conditions to the power component through input ports *Trans_cons*, *Proc_cons* and *Sensor_cons*, and the power component calculates power consumption according to these working conditions; meanwhile, the power component transfers its current state to other components through output port *Power_info*.

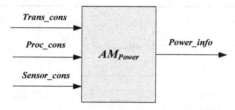

Fig. 5. Architecture of AM_{Power}

The formal description of atomic model AM_{Power} is as follows

$$AM_{Power} = <X, Y, S, \delta_{ext}, \delta_{int}, \lambda, ta >$$
$$X = \{"Trans\_cons", "Proc\_cons", "Sens\_cons"\}$$
$$Y = \{"Power\_info"\}$$
$$S = \{RUN, DEAD\}$$
$$\delta_{ext} : RUN \times "Trans\_cons" \to RUN$$
$$RUN \times "Proc\_cons" \to RUN$$
$$RUN \times "Sens\_cons" \to RUN$$
$$\delta_{int} : RUN \xrightarrow{if\ capacity\ >0} RUN$$
$$RUN \xrightarrow{if\ capacity\ >0} DEAD$$
$$\lambda : RUN \xrightarrow{if\ capacity\ >0} "Power\_info"$$
$$ta : RUN \to t_{Run}$$
$$DEAD \to \infty$$

2.5 WSN Node Modeling

One of essential advantages of DEVS is that it could achieve modular and hierarchical model design employing the I/O ports and coupling relations [6]. The component atomic models mentioned above could be coupled into sensor node model CM_{Node} using internal and external coupling relations, which is shown in Fig. 6.

The formal description of coupling model CM_{Node} is as follows

$$CM_{Node}=<X,Y,D, \{ M_d \} , EIC,EOC,IC, select>$$
$$X = \{"Mess\_rec", "Env\_info"\}$$
$$Y = \{"Mess\_forw"\}$$
$$\{M_d\} = \{Sensor, Processor, Transceiver, Power\}$$
$$EIC = \{(CM_{Node}.Mess\_rec, Transceiver.Mess\_rec),$$
$$(CM_{Node}.Env\_info, Sensor.Env\_info)\}$$

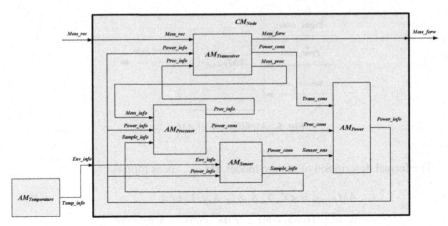

Fig. 6. Architecture of CM_{Node}

$$EOC = \{(Transceiver.Mess\_forw, CM_{Node}.Mess\_forw)\}$$
$$IC = \{(Transceiver.Power\_cons, Power.Trans\_cons)\},$$
$$(Transceiver.Mess\_proc, Processor.Mess\_info),$$
$$(Processor.Proc\_info, Transceiver.Proc\_info),$$
$$(Processor.Power\_cons, Power.Proc\_cons),$$
$$(Sensor.Sample\_info, Processor.Sample\_info),$$
$$(Sensor.Power\_cons, Power.Sens\_cons),$$
$$(Power.Power\_info, Transceiver.Power\_info),$$
$$(Power.Power\_info, Processor.Power\_info),$$
$$(Power.Power\_info, Sensor.Power\_info)\}$$
$$select(\{Sensor, Processor, Transceiver, Power\}) = Power$$

Coupling model CM_{Node} could communicates with other nodes in network topology only through ports *Mess_rec* and *Mess_forw* without extending ports as required while constructing network topology. That ensures a node coupling model is reusable in constructing WSN simulation scenarios.

2.6 WSN Modeling

A WSN can be established by setting coupling relations between WSN nodes according to network topology, where modeling of homogeneous nodes manifests the modular and reusable advantages of DEVS. The entire WSN model will be established based on the topology described in the simulation experiment in the next section.

3 Simulation Experiment and Analysis

3.1 Routing Protocol

Minimum Hop Count (MHC) routing protocol [11] is chosen as the routing protocol implementation in WSN node models in this paper. This protocol maintains a minimum hop count from current node to a sink node within each node, and nodes perform message flooding according to MHC, which achieves directional data flow to the sink node.

3.2 Simulator Introduction and Simulation Analysis

Simulation packet CD++ [12] is employed to execute WSN modeling and simulation based on DEVS in this paper. CD++ is composed of a series of models with hierarchical structure, and each model associates with a simulation entity. Atomic models are established with C++ language, and coupling models are established with the built-in formal definition language which describes the underlying member models and the coupling relations between these models.

The network topology of the simulation experiment is shown as Fig. 7. The topology of a WSN can be represented via describing coupling relations between nodes employing the formal definition language built in CD++. For establishing the integrated simulation scenario, an external temperature model $AM_{Temperature}$ is designed to provide external input for the WSN. The simulation scenario is set as single source and single sink, where node 8 is chosen as the source node to perceive temperature. The sensor component of node 8 samples the environment temperature, and when the temperature reaches the minimum measurement limit, the sensor component will transmit the real-time temperature to the processor component; when the temperature exceeds the maximum measurement limit, the sensor component will transmit the ultralimit signal to the processor component.

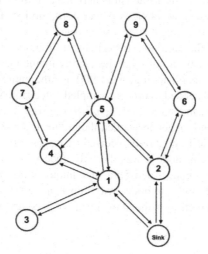

Fig. 7. Topology of wireless sensor network

Configurations for other relevant parameters in the simulation are as follows: the power capacity of each sensor node is set as 120000 units; the sensor component consumes 5 units per sampling cycle; the processor consumes 8 units per processing operation; the wireless communication component consumes 12 units per message forwarding. The total delay of message forwarding and processing is 100 ms.

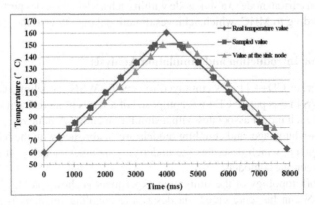

Fig. 8. Temperature variation curves during simulation

The experimental results are shown as Fig. 8, Fig. 9 and Fig. 10. Figure 8 manifests the temperature variation of the external temperature model, the sampling and processing of node 8 and the temperature data reception of the sink node. We can draw from Fig. 8 that the sink node started to receive temperature data 200 ms after node 8 forwarded data for the first time.

Figure 9 manifests the power consumption of each node. From 0 ms to about 500ms, the entire WSN was in the network establishment stage of MHC protocol. In this stage, query messages that serve for the establishment of the minimum hop gradient field were flooding, and all the nodes had obvious power consumption. From 500 ms to about 900 ms, the network establishment stage had finished, but node 8 did not perceive that the temperature reached the minimum measurement limit, and there were no messages flooding in the WSN, so powers of all nodes merely maintained basic running of sensor components and channel listening. After 900 ms, the entire WSN was in message forwarding stage: node 8 needed to process sampled value from sensor component continuously and to flood data messages; node 3, 6, 7, 9 could only receive flooding messages from a single node and no longer forwarded them, thus these nodes had the minimum power consumption; node 1 and 2 only received messages forwarded from node 5, and continued to forward them; node 4 was the only node which received messages forwarded from 2 nodes (i.e. node 1 and 5); as the unique node through which messages flooded from nodes with hop 3 to nodes with hop 1, node 5 had the maximum power consumption. Figure 10 manifests the amounts of received and forwarded messages of nodes in the WSN.

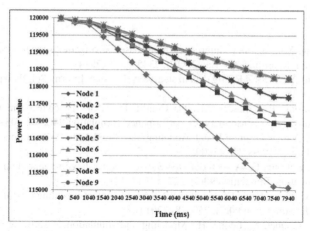

Fig. 9. Power consumption curves of each node

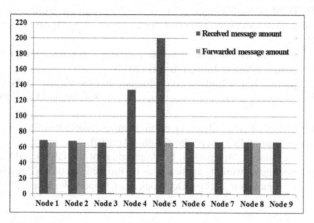

Fig. 10. Histogram of amount of received and forwarded messages of each node

4 Conclusion

A modeling and simulation method for WSN based on DEVS was proposed in this paper. The method was utilized to model sensor node components, sensor nodes, WSN and external environment, and ultimately a simulation model which can completely describe internal structure of sensor nodes, WSN topology and customized external environment was acquired. MHC was chosen as the WSN routing protocol in the simulation experiment to analyze the WSN performance, and the experiment results verified the modeling and simulation feasibility of the proposed DEVS- based method. A few nodes and a simple temperature variation were set as the simulation scenario in our first experiment; more elaborate models will be established to fulfill simulation with larger scale and more complicated scenario.

References

1. Xu, Y., Wang, X.F., He, Q.Y.: Internet of things based information support system for multi-agent decision. Ruan Jian Xue Bao **25**(10), 2325–2345 (2014)
2. Li, D.R., Yao, Y., Shao, Z.F.: Big data in smart city. Geomat. Inform. Sci. Wuhan Univ. **39**(6), 631–640 (2014)
3. Antoine-Santoni, T., Santucci, J.F., De Gentili, E., et al.: Discrete event modeling and simulation of wireless sensor network performance. Simulation **84**(2–3), 103–121 (2008)
4. Gabriel, A.: Wainer Discrete-Event Modeling and Simulation: A Practitioner's Approach. Taylor and Francis Press, UK (2009)
5. Bergero, F., Kofman, E.: A vectorial DEVS extension for large scale system modeling and parallel simulation. Simulation **90**(5), 522–546 (2014)
6. Seo, K.M., Choi, C., Kim, T.G., et al.: DEVS-based combat modeling for engagement-level simulation[J]. Simulation **90**(7), 759–781 (2014)
7. Bogado, V., Gonnet, S., Leone, H.: Modeling and simulation of software architecture in discrete event system specification for quality evaluation. Simul. Trans. Soc. Model. Simul. Int. **90**(3), 290–319 (2014)
8. Kapos, G.D., Dalakas, V., Nikolaidou, M., et al.: An integrated framework for automated simulation of SysML models using DEVS. Simul. Trans. Soc. Model. Simul. Int. **90**(717), 717–744 (2014)
9. Wang, R.Z., Shi, T.X., Jiao, W.P.: Collaborative sensing mechanism for intelligent sensors based on tuple space. J. Softw. **26**(4), 790–801 (2015)
10. Yang, C., Li, Q., Liu, J.: Dualsink-based continuous disaster tracking and early warning in power grid by wireless sensor networks. Geomat. Inf. Sci. Wuhan Univ. **38**(3), 303–306 (2013)
11. Duan, W., Qi, J., Zhao, Y., et al.: Research on minimum hop count routing protocol in wireless sensor network. Comput. Eng. Appl. **46**(22), 88–90 (2010)
12. Wainer, G.A., Tavanpour, M., Broutin, E.: Application of the DEVS and Cell-DEVS formalisms for modeling networking applications. In: Proceedings of the 2013 Winter Simulation Conference: Simulation: Making Decisions in a Complex World, pp. 2923–2934. IEEE Press (2013)

Spatial Reciprocity Aided CSI Acquirement for HST Massive MIMO

Kaihang Zheng[1(✉)], Yinglei Teng[1], An Liu[2], and Mei Song[1]

[1] School of Electronic Engineering, Beijing University of Posts and Telecommunications, Beijing, China
214882388@qq.com
[2] College of Information Science and Electronic Engineering, Zhejiang University, Hangzhou, China

Abstract. This paper proposes a channel estimation scheme for large-scale multiple-input multiple-output (MIMO) systems in high-speed train (HST) scenarios. On the premise that the priori velocity is accurate, we introduce fuzzy prior spatial knowledge and design a sparse received signal model with dynamic grids. After reconstructing the channel estimation into a sparse Bayesian learning (SBL) parameter estimation problem, the maximization-minimization (MM) algorithm is adopted to solve the problem, and a fast searching algorithm based on significant gradient is proposed to solve the multi-peak optimization problem of the surrogate function. Finally, the simulation verifies that the scheme can converge quickly and has accurate estimation results.

Keywords: Joint channel estimation · Spatial reciprocity · Massive MIMO · Sparse Bayesian learning (SBL)

1 Introduction

In the past 50 years, the rapid development of high-speed trains (HST) brings severe challenges to wireless communication. The main problem is the high Doppler frequency shift (DFS) introduced by the fast velocity, which directly leads to an increase in inter-carrier interference (ICI), and makes it difficult to obtain high precision real-time channel state information (CSI) with the rapid time-varying channel. Moreover, the difficulty of the handover scheme between base stations (BS) and HST channels modeling make the problem more serious [1].

For channel estimation schemes in high mobility scenarios, some research is devoted to estimating and compensating DFS to obtain CSI [2–5]. [2, 3] adopt a beamforming network to separate multiple DFSs mixed in multiple channel paths, and then the conventional compensation and channel estimation methods for single DFS can be carried out for each beamforming branch. However, this solution has very strict requirements on the number of antennas and the setting of beamforming networks, which is difficult to obtain a good trade-off between the high angular resolution and the blurring of adjacent grids. Although [4, 5] has a relative high estimation accuracy, their assumption of the

© Springer Nature Switzerland AG 2021
Q. Zu et al. (Eds.): HCC 2020, LNCS 12634, pp. 31–42, 2021.
https://doi.org/10.1007/978-3-030-70626-5_4

communication environment is too ideal and ignores the influence of non-line-of-sight (NLOS) channels. Even if line-of-sight (LOS) channel is dominate in most of HST channels, NLOS channels still cannot be ignored [6].

High-mobility scenarios, including HST, usually have a fixed trajectory for vehicles, which means that the location and direction of the train can be predicted with the assist of technologies such as Global Position System (GPS). For the LOS dominant scenario, the location will especially play a decisive role in the performance of channel estimation. [7, 8] studied the application of train position in HST scenarios, and they presumed that train passing through the same position experience similar CSI. Therefore, when the train subsequently passes through a specific position, the initial channel estimation result or codebook can be used to assist the current estimation. However, this requires a longer environmental coherence time than milliseconds (the HST ECT) and initial estimation scheme is not clear. In the HST scenario, the ECT is much longer than the channel coherence time, and most channel parameters can be considered constant in ECT. In response to these slow time-varying items, [9–12] carried out relevant research. [9] and [10] focused on LoS estimation but lack the consideration of dense scatterer scenario such as tunnel and station. [11, 12] models the time-domain variation of each channel parameter as a Markov chain, which is more suitable for channels with relatively random changes in the traditional environment. In addition, these schemes didn't take into account the problem that the channel estimation frequency should be increased due to the excessively fast channel aging, resulting in a decrease in frequency band utilization.

In this paper, we aim at a joint uplink/downlink (UL/DL) channel estimation method under HST environment in frequency division duplex (FDD) system. We first consider a dynamic grid model of received signals according to the sparse representation of massive MIMO channels and transform the channel parameter estimation problem into a sparse Bayesian learning (SBL) problem. Then, we utilize a block majorization-minimization (MM) algorithm to update the parameters estimation value in turn. It should be noted that because we focus on the estimation of the channel parameters instead of the channel itself, the estimated frequency requirements can be reduced to the order of milliseconds or even lower.

The rest of this paper is organized as follows. In Sect. 2, we describe the system model. In Sect. 3, we propose a sparse received signal model with dynamic grids and construct sparse Bayesian learning problem. In Sect. 4, we present the block MM algorithm and a searching algorithm for angle to solve the multi-peak optimal problem. The simulation results and conclusions are given in Sect. 5 and Sect. 6, respectively.

2 System Model

Considering a massive MIMO HST system with an N_{BS}-antenna equipped BS and an N_{MS}-antenna HST terminal. According to [6], the Rice channel model can fit different HST communication environments by adjusting Rice factor. Without loss of generality, we define the HST channel as a multipath flat fading model, and assume the train travels in the positive direction as Fig. 1 illustrates. The DL channel matrix and the steering vector is given by,

$$\mathbf{a}^d(\theta) = [1, e^{j2\pi \frac{d_0 \cos\theta}{\lambda^d}}, \cdots, e^{j2\pi \frac{d_0 \cos\theta}{\lambda^d}(N_{MS}-1)}]^T, \tag{1}$$

$$\mathbf{H}_n^d = \sum_{i=1}^{L} h_i^d D_{n,i}^d \mathbf{a}^d(\theta_{MS,i})(\mathbf{a}^d(\theta_{BS,i}))^H. \tag{2}$$

In the steering vector, λ^d is the wavelength, θ is the angles arrival or departure from BS or train respectively. d_0 stands for the separation distance between the antennas. In the channel, L denotes the number of paths, f_d is the carrier frequency, h_i^d stands for the UL/DL channel coefficient of the i-th path, and in particular the first path h_1^d represents the LoS channel. $D_{i,n}^d = e^{j2\pi f_D^d \cos\theta_{MS,i} n T_s}$, it denotes the DFS of the i-th path at the n-th symbol index, and T_s is sampling interval. $f_D^d = f_d v/c$ is the maximum DFS, where v and c denote the velocity of the train and light. The UL channel matrix has similar structure by changing the superscript d to u.

As illustrated in Fig. 1, we assume that the travel tack is horizontally coordinated with BS, and the antenna array of the train terminals are arranged parallel to the track. The height of the BS is H_0, the vertical distance from the track is D, and the real-time location of the train is s. Please note that, usually the pre-obtained v can be regarded as accurate but the s obtained by GPS is inaccurate. It will be adopted in Sect. 3.2 as priori knowledge to assist LoS angle estimation.

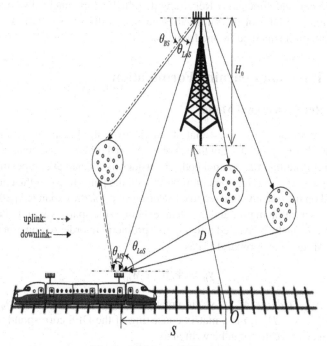

Fig. 1. Multipath HST model.

In the FDD massive MIMO system, the transmission time is divided into frames, and each frame covers N_p DL and UL subframes. As illustrated in Fig. 2, the DL pilot $\mathbf{x}^d \in \mathbb{C}^{N_{BS}}$ and UL pilot $\mathbf{x}^u \in \mathbb{C}^{N_{MS}}$ are send at the beginning of each subframe, and they are consistent within one frame.

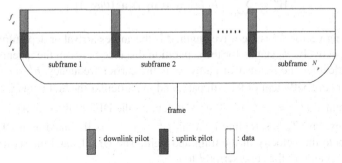

Fig. 2. The frame structure.

Correspondingly, the received signal of the n-th pilot at UL/DL can be expressed as:

$$\mathbf{y}_n^* = \mathbf{H}_n^* \mathbf{x}^* + \mathbf{w}_n^* \tag{3}$$

where $\mathbf{y}_n^*$ is the received signal, $\mathbf{w}_n^*$ is a Gaussian noise vector with zero mean and variance $\sigma_*^2/2$ per-real dimension. Regarding that the HST channel is fast time-varying, we can presume the DFS of the pilot varies between different subframes, while other parameters remain unchanged in one frame period.

3   Sparse Bayesian Learning Formulation

3.1   Sparse Received Signal Model

The goal of obtaining CSI is to estimate $\mathbf{H}$ or $\{\mathbf{h},\ \boldsymbol{\theta}_{MS},\ \boldsymbol{\theta}_{BS}\}$ from the received signal, where $\mathbf{h} = \mathbf{h}^d$, $\mathbf{h}^u$, $\boldsymbol{\theta}_{MS}$ and $\boldsymbol{\theta}_{BS}$ are the set of the UL/DL channel coefficients indicating the angular information of L channel paths. Considering the actual composition channel paths are sparse in the angular domain, if we extend the paths to cover the entire angular domain based on the transmit and received steering vectors, the expanded path set can be taken as the observation matrix and the channel gain as the sparse vector to be observed, then the problem can be converted into a sparse problem. In order to construct the sparse received signal model, we rebuild (3) as:

$$\mathbf{y}_n^* = \boldsymbol{\Phi}_n^* \mathbf{h}^* + \mathbf{w}_n^*, \tag{5}$$

where $\boldsymbol{\Phi}_n^* = \mathbf{A}_{MS}^* \mathbf{D}_n^* diag((\mathbf{A}_{BS}^*)^H \mathbf{x}^*)$ is a matrix composed by θ_{MS}, θ_{BS}, D_n, and $\mathbf{D}_n^* = diag(D_{n,1}^*, D_{n,2}^* \cdots, D_{n,M}^*)$ is a matrix consisting of the DFS correspond to θ_{MS}. The stacking of steering vectors can be written as:

$$\mathbf{A}_{MS}^* = [\mathbf{a}^*(\theta_{MS,1}), \mathbf{a}^*(\theta_{MS,2}), \cdots, \mathbf{a}^*(\theta_{MS,M})], \tag{6}$$

$$A_{BS}^* = [a^*(\theta_{BS,1}),\, a^*(\theta_{BS,2}),\, \cdots,\, a^*(\theta_{BS,M})]. \tag{7}$$

Here, M is the number of the grids, where $M = N_1 * N_2$. Define the initial angle sets of trains and BSs as $\breve{\theta}_{MS} = \left[\breve{\theta}_{MS,1}, \breve{\theta}_{MS,2}, \cdots, \breve{\theta}_{MS,N_1}\right]$ and $\breve{\theta}_{BS} = \left[\breve{\theta}_{BS,1}, \breve{\theta}_{BS,2}, \cdots, \breve{\theta}_{BS,N_2}\right]$, respectively. These angle sets are initiated uniformly in the range of $(0, \pi]$ to guarantee they can cover the whole angular domain. Then, we have the 2D angular domain discrete grids, and $\boldsymbol{\theta}_{BS} = \left[\breve{\theta}_{MS,1}, \cdots, \breve{\theta}_{MS,1}, \breve{\theta}_{MS,2}, \cdots \breve{\theta}_{MS,2}, \cdots, \breve{\theta}_{MS,N_1}\right]$, $\boldsymbol{\theta}_{BS} = [\breve{\theta}_{BS}, \breve{\theta}_{BS}, \cdots, \breve{\theta}_{BS}]$. The channel vector $\mathbf{h}^* = [h_1^*, h_2^*, \cdots, h_M^*]^T$ is defined as the path coefficients. Stacking N_P observation $\mathbf{y}_n^*$ along n, we have

$$\mathbf{y}_* = \boldsymbol{\Phi}_* \mathbf{h}^* + \mathbf{w}_* \tag{8}$$

where $\boldsymbol{\Phi}_* = [\boldsymbol{\Phi}_1^{*H}, \boldsymbol{\Phi}_2^{*H}, \cdots \boldsymbol{\Phi}_{N_P}^{*H}]^H$, $\mathbf{w}_* = [\mathbf{w}_1^{*H}, \mathbf{w}_2^{*H}, \cdots \mathbf{w}_{N_P}^{*H}]^H$, φ_i^* is defined as the i-th column in $\boldsymbol{\Phi}_*$ which contains the same angle information of a path in different transmission moments. Ideally, the sparse vector $\mathbf{h}^*$ has L non-zero value which is much smaller than M. Once we get $\mathbf{h}^*$ or the support of $\mathbf{h}^*$, the channel parameters can be obtained from it.

3.2 Prior of Parameters and Sparse Problem Formulation

Since we adapt dynamic grids to compensate the estimation errors between grids, the observation matrix $\boldsymbol{\Phi}_*$ is not static, making traditional compressed sensing algorithms can hardly work. We choose the SBL to solve the problem. Under the assumption of complex Gaussian noise, we have

$$p(\mathbf{y}^*|\mathbf{h}^*, \boldsymbol{\theta}_{MS}, \boldsymbol{\theta}_{BS}) = \mathcal{CN}(\mathbf{y}^*|\boldsymbol{\Phi}_*(\boldsymbol{\theta}_{MS}, \boldsymbol{\theta}_{BS})\mathbf{h}^*, \omega_*^{-1}\mathbf{I}) \tag{9}$$

where $\omega_* = 1/\sigma_*^2$ stands for the noise precision. On account of the unknown ω, it is modeled as a Gamma hyperprior $p(\omega_*) = \gamma(\omega|1 + a_{\omega_*}, b_{\omega_*})$, where we set $a_{\omega_*}, b_{\omega_*} \to 0$ to keep the noise priors non-informative. Assume the prior for each variable h_i^* of sparse channel vector is a Gaussian prior distribution with a variance α_i^{-1}. Letting $\boldsymbol{\alpha} = [\alpha_1, \alpha_2 \cdots, \alpha_M]$, we have

$$p(\mathbf{h}_d|\boldsymbol{\alpha}) = \mathcal{CN}(\mathbf{h}_d|0,\, diag(\boldsymbol{\alpha}^{-1})). \tag{10}$$

For each element α_i, we model it as the same Gamma hyperprior $p(\alpha_i) = \gamma(\alpha_i|1 + a_\alpha, b_\alpha)$, where we set $b_\alpha \to 0$ to promote the sparsity of channel vector. Accordingly, the marginal distribution of $\mathbf{h}_d$ gives,

$$p(\mathbf{h}_d) = \int \mathcal{CN}(\mathbf{h}_d|0,\, diag(\boldsymbol{\alpha}^{-1}))p(\boldsymbol{\alpha})d\boldsymbol{\alpha} \propto \prod_{i=1}^M (b_\alpha + h_i^d)^{-(a_\alpha + 3/2)}. \tag{11}$$

Since the value of $\boldsymbol{\alpha}$ indicates the possibility of h_i^d being zero, it can be regarded as the support of $\mathbf{h}_d$. We can confirm the support by choosing the paths which have small α_i, and then recover $\mathbf{h}_d$ by least square (LS) method easily. It's worth noting that although channel disparity does not exist in FDD, FDD still has spatial disparity when the duplex frequency is not much different even in high mobility scenario [13]. It means that the same angle information is shared by UL and DL, and their DFSs are only differs in the carrier frequency. In the proposed sparse received signal model, the spatial reciprocity is reflected in the same support shared by Ul and DL. In order to strictly guarantee spatial reciprocity, we introduce parameter $\boldsymbol{\gamma} = [\gamma_1, \gamma_2, \cdots \gamma_M]^T$ and model h_i^u as a Gaussian distribution with a variance $\alpha_i \gamma_i$, and γ_i is set as a positive number to control the joint sparse structure. Then we have

$$p(\mathbf{h}_u|\boldsymbol{\alpha}, \boldsymbol{\gamma}) = \mathcal{CN}(\mathbf{h}_u|0, \, diag((\boldsymbol{\alpha} \cdot \boldsymbol{\gamma})^{-1})). \tag{12}$$

This assumption implies that only when both $\theta_{BS,i}$ and $\theta_{MS,i}$ are near to the actual angles in the same path, the optimal solution to both α_i and $\alpha_i \gamma_i$ are of a lower order of magnitude.

Moreover, we adopt the fuzzy priori knowledge of HST location to improve the estimation precision of the parameters in LoS path. Assume the real-time location x follows the Gaussian distribution with a variance σ_x^2 and expectation x_0 obtained by GPS. As illustrated in Fig. 1, the relationship between train location and θ_{LoS} can be obtained through $\frac{\sqrt{H_0^2+D^2}}{x} = \frac{H}{x} = \tan\theta_{LoS}$. We can further obtain the priori LoS angle distribution:

$$p(\theta_{LoS}) = p(\cot\theta_{LoS})\frac{\partial\cot\theta_{LoS}}{\partial\theta_{LoS}} = \frac{1}{\sqrt{2\pi\sigma_x^2}}e^{-\frac{(\frac{H}{\tan\theta}-x_0)^2}{2\sigma_x^2}}\frac{H}{\sin^2\theta}. \tag{13}$$

Adding the prior LoS angle into the grid, we have $\widetilde{\boldsymbol{\theta}}_{MS} = [\theta_{LoS}, \theta_{MS,1}, \theta_{MS,2}\cdots,$ $\theta_{MS,M}] \in \mathbb{C}^{M+1}$, $\widetilde{\boldsymbol{\theta}}_{BS} = [\theta_{LoS}, \theta_{BS,1}, \theta_{BS,2}\cdots, \theta_{BS,M}] \in \mathbb{C}^{M+1}$. Due to the difficulty of estimating $\mathbf{h}^*$ directly, our goal shifts to solving the other parameters including $\{\omega, \boldsymbol{\alpha},$ $\boldsymbol{\gamma}, \widetilde{\boldsymbol{\theta}}_{MS}, \widetilde{\boldsymbol{\theta}}_{BS}\}$. After that, $\mathbf{h}$ can be obtained via the least square (LS) solution. Therefore, we aim to estimate the most likely values of the remaining parameters through solving the maximum posterior likelihood function of $p(\omega, \boldsymbol{\alpha}, \boldsymbol{\gamma}, \widetilde{\boldsymbol{\theta}}_{MS}, \widetilde{\boldsymbol{\theta}}_{BS}|\mathbf{y})$, i.e.

$$(\widehat{\omega}, \widehat{\boldsymbol{\alpha}}, \widehat{\boldsymbol{\gamma}}, \widetilde{\boldsymbol{\theta}}_{MS}, \widetilde{\boldsymbol{\theta}}_{BS}) = \operatorname{argmax} \ln p(\mathbf{y}, \omega, \boldsymbol{\alpha}, \boldsymbol{\gamma}, \widetilde{\boldsymbol{\theta}}_{MS}, \widetilde{\boldsymbol{\theta}}_{BS}), \tag{14}$$

where $\omega = \{\omega_d, \omega_u\}$.

3.3 Block MM Algorithm

To solve problem (14), we propose a block MM algorithm to find the optimal solution. In each iteration, we construct and solve a surrogate function of each parameter which should satisfied some principles [14]. Let $z^{(k)}$ denote the optimization variables at the beginning of the k-th iteration, and all parameters will update by solving the optimal problem $\dot{z}^{k+1} = \text{argmax}\mathcal{U}(z|\dot{z}^k)$ in turn. We construct the surrogate function at any fixed point $\dot{z}$ as follow inspired by:

$$\mathcal{U}(z|\dot{z}) = \int p(\mathbf{h}_d|\mathbf{y}, \dot{z})ln\frac{p(\mathbf{h}_d, \mathbf{y}, z)}{p(\mathbf{h}_d|\mathbf{y}, \dot{z})}d\mathbf{h}_d. \tag{15}$$

We can obtain the update rule of ω, α and γ easily by calculating their zero points of derivations. However, the optimal problem of θ_{MS} and θ_{BS} are non-convex and have numbers of stationary points, we use one-step gradient update to obtain $\theta_{MS}^{(k+1)}$, but for $\theta_{BS}^{(k+1)}$, we recruit a SGFS algorithm. Recalling (9)–(12), $p(\mathbf{h}_d|\mathbf{y}, z)$ obey the complex Gaussian:

$$p(\mathbf{h}_d|\mathbf{y}, z) = \mathcal{CN}(\mathbf{h}^d|\mu^d(z), \Sigma^d(z)), \tag{16}$$

$$p(\mathbf{h}_u|\mathbf{y}, z) = \mathcal{CN}(\mathbf{h}^u|\mu^u(z), \Sigma^u(z)) \tag{17}$$

where $\mu^d = w^d \Sigma^d (\tilde{\Phi}^d)^H \mathbf{y}^d$, $\Sigma^d = (w^d \tilde{\Phi}^d (\tilde{\theta}_{BS}, \tilde{\theta}_{MS})^H \tilde{\Phi}^d (\tilde{\theta}_{BS}, \tilde{\theta}_{MS}) + diag(\alpha))^{-1}$ and $p(\mathbf{h}_u|\mathbf{y}, z)$ has the same distribution but with different parameters, i.e. $\mu^u = w^u \Sigma^u (\tilde{\Phi}^u)^H \mathbf{y}^u$, $\Sigma^u = (w^u \tilde{\Phi}^u (\tilde{\theta}_{BS}, \tilde{\theta}_{MS})^H \tilde{\Phi}^u (\tilde{\theta}_{BS}, \tilde{\theta}_{MS}) + diag(\alpha \cdot \gamma))^{-1}$. Based on the above deductions, we can obtain the following update rules and algorithms.

The surrogate function of ω, α, γ has a unique solution in each iteration as follow:

$$\omega_d^{(k+1)} = (N_p N_{MS} + a_{\omega_d} - 1)/(b_{\omega_d} + \Psi^d(\omega_d^{(k)}, \alpha^{(k)}, \tilde{\theta}_{BS}^{(k)}, \tilde{\theta}_{MS}^{(k)})), \tag{18}$$

$$\omega_u^{(k+1)} = (N_p N_{BS} + a_{\omega_u} - 1)/(b_{\omega_u} + \Psi^u(\omega_u^{(k)}, \alpha^{(k)}, \tilde{\theta}_{BS}^{(k)}, \tilde{\theta}_{MS}^{(k)})) \tag{19}$$

$$\alpha_i^{(k+1)} = (a_\alpha + 1)/(b_\alpha + \alpha_i^{(k)}(\Xi_i^d + \gamma_i \Xi_i^u)) \tag{20}$$

$$\gamma_i^{(k+1)} = 1/(\alpha_i^{(k+1)}\gamma_i^{(k)} \Xi_i^u) \tag{21}$$

where $\Psi^d(\omega_d^{(k)}, \alpha^{(k)}, \tilde{\theta}_{BS}^{(k)}, \tilde{\theta}_{MS}^{(k)}) = tr(\tilde{\Phi}^d \Sigma^d (\tilde{\Phi}^d)^H) + ||\mathbf{y}^d - \tilde{\Phi}^d \mu^d||_2^2$, $\Psi^u(\omega_u^{(k)}, \alpha^{(k)}, \tilde{\theta}_{BS}^{(k)}, \tilde{\theta}_{MS}^{(k)}) = tr(\tilde{\Phi}^u \Sigma^u (\tilde{\Phi}^u)^H) + ||\mathbf{y}^u - \tilde{\Phi}^u \mu^d||_2^2$, and $\Xi_i^d = (\Sigma^d + \mu^d(\mu^d)^H)_{i,i}$, $\Xi_i^u = (\Sigma^u + \mu^u(\mu^u)^H)_{i,i}$.

The surrogate functions of the above parameters are convex and the optimal points in the current iteration are easy to calculate after obtaining the derivation. However, since the surrogate functions of $\tilde{\theta}_{BS}$ and $\tilde{\theta}_{MS}$ are non-convex with multi-peaks, which is not only difficult to obtain the extreme point, but also unable to obtain a series expansion with a better fitting condition. Fortunately, the surrogate function of $\tilde{\theta}_{MS}$ has relatively slow fluctuation in the main lobe. This observation inspired us choose a dynamic step setting which make θ_{MS} update in the main lope and cross other lopes to gradually approach optimal solution. Accordingly, we use the rule as below to update θ_{MS}:

$$\theta_{MS}^{(k+1)} = \theta_{MS}^{(k)} + \Delta_{\theta_{MS}}^{(k)} sign(\tau_{\theta_{MS}}) \tag{22}$$

where $\Delta_{\theta_{MS}}^{(k)}$ is the adjustable step size, $\tau_{\theta_{MS}}$ is the derivation of θ_{MS}, $sign()$ is the sign-function. However, GDM isn't sufficient to solve $\tilde{\theta}_{BS}$ due to such frequent fluctuation. Thus, we propose a SGFS algorithm to assist $\tilde{\theta}_{BS}$ to traverse the angle domain that approaches the optimal solution.

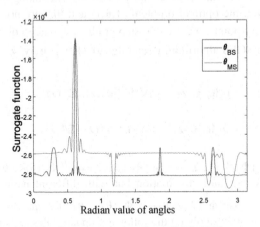

Fig. 3. Surrogate function of θBS, θMS.

Although every $\theta_{MS,i}$ changes with the iteration, the iteration result of $\theta_{MS,i}$ with the same initial value does not differ greatly because the step size is small. Therefore, we divide the M-items θ_{BS} into N_1 groups with N_2 items ensure that the $\theta_{BS,i}$ in each group covers the angular domain and corresponds to the similar $\theta_{MS,i}$ according to the initialization of the grids. Based on the particularly significant gradient near the optimal solution, we determine the update rule of each $\theta_{BS,i}$ in the same group by comparing their gradient values. Only when there exist a $\theta_{BS,i}$ has a higher gradient than others in the group by at least an order of magnitude, it will update with gradually shorter steps. Otherwise, all $\theta_{BS,i}$ will updated in the same direction with long step to quickly traverse the entire angular domain to search for the optimal solution neighborhood. After the traversal, there exit one $\theta_{BS,i}$ in each group approach the optimal solution, and the support of α and γ will indicate its index.

4 Simulation

To simulate the actual scenario, we set the coverage radius of a BS as 1 km, the location of HST is in the region of $[-1000, 1000]$. Parameter D and H_0 are set to be 50 m and 30 m, and the expected HST velocity v is 360 km/h. In the simulations, we consider $f_d = 2.7$ GHz and $f_u = 3$ GHz carrier frequency, $N_{BS} = 128$ and $N_{MS} = 64$ antennas, $L = 4$ channel paths, and half DL wavelength antenna spacing $d_0 = 0.5\lambda_d$. The grid parameters are set as $N_1 = 20$ and $N_2 = 11$, and has N_c subframes in a frame.

First, we'll focus on the convergence speed of proposed scheme. Figure 4 show the root mean square error (RMSE) of the angles versus iterations with 10 dB SNR

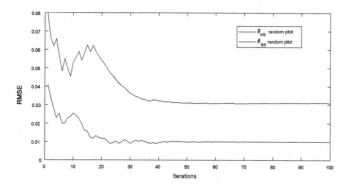

Fig. 4. RMSE of θ_{BS}, θ_{MS}.

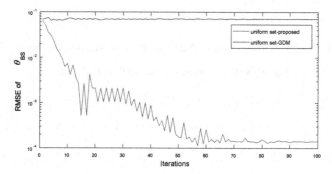

Fig. 5. RMSE of θ_{BS} adopting SGFS algorithm and GDM.

and the number of pilots $N_p = 7$, which is also the default parameters settings in the following simulations. From this figure, it can be seen that the proposed algorithm can gain accurate estimated value and converge within 30 steps, which is a sufficient fast convergence speed for HST scenario. Figure 5 illustrates the superiority of the proposed SGFS algorithm. As can be seen, the performance of GDM is not ideal, it is easy to converge to the local optimal solution due to too many stagnation points of the surrogate function about the angle which can be seen in Fig. 3. But the SGFS algorithm can obtain a better convergence performance, $\boldsymbol{\theta}_{BS}$ can converge near the optimal solution at a fast speed and gradually approach the optimal solution.

Then, we will demonstrate the superiority of the proposed scheme in the accuracy. The comparison performance of angle estimation with different algorithms versus SNR and antenna number is shown in Fig. 6. As can be seen, the proposed scheme can adapt to various SNR environments and antenna configurations, and there is no need to make major changes to the scheme according to the environment. It is worth noting that we use the DL signal to compare the effects of the single-path estimation and the joint estimation, which can prove that the joint estimation has much better performance. This is due to the lack of spatial reciprocity constraints in the case of a single path, and it will only have a relative better estimation effect on one side angle, i.e. θ_{MS}, even cost more pilot resource.

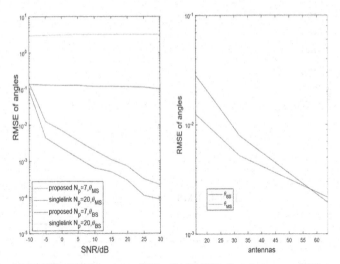

(a) RMSE of angles versus SNR   (b) RMSE of angles versus SNR

Fig. 6. RMSE of angles versus SNR and antennas.

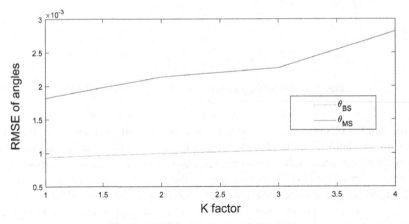

Fig. 7. RMSE of angles versus K factor.

Finally, we prove the robustness of the proposed scheme in different communication environments. Since Rice factor K represents the proportion of LOS channel in HST channel, it can be used to abstract different communication environments. Figure 7 illustrates the estimation performance of angles under different Rice factor K. It can be seen that the proposed scheme can adapt to various communication environments. Although the performance deteriorates with the increase of the K factor, that is because the energy of the NLOS channel becomes smaller gradually and even difficult to distinguish from noise, but this is acceptable. On one hand, even in this case, there are still relatively accurate estimation results. On the other hand, the influence of the NLOS channel in a high-K factor environment is negligible, and LOS still has a good estimation effect at this moment.

5 Conclusion

In this paper, we study the problem of channel estimation. The problem is first transformed into the SBL channel parameter estimation problem which is obtained by adopt the channel sparsity of the angular domain. Then the idea of MM algorithm is learned to solve the surrogate function of each parameter to gradually approach the optimal solution, and the challenge brought by the multi-peak optimization problem is solved by SGFS. Simulation proves that our proposed scheme can adapt to the influence of various antenna numbers and SNR. The comparison with the single-path scheme verifies the superiority of joint UL/DL scheme.

Acknowledgement. This work was supported in part by the National Key R&D Program of China (No. 2018YFB1201500), the National Natural Science Foundation of China under Grant No. 61771072.

References

1. Ren, X., Chen, W., Tao, M.: Position-based compressed channel estimation and pilot design for high-mobility OFDM systems. IEEE Trans. Veh. Technol. **64**(5), 1918–1929 (2014)
2. Guo, W., Zhang, W., Mu, P., Gao, F.: High-mobility OFDM downlink transmission with large-scale antenna array. IEEE Trans. Veh. Technol. **66**(9), 8600–8604 (2017)
3. Guo, W., Zhang, W., Mu, P., Gao, F., Lin, H.: High-mobility wideband massive MIMO communications: doppler compensation, analysis and scaling law. IEEE Transactions on Wireless Communications (2019)
4. Fan, D., Zhong, Z., Wang, G., Gao, F.: Doppler shift estimation for high-speed railway wireless communication systems with large-scale linear antennas. In: 2015 International Workshop on High Mobility Wireless Communications (HMWC), pp. 96–100. IEEE (2015)
5. Hou, Z., Zhou, Y., Tian, L., Shi, J., Li, Y., Vucetic, B.: Radio environment mapaided doppler shift estimation in LTE railway. IEEE Trans. Veh. Technol. **66**(5), 4462–4467 (2016)
6. He, R., Zhong, Z., Ai, B., Wang, G., Ding, J., Molisch, A.F.: Measurements and analysis of propagation channels in high-speed railway viaducts. IEEE Trans. Wireless Commun. **12**(2), 794–805 (2012)
7. Li, T., Wang, X., Fan, P., Riihonen, T.: Position-aided large-scale MIMO channel estimation for high-speed railway communication systems. IEEE Trans. Veh. Technol. **66**(10), 8964–8978 (2017)
8. Garcia, N., Wymeersch, H., Strö̈m, E.G., Slock, D.: Location-aided mm-wave channel estimation for vehicular communication. In: 2016 IEEE 17th International Workshop on Signal Processing Advances in Wireless Communications (SPAWC), pp. 1–5. IEEE (2016)
9. Zhang, C., Zhang, J., Huang, Y., Yang, L.: Location-aided channel tracking and downlink transmission for HST massive MIMO systems. IET Commun. **11**(13), 2082–2088 (2017)
10. Gong, Z., Jiang, F., Li, C.: Angle domain channel tracking with large antenna array for high mobility V2I millimeter wave communications. IEEE J. Selected Topics Signal Process. **13**(5), 1077–1089 (2019)

11. Liu, G., Liu, A., Zhang, R., Zhao, M.-J.: Angular-Domain Selective Channel Tracking and Doppler Compensation for High-Mobility mmWave Massive MIMO, arXiv preprint arXiv: 1911.08683 (2019)
12. Han, Y., Liu, Q., Wen, C.-K., Matthaiou, M., Ma, X.: Tracking fdd massive mimo downlink channels by exploiting delay and angular reciprocity. IEEE J. Sel. Topics Signal Process. **13**(5), 1062–1076 (2019)
13. Hugl, K., Kalliola, K., Laurila, J.: Spatial reciprocity of uplink and downlink radio channels in FDD systems. In: Proceedings of the COST, vol. 273, no. 2. Citeseer, p. 066 (2002)
14. Dempster, A.P., Laird, N.M., Rubin, D.B.: Maximum likelihood from incomplete data via the em algorithm. J. Roy. Stat. Soc.: Ser. B (Methodol.) **39**(1), 1–22 (1977)

A Flexible Film Thermocouple Temperature Sensor

Yulong Bao, Bin Xu, Huang Wang, Dandan Yuan, Xiaoxiao Yan, Haoxin Shu[✉], and Gang Tang[✉]

Jiangxi Province Key Laboratory of Precision Drive and Control, Nanchang Institute of Technology, Nanchang 330099, China

`2916579466@qq.com, tanggangnit@163.com`

Abstract. This article introduces a thin-film thermocouple temperature sensor with symmetrical electrode structure. It uses PI film as a flexible substrate. Cu film and CuNi film made by MEMS manufacturing process as positive and negative electrodes. The device itself has the advantages of miniature, bendable and fast response speed. In order to reduce the resistance value of the film, an experiment was conducted to optimize the thickness of the metal film and the temperature of the sputtering substrate. The critical dimensions of Cu/CuNi film are 650 nm and 400 nm respectively. The best sputtering substrate temperature for Cu/CuNi films are 100 °C and 150 °C. Testing the adhesion of thin film thermocouples using the peel-off method. The test result is 9.4 N. Finally, the film thermocouple temperature sensor is subjected to a temperature static calibration experiment. The result shows that the actual potential difference error is within ± 1 °C. It belongs to the second class standard in the formulation of thermocouple standards in China. Through curve fitting, the corresponding relationship between temperature and potential difference is more accurate.

Keywords: Thin film thermocouple · Sensor · MEMS process · Static calibration

1 Introduction

With the rapid development of science and technology, various types of film materials are widely used in production and life [1–4]. The thin film thermocouple temperature sensor is a new type of sensor born with the development of thin film technology [5–7]. Compared with the traditional bulk thermocouple, due to its small device and the thickness of the thermal junction at the micro-nano level, it has the advantages of fast response and small heat capacity [8, 9].

In this study, we propose a thin-film thermocouple temperature sensor based on PI flexible substrate to achieve flexibility and rapid temperature measurement. Design the appropriate dimensional structure, prepare the finished product using MEMS technology and conduct experimental testing to verify the performance of the finished product.

© Springer Nature Switzerland AG 2021
Q. Zu et al. (Eds.): HCC 2020, LNCS 12634, pp. 43–54, 2021.
https://doi.org/10.1007/978-3-030-70626-5_5

2 Theoretical Analysis

2.1 Working Principle

Thermocouple temperature sensor is a widely used temperature measurement device, and its temperature measurement principle is thermoelectric effect [10]. The thermoelectric effect refers to the thermoelectric phenomenon that occurs due to the mutual contact of different types of metals, that is, two different metals form a closed loop. When there is a temperature difference between the two connectors, current will be generated in the loop.

A closed loop is composed of two conductors A and B with different properties, as shown in Fig. 1. When the nodes (1) and (2) are at different temperatures (T, To), a thermoelectric potential will be generated between them and a certain current will be formed in the loop. This phenomenon is called thermoelectric effect.

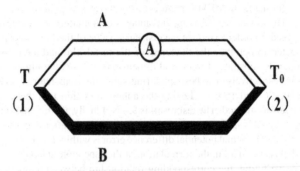

Fig. 1. Thermocouple temperature sensor schematic

Experiments have proved that the thermoelectric potential generated by the thermoelectric effect consists of two parts: contact potential (Peltier potential) and thermoelectric potential (Thomson potential). When two different metals are in contact with each other, since the free electron density of each metal conductor is different, electron migration and diffusion will occur at the junction. The metal that loses free electrons is positively charged, and the metal that gets free electrons is negatively charged. When the electron diffusion reaches equilibrium, an electric potential is formed at the contact of the two metals, which is called the contact potential; Its size is not only related to the properties of the two metals, but also related to the metal junction temperature. The expression is:

$$E_{AB}(T) = \frac{kT}{e} \ln \frac{N_A}{N_B} \tag{1}$$

$E_{AB}(T)$——Contact potential of A and B metals at temperature T;
k——Boltzmann constant, $k = 1.38 \times 10^{-23} (\text{J/K})$;
e——Electronic charge, $e = 1.6 \times 10^{-19} (\text{C})$;
N_A, N_B——Free electron density of metals A and B;
T——The temperature value at the node.

For a single metal, if the two ends have different temperatures, the free electrons at the higher temperature end will migrate to the lower end, so that the two ends of the single metal will have different potentials, thus forming an electric potential, which we call thermoelectric potential. Its size is related to the nature of the metal and the temperature difference between the two ends, the expression is:

$$E_A(T, T_0) = \int_{T0}^{T} \sigma_A dT \tag{2}$$

$E_A(T,T_0)$——The thermoelectric potential when the temperature at both ends of the metal A is T and T_0;
σ_A——Coefficient of temperature difference;
T, T_0——Absolute temperature of high and low end;
A and B closed circuit composed of two metal conductors, the total temperature difference potential:

$$E_A(T, T_0) - E_B(T, T_0) = \int_{T0}^{T} (\sigma_A - \sigma_B) dT \tag{3}$$

In summary, the total thermoelectric potential of the loop:

$$E_{AB}(T, T_0) = E_{AB}(T) - E_{AB}(T_0) + \int_{T0}^{T} (\sigma_A - \sigma_B) dT \tag{4}$$

2.2 Four Important Laws of Thermocouples

The Law of Homogeneous Conductors. If the thermocouple loop has the same two metal conductors, both of which are closed loops of homogeneous material, there is no current flowing in the loop and no potential, independent of the thermal electrode length, diameter and node temperature. If there is a current in the closed circuit, the metal conductor must not be homogeneous.

If the two metallic conductors of the thermocouple circuit are composed of two different homogeneous materials, the thermopotential of the thermocouple is only related to the temperature at the two nodes and is independent of the temperature of the other distributions along the way. If the two metals are non-homogeneous conductors, the hot electrode will generate additional potential when the temperature changes, causing errors in measurement. Therefore, conductor homogeneity is an important part of measuring the quality of thermocouple temperature sensor.

Intermediate Conductor Law. In the thermocouple loop composed of conductors A and B, the introduction of a third conductor C has no effect on the total potential of the closed loop as long as the temperature at both ends of conductor C is maintained the same. It can be seen that the thermocouple temperature sensor access to the measuring instrument as long as to ensure access to the same material, the same temperature, the measuring instrument on the thermocouple potential difference will not have an impact.

The Law of Intermediate Temperature. The thermal potential between the two junctions of the thermocouple loop is equal to the thermal potential of the thermocouple at a temperature of T, T_n and the algebraic sum of the thermal potentials at a temperature of T, T_0. T_n is called the intermediate temperature and the thermal potential of the loop is:

$$E_{AB}(T,T_0) = E_{AB}(T,T_n)+E_{AB}(T_n,T_0) \tag{5}$$

The international unified standard thermocouple indexing table to describe the cold end of the temperature of 0 °C when the corresponding relationship between the thermal potential and the temperature of the hot end. Experimental process when the reference end or cold end temperature is not 0 °C, you can use this law against the index table to obtain the temperature of the working end T, thermocouple in the use of compensation wire is also to follow the intermediate temperature law.

Standard Electrode Law. If the thermocouple thermal potentials $E_{AC}(T,T_0)$ and $E_{BC}(T,T_0)$ are known for the thermocouple composed of two electrodes A and B and another electrode C, respectively. At the same nodal temperature (T, T_0), the thermocouple thermal potential $E_{AB}(T,T_0)$ consisting of electrodes A and B is:

$$E_{AB}(T,T_0) = E_{AC}(T,T_0) - E_{BC}(T,T_0) \tag{6}$$

The law is called the standard electrode law and the electrode C is called the standard electrode. In actual measurement, because of the advantages of stable physical and chemical properties and high melting point of platinum wire, platinum wire is often used as the standard electrode.

2.3　Thermoelectric Performance

The physical quantity that characterizes a thermocouple is the thermoelectric potential rate, which is essentially the relative thermoelectric properties of the two different materials that make up the thermocouple. The thermoelectric properties of a single material are called absolute thermoelectric potential rate or absolute Seebeck coefficient [11]. Defined as:

$$S = \int_0^T \frac{\mu}{T}dT \tag{7}$$

in the formula, μ ——Thomson coefficient of the material.

T ——absolute temperature.

This definition of absolute thermoelectric potential rate is derived from the Kelvin relational formula, which is collated as:

$$S_{AB} = \int_0^T \frac{\mu_A}{T}dT - \int_0^T \frac{\mu_B}{T}dT \tag{8}$$

Putting Eq. (1) into the above equation:

$$S_{AB} = S_A - S_B \tag{9}$$

It can be seen from formula (3) that the Seebeck coefficient S_{AB} of a thermocouple temperature sensor is equal to the algebraic sum of the absolute thermoelectric potential rates of the two hot electrode materials that make up the thermocouple.

3   Device Design and Manufacturing

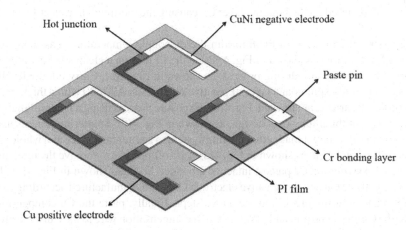

Fig. 2. Schematic diagram of the structure of thin film thermocouple temperature sensor

The designed thin film thermocouple temperature sensor is shown in Fig. 2. The structure includes PI film substrate, bonding layer Cr, positive electrode Cu, negative electrode CuNi, and a thermal junction composed of two electrodes in contact. The overall size of the device is 24 mm × 24 mm, and there are 4 thin-film thermocouple temperature sensors of the same size, each of which is relatively independent.

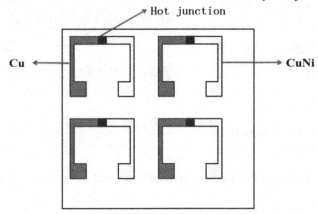

Fig. 3. A two-dimensional diagram of a thermocouple temperature sensor

The two-dimensional diagram of the thermocouple temperature sensor is shown in Fig. 3. The Cu and CuNi film structure is designed to be a symmetrical structure, and the size of the thermal junction is 1 mm × 1 mm. The size of the compensation wire pin is 2 mm × 2 mm, which is convenient for bonding the lead. The bonding layer Cr is located between the PI film substrate and the electrode layer, the width at the hot junction is increased by 0.5 mm, and the connection pin is increased by 0.5 mm. The purpose is to prevent deviations during preparation, so that the Cu and CuNi electrodes are not completely sputtered on the bonding layer Cr, causing the thermocouple electrode to fall off.

The process flow chart of the thin-film thermocouple temperature sensor prepared by MEMS technology is shown in Fig. 4. Paste the PI film on a clean silicon wafer and perform the cleaning and drying process, as shown in Fig. 4(a); Carry out the leveling process, and evenly spin-coat a layer of positive photoresist AZ4620 with a thickness of 1 μm on the PI surface, as shown in Fig. 4(b); Exposure process using a photolithography machine, photolithography on a design, as shown in Fig. 4(c); Sputtering process using high vacuum triple-target magnetron coating system to sputter the device as a whole with a layer of Cr metal film, as shown in Fig. 4(d); Lift-off process to remove the remaining photoresist, leaving the Cr pattern intact as a bonded layer, as shown in Fig. 4(e). The positive electrode Cu and the negative electrode CuNi are manufactured according to the MEMS manufacturing process in the above steps. Finally, paste the Cu compensation wire to the Cu electrode pin, and paste the CuNi compensation wire to the CuNi electrode pin.

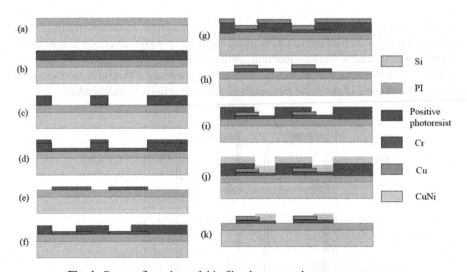

Fig. 4. Process flow chart of thin film thermocouple temperature sensor

4 Experimental Testing and Analysis

4.1 Experimental Test Platform

The temperature static calibration test platform of the thermocouple temperature sensor is shown in Fig. 5. The instruments include: a constant temperature heating platform and a KEITHLEY electrometer. The constant temperature heating table can control the temperature stably, the temperature control range is between normal temperature and 350 °C, the control accuracy is accurate and the performance is stable, and the error is within ±1 °C. The KEITHLEY electrometer has voltage measurement parameters, which can be used for potential difference testing to obtain the relationship between temperature and potential difference.

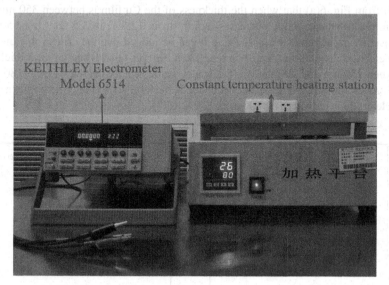

Fig. 5. Temperature static calibration experiment platform

4.2 Metal Film Thickness Optimization Experiment

When a metal conductor has a temperature gradient, it will cause the number of electrons near the Fermi energy at both ends of the conductor to change with energy, with more ultra-high-energy electrons on the high-temperature side and more ultra-low-energy electrons on the low-temperature side. When conducting electrons diffuse in a conductor, their diffusion rate is related to energy: the higher the energy of electrons, the less scattering they will receive during diffusion, and the greater the diffusion rate. A net diffused electron flow will be formed in the conductor, so that conduction electrons will accumulate at one end of the conductor, generating an electromotive force. The thermoelectromotive force generated by the diffusion of electrons due to the existence of the conductor temperature difference is called "diffusion thermoelectromotive force". The derivative of temperature is called absolute thermoelectric potential rate [12].

When the thickness of the metal thin film or the alloy thin film is small to a certain extent, the scattering probability of the conductive electrons by the boundary increases, which affects its diffusion thermoelectromotive force rate. As a result, the thermoelectric properties of the thin film are different from solid materials. This phenomenon is called the "film size effect".

The thickness of the sputtered metal film is different, which will affect the film quality and resistivity of the thermocouple temperature sensor. The "film size effect" shows that the resistivity of the metal film will increase sharply when the thickness is less than a certain thickness. The rate will be basically stable, and this thickness is called the critical dimension.

The experimental results of film thickness optimization are shown in Fig. 6. It can be seen from Fig. 6(a) that when the thickness of the Cu film is between 350 nm and 650 nm, the resistance value decreases rapidly from 12.8 Ω to 3.5 Ω. After 650 nm, the resistance value tends to stabilize, so it can be judged that the critical thickness of the Cu electrode is 650 nm. It can be seen from Fig. 6(b) that when the film thickness is between 250 nm and 400 nm, the resistance value rapidly decreases from 180 Ω to 53.2 Ω. After 400 nm, the resistance value tends to stabilize, so the critical thickness of CuNi electrode is judged to be 400 nm.

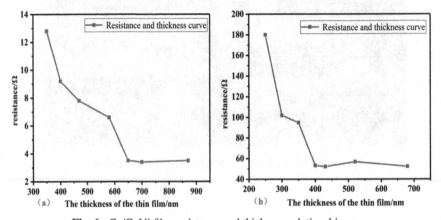

Fig. 6. Cu/CuNi film resistance and thickness relationship curve

4.3   Temperature Optimization Experiment of Metal Film Sputtering Substrate

Properly increasing the temperature of the substrate during sputtering can not only reduce the film stress and enhance the bonding force of the film substrate, but also reduce the film resistivity and increase the thermoelectromotive force.

The experimental results of the temperature optimization of the metal thin film sputtering substrate are shown in Fig. 7. It can be seen from Fig. 7(a) that the resistance value drops rapidly from 3.5 Ω to 1.1 Ω when the temperature is 30 °C to 100 °C, and the resistance value between 100 °C to 200 °C is relatively stable. It is inferred that the optimal Cu film sputtering. The substrate temperature is 100 °C. It can be seen from Fig. 7(b) that the resistance value drops rapidly from 53.2 Ω to 32.4 Ω at a temperature of 30 °C to 150 °C. Although the resistance value is also falling at 150 °C to 200 °C, the overall resistance value tends to be stable, so Inferred that the optimal CuNi film sputtering substrate temperature is 150 °C.

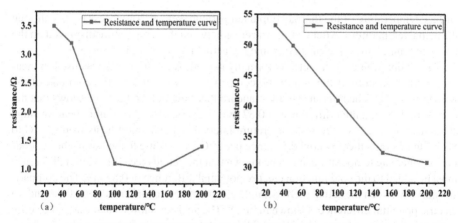

Fig. 7. Cu/CuNi film resistance and temperature curve

4.4 Adhesion Test Experiment

The adhesion between the film and the substrate directly affects the performance of the film, and films with poor adhesion cannot be applied to flexible devices. Internal stress is generated during the film preparation process, and internal stress is also generated between the flexible substrate and the metal film. Excessive internal stress may cause the film to curl or crack, or the film will fall off when the flexible substrate is torn. Therefore, the adhesion of the film determines the possibility and reliability of the application of flexible devices.

Adhesion is one of the important indicators to describe the performance of the film. The adhesion test methods include two types: one is the adhesive method, including the pull method, the peel method, and the pull method; the other is the direct method, including the scratch method, the friction method, Centrifugation etc. This paper uses the peeling method to test the adhesion between the film and the flexible substrate. A length of PI tape is torn out, one end is fixed with a clip, and the other end is pasted on the Cu/CuNi electrode of the thermocouple temperature sensor to cover the entire electrode layer. Pull the tension sensor of the push-pull dynamometer upward, observe the change of the reading and the peeling of the film by the tape, and stop if there is a small amount of film peeling off. After testing, it is concluded that when the applied pressure is about 9.4N, the metal film begins to fall off slightly, which shows that the adhesion force is 9.4N.

4.5 Thermocouple Static Calibration Experiment

According to the temperature static calibration test platform built in Fig. 5, the thermocouple temperature sensor is statically calibrated. First, paste the thermocouple temperature sensor on the silicon chip, put it on the constant temperature heating table and fix it with PI tape. The test cable of the KEITHLEY electrometer clamps the compensation lead of the positive and negative electrodes respectively. Because there may be an error between the temperature displayed by the constant temperature heating station and the surface temperature of the thermocouple temperature sensor, it is necessary to calibrate the actual temperature. After testing, the error between the temperature displayed on the constant temperature heating station and the actual temperature is about 5 °C.

The calibration experiment was formally started. A thermometer was used to measure the temperature of the end of the compensating wire, that is, the cold end of the wire to be 22 °C. The reference to the standard thermocouple index table shows that the corresponding potential difference is 0.87 mV. The constant temperature heating stage starts heating at 22 °C, the set temperature interval is 1 °C, and it heats from 22 °C to 80 °C and records the potential difference per 1 °C. According to the law of thermocouple intermediate temperature, it can be known that the actual electric potential difference must be added to the cold junction electric potential difference of 0.87 mV. The comparison curve between the actual thermocouple electric potential difference and the standard electric potential difference is shown in Fig. 8. The analysis shows that the actual electric potential difference error is ±1 °C, which is in the standard of thermocouples in China. It is a level II standard under development.

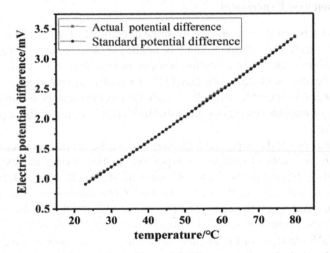

Fig. 8. Comparison curve between actual potential difference and standard potential difference

Because the temperature interval set in the thermocouple static calibration experiment is 1 °C, the potential difference value cannot further express more accurate temperature changes, so this article fits the actual thermocouple potential difference and the temperature change curve to make it at 23 °C to 80 °C any potential difference value within the range corresponds to a temperature value. The fitting curve is shown in Fig. 9. The fitting relationship is a linear equation, and the fitting formula is:

$$y = 0.04309x - 0.09827 \tag{10}$$

$$R^2 = 0.99963$$

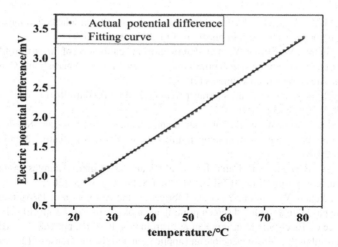

Fig. 9. Fitting curve of relationship between actual potential difference and temperature change

5   Conclusion

In this paper, a T-type thermocouple temperature sensor with a symmetrical electrode structure is designed. It consists of PI film material as a flexible substrate, which is flexible and can be bent for use. Manufactured by MEMS manufacturing process, the device is more miniature and precise. In order to reduce the resistance value of the metal film, an optimization experiment for the thickness of the metal film and the temperature of the sputtering substrate was set to obtain the critical size and the best sputtering substrate temperature. At the same time, the adhesion test experiment was added, and the adhesion of the film thermocouple temperature sensor was about 9.4N. Finally, through temperature static calibration experimental research and analysis, the results show that the actual potential difference error is ± 1 °C, which belongs to the second-level standard in the formulation of thermocouple standards in China. In order to be more precise in the experimental test results, curve fitting was performed on the relationship between the actual thermocouple potential difference and the temperature change.

Acknowledgments. This work was supported in part by Science Foundation of the department of education of Jiangxi province (GJJ170987, GJJ180936, GJJ180938). Open project of Shanxi Key Laboratory of Intelligent Robot (SKLIRKF2017004).

References

1. Nur, M.E.A., Lonsdale, W., Vasiliev, M., et al.: Application-specific oxide-based and metal-dielectric thin film materials prepared by RF magnetron sputtering (2019)
2. Hora, J., Hall, C., Evans, D., et al.: Inorganic thin film deposition and application on organic polymer substrates. Adv. Eng. Mater. **20**(5), 1700868 (2018)
3. Langford, C.L.: Application of molecular layer deposition for graphite anodes in lithium-ion batteries and porous thin-film materials (2016)

4. Szindler, M.M.: Polymeric electrolyte thin film for dye sensitized solar cells application. Solid State Phenomena. Trans Tech Publications Ltd. **293**, 73–81 (2019)

5. Jiong, D., Jichen, W., Suijun, Y., et al.: Fabrication and calibration of Pt/Au thin-film thermocouple based on a modified screen printing technology. In: IEEE International Conference on Electronic Measurement Instruments (2017)

6. Kumar, S.R., Kasiviswanathan, S.: Transparent ITO-Mn: ITO thin-film thermocouples. IEEE Sens. J. **9**(7), 809–813 (2009)

7. Liu, D., Shi, P., Ren, W., et al.: Fabrication and characterization of La0.8Sr0.2CrO3/In2O3 thin film thermocouple for high temperature sensing. Sensors Actuators A-physical **2018**, 459–465 (2018)

8. Albrecht, A., Bobinger, M., Calia, J.B., et al.: Transparent thermocouples based on spray-coated nanocomposites. In: 2017 IEEE SENSORS, pp. 1–3. IEEE (2017)

9. Jiong, D., Jichen. W., Suijun, Y., et al.: Fabrication and calibration of Pt/Au thin-film thermocouple based on a modified screen printing technology. In: 2017 13th IEEE International Conference on Electronic Measurement & Instruments (ICEMI), pp. 168–174. IEEE (2017)

10. Lifan, M., Jinhui, L.: Sensor principle and application. Publishing House of Electronics Press, Beijing (2011)

11. Zexian, H.: Principle and identification of thermocouple. China Metrology Press, Beijing (1993)

12. Zhengfei, H., Biao, Y.: Introduction to Material Physics. Chemical Industry Press, Beijing (2009)

Design of a Morphing Surface Using Auxetic Lattice Skin for Space-Reconfigurable Reflectors

Bin Xu[1], Houfei Fang[2(✉)], Yangqin He[2], Shuidong Jiang[2], and Lan Lan[2]

[1] China Academy of Space Technology (Xi'an), Xi'an 710000, China
[2] Shanghai YS Information Technology Co., Ltd., Shanghai, People's Republic of China
houfei_fang@yahoo.com

Abstract. The effect of Poisson's ratio to the reflector reshaping is investigated through mechanical study of reconfigurable reflectors in this paper. The value of Poisson's ratio corresponding to the minimum deforming stress is given and an auxetic lattice is proposed for the reflector surface. The parameters of the auxetic lattice are investigated for vary Poisson's ratio. A case of reconfigurable reflector is studied, the curvature change and strain are calculated by surface geometry analyse, and the negative Poisson's ratio is established for vary thickness. According to RMS calculation by the FEM structure analyse, the thickness can finally be established.

Keywords: Reflector · Auxetic lattice · Reconfigurable · Negative Poisson's ratio

Symbols

E	Young's Modulus
ν	Poisson's ratio
σ	Stress
σ_B	Bending stress
σ_M	Membrane stress
τ	Shear stress
ε	strain
γ	Shear strain
x, y, z	Crosswise, length wise direction, vertical direction
K_x, K_y, K_g	Curvature in x, y dimension and Guassian curvature
ν_{Bmin}	Poisson's ratio corresponding to minimal bending stress
ν_{Mmin}	Poisson's ratio corresponding to minimal membrane stress
ν_{min}	Poisson's ratio corresponding to minimal stress

© Springer Nature Switzerland AG 2021
Q. Zu et al. (Eds.): HCC 2020, LNCS 12634, pp. 55–67, 2021.
https://doi.org/10.1007/978-3-030-70626-5_6

1 Introduction

Conventional reflecting antennas with fixed reflector surface can not adapt the radiation patterns. In contrast, the in-orbit space antenna with mechanically reconfigurable reflector (MRR) can adapt its radiation pattern by reshaping the reflector, and cover several different areas during lifetime.

In past few decades, many investigations have been performed in MRR research [1–18]. As a pioneer, Prof. P.J.B. Clarricoats [1–4] investigated the reshaping of a metal tricot mesh with a number of actuators distributed over the surface. Leri Datashvili etc. [16, 17] designed a reflector morphing skin using flexible fiber composites, which have good mechanical and radio frequency performances, and a prototype driven by 19 actuators was fabricated, see Fig. 1(a). Shubao Shao with coworkers [18] employed a sandwich structure material to fabricate a reflector surface. That sandwich structure is composed of three layers: BWF CFRS layer with orthogonally woven fibers on the top and bottom, aluminum honeycomb as the core layer, as shown in Fig. 1(b).

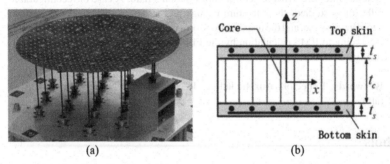

(a)   (b)

Fig. 1. The morphing skin of MRR (a) reflector surface fabricated by CFRS (b) reflector surface fabricated by aluminum honeycomb

The previous researches of mechanical properties needed for the reconfigurable surface are most about tensile modulus, shear modulus and bending stiffness. As Poisson's ratio is unchangeable for a certain material, most of which are positive, researchers have little interest in the investigation of the effect of Poisson's ratio on surface reshaping. During the last few years, so-called "designer materials" with arbitrarily complex nano/micro-architecture have attracted increasing attention to the concept of mechanical metamaterials [19]. Owing to their rationally designed nano/micro-architecture, mechanical metamaterials exhibit unusual properties at the macro-scale, such as various Poisson's ratio, including negative Poisson's ratio (NPR).

In the present paper, the stress analyses on reflector reshaping is performed, and the relationship between the Poisson's ratio and main principal stress on reflector surface during reshaping is investigated. In the situation of large curvature deformation, material with negative Poisson's ratio has advantage for its small stress, and a NPR lattice skin is introduced to form the reflector surface.

2   Mechanical Investigation of Surface Reshaping

The reflector reshaping is achieved by the actuation of numbers of actuator under the surface. An elastic out-of-plane deformation from original shape is achieved through a combination of a change in curvature and in-plane strain. And the stress is found by the superposition of bending (σ_B) and membrane (in plane) stress (σ_M). According to the theory of elastic mechanics, the following formulas are used to calculate the stress of the morphing skin in the x, y direction and shear stress.

$$\sigma_x = \frac{E}{1 - v^2}(\varepsilon_x + v\varepsilon_y), \sigma_y = \frac{E}{1 - v^2}(\varepsilon_y + v\varepsilon_x), \tau_{xy} = \frac{E}{2(1 + v)}\gamma_{xy} \qquad (1)$$

2.1   Stress Analysis of Bending Deformation Dominated

For the case of a reflector surface, K_x and K_y are vary with position or with actuation. And $K_x \times K_y$ may be either positive which appears a bowl shaped surface, or negative, which corresponds to a saddle shape. In the situation of bending deformation dominated, the membrane stress is neglected, and consider the principal curvature directions as x and y dimension. Thus, the major principal stress is the maximum of σ_{Bx} and σ_{By}. The bending strain on surface in two direction can be calculated by the change of curvature (K_x, K_y) and thickness t:

$$\varepsilon_{Bx} = \delta K_x \times \frac{t}{2}, \varepsilon_{By} = \delta K_y \times \frac{t}{2} \qquad (2)$$

So the bending stress in x and y dimension are:

$$\sigma_{Bx} = \frac{Et}{2(1 - v^2)}(\delta K_x + v\delta K_y), \sigma_{By} = \frac{Et}{2(1 - v^2)}(\delta K_y + v\delta K_x) \qquad (3)$$

In order to investigate the effect of Poisson's ratio on stress, Assume:

$$\beta = \frac{\delta K_y}{\delta K_x}, \delta K_x \neq 0 \qquad (4)$$

And Eq. (3) can be written as:

$$\sigma_{Bx} = \frac{Et\delta K_x}{2}\frac{1 + \beta v}{(1 - v^2)}, \sigma_{By} = \frac{\alpha\delta K_x}{2}\frac{\beta + v}{(1 - v^2)} \qquad (5)$$

The maximum principal bending stress:

$$\sigma_B = \max\left\{\frac{Et\delta K_x}{2}\frac{1 + \beta v}{1 - v^2}, \frac{Et\delta K_x}{2}\frac{\beta + v}{1 - v^2}\right\} \qquad (6)$$

$Et\delta K_x/2$ can be considered as a proportionality coefficient K, and the investigation on the relationship between the bending stress with Poisson's ratio come down to $\frac{1+\beta v}{(1-v^2)}$

and $\frac{\beta+v}{(1-v^2)}$. As the domain of Poisson's ratio is $(-1, 1)$, σ_B can also be expressed as a piecewise function:

$$\sigma_B = \begin{cases} K\frac{-\beta-v}{1-v^2}, & \beta \in (-\infty, -1) \\ K\frac{1+\beta v}{1-v^2}, & \beta \in [-1, 1] \\ K\frac{\beta+v}{1-v^2}, & \beta \in (1, +\infty) \end{cases} \tag{7}$$

In order to find the maximum bending stress at vary Poisson's ratio, derivative of σ_B is calculated as followed:

$$\frac{d\sigma_B}{dv} = \begin{cases} K\frac{-v^2-2\beta v-1}{(1-v^2)^2}, & \beta \in (-\infty, -1) \\ K\frac{\beta v^2+2v+\beta}{(1-v^2)^2}, & \beta \in [-1, 1] \\ K\frac{v^2+2\beta v+1}{(1-v^2)^2}, & \beta \in (1, +\infty) \end{cases} \tag{8}$$

Assuming $\frac{d\sigma_B}{dv} = 0$, the value of v corresponding to the minimum bending stress (short for "v_{Bmin}") is followed. And when $\beta = \pm 1$, the bending stress is monotonic about Poisson's ratio. Taking the domain of v into consideration, the value over the range of $(-1, 1)$ should be excluded.

$$v_{B\min} = \begin{cases} -\beta - \sqrt{\beta^2 - 1} > 0, & \beta \in (-\infty, -1) \\ \frac{-1+\sqrt{1-\beta^2}}{\beta} > 0, & \beta \in (-1, 0) \\ 0, & \beta = 0 \\ \frac{-1+\sqrt{1-\beta^2}}{\beta} < 0, & \beta \in (0, 1) \\ -\beta + \sqrt{\beta^2 - 1} < 0, & \beta \in (1, +\infty) \end{cases} \tag{9}$$

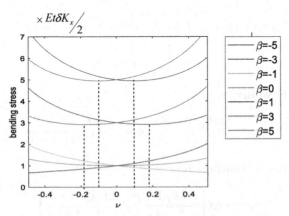

Fig. 2. The curves of bending stress vs. Poisson's ratio

According to the above analysis, when $\beta < 0, v_{Bmin}$ is positive. On the opposite, v_{Bmin} is negative. And v_{Bmin} is zero when $\beta = 0$. To verify the above conclusions, the discrete values of β investigated are given below:

$$\beta = \{-5, -3, -1, 0, 13, 5\}$$

The curves of bending stress vs. Poisson's ratio at different β are shown in Fig. 2. When $\beta = -5, -3, 0, 3, 5$, v_{Bmin} are $0.101, 0.172, 0, -0.172, -0.101$ respectively. When $\beta = 1$, the bending stress is increase with v, and when $\beta = -1$, the bending stress decreases with v.

2.2 Membrane Stress vs. Poisson's Ratio

The membrane stress in x and y dimension are as followed:

$$\sigma_{Mx} = \frac{E}{1 - v^2}(\varepsilon_{Mx} + v\varepsilon_{My}), \sigma_{My} = \frac{E}{1 - v^2}(\varepsilon_{My} + v\varepsilon_{Mx}) \tag{10}$$

The main principal stress for the situation of only in-plane deformation is

$$\sigma_M = \max(\sigma_{Mx}, \sigma_{My}) \tag{11}$$

Similarly, as the investigation of bending stress, when $\varepsilon_{xm}/\varepsilon_{ym} < 0, \neq -1$, the Poisson's ratio corresponding to the minimum membrane stress (short for "v_{Mmin}") is positive, when $\varepsilon_{xm}/\varepsilon_{ym} > 0, \neq 1$, v_{Mmin} is negative. When $\varepsilon_{xm}/\varepsilon_{ym} = \pm 1$, the membrane stress is monotonic increase or decrease with Poisson's ratio. And if one of $\varepsilon_{xm}, \varepsilon_{ym}$ is zero, the Poisson's ratio corresponding to the minimum membrane stress is zero.

2.3 Stress Analysis of Reflector Reshaping

The deformation of reflector surface under point actuation mostly including bending and extension. Total strain in x and y dimension combined bending strain and membrane strain can be add together in these form:

$$\varepsilon_x = \varepsilon_{Bx} + \varepsilon_{Mx}, \varepsilon_y = \varepsilon_{By} + \varepsilon_{My} \tag{12}$$

Total stress in x, y dimension and shear stress can be calculated by following matrix:

$$\begin{bmatrix} \sigma_x \\ \sigma_y \\ \tau_{xy} \end{bmatrix} = \begin{bmatrix} \frac{E}{1-v^2} & \frac{vE}{1-v^2} & 0 \\ \frac{vE}{1-v^2} & \frac{E}{1-v^2} & 0 \\ 0 & 0 & \frac{E}{2(1+v)} \end{bmatrix} \begin{bmatrix} \varepsilon_{Mx} + \frac{t}{2}\delta k_x \\ \varepsilon_{My} + \frac{t}{2}\delta k_y \\ \gamma_{xy} \end{bmatrix} \tag{13}$$

Main principal stress is obtained by bring the value of σ_x, σ_y and τ_{xy} into Eq. (14):

$$\sigma_1 = \frac{1}{2}|\sigma_x + \sigma_y| + \frac{1}{2}\sqrt{(\sigma_x - \sigma_y)^2 + 4\tau_{xy}^2} \tag{14}$$

3   Geometry Analysis of Reflector Surfaces

The reflector prototype presented in this paper has two shapes which are corresponding to two different coverage area (area 1, area 2), as shown in Fig. 3. The curvature in x and y dimension and Gaussian curvature are calculated by Eq. (15). The curvature change of surface reshaping from shape 1 to shape 2 are shown in Fig. 4.

$$K_x(x, y) = \frac{\partial^2 z/\partial x^2}{\left(1 + (\frac{\partial z}{\partial x})^2\right)^{3/2}}, K_y(x, y) = \frac{\partial^2 z/\partial y^2}{\left(1 + (\frac{\partial z}{\partial y})^2\right)^{3/2}}, Kg = K_x \times K_y \qquad (15)$$

Original shape:

Target shape:

Fig. 3. Reflector surface

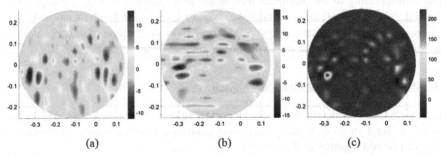

(a)   (b)   (c)

Fig. 4. Curvature calculation (a) change of K_x (b) change of K_y (c) change of K_g

In order to reduce the stress at rim, the boundary defined for the proposed reconfigurable reflector is partially fixed rim with free radial displacements only. The maximum stress occurs at point (short for "point P") of the maximum Gaussian curvature change because the strain at this point is maximum. Reducing the maximum stress can increase the allowance provided by material yield stress.

As can be seen from the calculation results of Fig. 4, the curvature change at point P in x and y dimension is $\delta K_x = 12.67$, $\delta K_y = 17.36$.

Matlab is used to calculate the membrane and shear strain according to the surface geometry deformation. The results are shown in Fig. 5. The membrane strain in x and y dimension of point P are 0.022 and −0.005 respectively, and the shear strain of point P is 0.017. As can be seen from Eq. (3), the bending stress scales with thickness. Assume

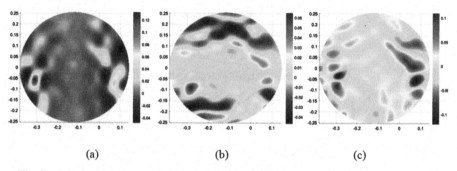

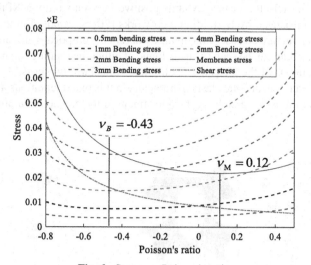

Fig. 5. Calculated strain (a) strain in x dimension (b) strain in y dimension (c) shear strain

the thickness is 0.5 mm, 1 mm, 2 mm, 3 mm, 4 mm, the bending stress, the membrane stress and the shear stress are calculated by Eq. (1), (6), (11), the results is shown in Fig. 6. The Poisson's ratio corresponding to the minimum bending stress and membrane stress are -0.43, 0.12 respectively, and the shear stress decreases with Poisson's ratio increasing. The main principal stress calculated by Eq. (14) is shown in Fig. 7. When the thickness is 0.5 mm, the Poission's ratio corresponding to the minimum main principal stress (short for "ν_{min}") is 0.06, which is closer to ν_{Mmin}, because the membrane stress is dominated. Along with the thickness increasing, the bending stress play more role in the main principle stress, and ν_{min} moves in negative direction and closer to $\nu_{B\,min}$.

Fig. 6. Stress vs. Poisson's ratio

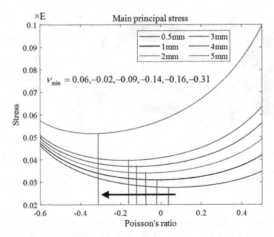

Fig. 7. Main principal stress vs. Poisson's ratio

4 Optimal Design of Reflector Surface

4.1 Lattice Based Structure

For the reflector surface with thickness exceed 1 mm, Poisson's ratio ν is negative for minimum stress according to Fig. 7. Traditional honey comb structure material is common used in reflector surface exhibits positive Poisson's ratio, a NPR lattice with elliptic voids introduced by Bertoldi and co-works [20] is proposed in this article, see Fig. 8. The 2D structure consist of the matrix wherein the elliptical voids are arranged in periodic manner, and can be seen as an analogue of either the rotating square model with square-like elements of the matrix. The elliptical voids have two possible orientations, horizontal and vertical, and the elastic parameters in this two orientations are same [21]. The Poisson's ratio is designable by the adjustment of the geometry parameters a, b, g.

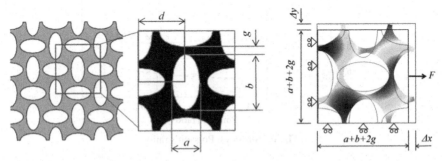

Fig. 8. The elliptic voids structure NPR material   **Fig. 9.** FE simulation model

During the optimization, the transmission loss should be taken into consideration, the diameter should be smaller than $\lambda/8$. As an example, for the X band, the wave length is about 30 mm, the parameter a determined as 3.65 mm. To lower the tensile stiffness, the parameter g should be small, but that will increase the manufacturing difficulty and cost. To compromise this paradox, g is supposed as 0.3 mm. A FE model, as shown in Fig. 9, is used to simulate the deformation and the effective Poisson's ratio v_{eff} and the ratio of effective Young's modulus (E_{eff}) to the Young's modulus of material (E_{mat}) are calculated by the following way:

$$v_{eff} = -\frac{\varepsilon_y}{\varepsilon_x} = -\frac{\Delta y/(a+b+2g)}{\Delta x/(a+b+2g)} = -\frac{\Delta y}{\Delta x} \tag{16}$$

The material investigated is carbon fibre reinforced silicon (CFRS), for example, Young's modulus $E_{mat} = 500$ MPa, Poisson's ratio is 0.4. The simulation results is shown in Fig. 10. With the ratio of b/a increasing, the effective Poisson's ratio v_{eff} increase.

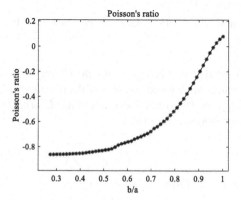

Fig. 10. Effective Poisson's ratio vs. b/a

4.2 Thickness of Reflector Surface

The thickness of reflector surface is an important parameters with an influence over out-of-plane deformation. As can be seen from Eq. (17), the bending stiffness is a cubic relationship of thickness. As mentioned above, the bending stiffness to tensile modulus should be large enough for achieving smooth reshaping. For this purpose, increasing thickness is an effective way which can significantly increase the bending stiffness, but it will lead the final displacement to lower local change in curvature and a more overall displacement.

$$D = \frac{Et^3}{12(1-v^2)} \tag{17}$$

To optimize the thickness, the discrete values of the variables investigated are given below:

$$t = \{0.5\,\text{mm}, 1\,\text{mm}, 2\,\text{mm}, 3\,\text{mm}, 4\,\text{mm}, 5\,\text{mm}\}$$

According to stress analysis for vary thickness above in Fig. 7, the Poisson's ratio corresponding to the minimum stress can be established. As the parameter a = 3.65 mm and g = 0.3 mm of lattice are determined, the value of parameter b can be determined according to the simulation results in Fig. 10. So, the parameters list as followed (Table 1):

Table 1. The parameters of reflector surface

t(mm)	0.5	1	2	3	4	5
v_{eff}	0.06	−0.02	−0.09	−0.14	−0.16	−0.31
a(mm)	3.65					
g(mm)	0.3					
b(mm)	3.6	3.45	3.4	3.35	3.3	3.16

ANSYS is used to simulate the reconfigured reflector surface by the actuation of 225 actuators. And the actuators are located at points of the maximum Gaussian curvature in each even distribution region. The surface accuracy of the deformed reflector is evaluated by the root mean square (RMS) error (18):

$$RMS = \sqrt{\frac{\sum\limits_{i=1}^{n} \left(Z_i^* - Z_i\right)^2}{n}} \qquad (18)$$

RMS of reconfigured reflectors with vary thickness is calculated and shown in Fig. 11, RMS decreases with thickness increasing firstly, and when thickness beyond a value (t = 5 mm in this case), the RMS decrease, see Fig. 11(f). As the actuation force accompanying is also increasing significantly when thickness increase. Thus, at the constraint of the reconfiguration accuracy, small thickness is desirable.

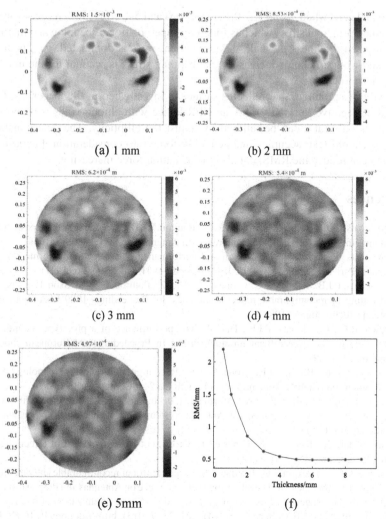

Fig. 11. Error distribution between simulated reconfigured shape with thickness and original shape (a) 1 mm (b) 2 mm (c) 3 mm (d) 4 mm (e) 5 mm (f) Curve of RMS vs. thickness

5   Conclusion

The investigation of Poisson's ratio for the material of mechanical reconfigurable reflector surface is given in this paper. The reshaping deformation is achieved though a combination of a change of curvature and in-plane strain. Based on the stress analysis, Poisson's ratio corresponding to the minimum stress are related to the thickness, the ratio of curvature change and the ratio of in-plane strains in two principal dimensions. When curvature change in same direction (i.e. $\delta Ky/\delta Kx > 0$), the Poisson's ratio corresponding to the minimum bending stress (v_{Bmin}) is negative, otherwise, $vBmin$ is positive, and when $\delta Kx \times \delta Ky = 0$, v_{Bmin} is zero. Similarly, for the membrane stress, v_{Mmin} are negative, positive and zero when $\delta\varepsilon_y/\delta\varepsilon_x > 0$, $\delta\varepsilon_y/\delta\varepsilon_x < 0$ and $\delta\varepsilon_y \times \delta\varepsilon_x = 0$ respectively. And for the main principal stress which is found by superposition of bending and membrane stresses, when thickness is small, the membrane stress is dominated, v_{min} is close to v_{Mmin}, and while thickness increasing, the bending stress is account for the main part, v_{min} is close to v_{Bmin}.

A case of reconfigurable reflector surface is investigated in this paper. As for the demand of negative Poisson's ratio, an NPR lattice with elliptic voids is proposed, of which Poisson's ratio can be tailored by adjust the ellipticity. The determination of thickness should take accuracy and actuation force into consideration. Increasing the thickness can reduce the RMS, but also the actuation force increasing.

References

1. Clarricoats, P.J.B., Hai, Z., Brown, R.C., Poulton, G.T.: A reconfigurable mesh reflector antenna. In: Sixth International Conference on Antennas and Propagation (1989)
2. Clarricoats, P.J.B., Zhou, H.: Design and performance of a reconfigurable mesh reflector antenna. Part 1: antenna design. In: IEE Proceedings-H, vol. 13, no. 6 (1991)
3. Clarricoats, P.J.B., Brown, R.C., Crone, G.E., Hai, Z., Poulton, G.T., Willson, P.J.: The design and testing of reconfigurable reflector antennas. In: Proceedings 1989 ESA Workshop on Antenna Technologies (1989)
4. Brown, R.C., Clarricoats, P.J.B., Hai, Z.: The performance of a prototype reconfigurable mesh reflector for spacecraft antenna applications. In: Proceedings 19th European Microwave Conference (1989)
5. Pontoppidan, K., Boisset, J.P., Crone, G.A.E.: Reconfigurable reflector technology. In: IEE Colloquium on Satellite Antenna Technology in the 21st Century (1991)
6. Pontoppidan, K.: Light-weight reconformable reflector antenna dish. In: Proceedings of 28th ESA Antenna Workshop on Space Antenna Systems and Technologies, Noordwijk, The Netherlands (2005)
7. Pontoppidan, K., Boisset, J.P., Ginestet, P., Crone, G.: Design and test of a reconfigurable reflector antenna. JINA, Nice (1992)
8. Rodrigues, G., Angevain, J.C., Santiago, J.: Shape control of reconfigurable antenna reflector: concepts and strategies. In: The 8th European Conference on Antennas and Propagation (2014)
9. Rodrigues, G., Angevain, J.-C., Santiago-Prowald, J.: Shape optimization of reconfigurable antenna reflectors. CEAS Space J. 5(3–4), 221–232 (2013). https://doi.org/10.1007/s12567-013-0038-5
10. Sakamoto, H., et al.: Shape control experiment of space reconfigurable reflector using antenna reception power. In: 3rd AIAA Spacecraft Structures Conf. AIAA 2016-0703 (2016)
11. Tanaka, H., Natori, M.C.: Study on a reconfigurable antenna system consisting of cable networks. Trans. Japan Soc. Aeronaut. Space Sci. 50, 48–55 (2007)
12. Susheel, C.K., Kumar, R., Chauhan, V.S.: An investigation into shape and vibration control of space antenna reflectors. Smart Mater. Struct. 25, 125018 (2016)
13. Viskum, H., Pontopiddan, K., Clarricoats, P.J.B., Crone, G.A.E.: Coverage flexibility by means of a reconformable subreflector. In: Proceedings of the APS International Symposium, pp. 13–18. IEEE (1997)
14. Washington, G.N., Angelino, M., Yoon, H.T., Theunissen, W.H.: Design, modeling, and optimization of mechanically reconfigurable aperture antennas. IEEE Trans. Antennas Propag. 50(5), 628–637 (2002)
15. Cappellin, C., Pontoppidan, K.: Feasibility study and sensitivity analysis for a reconfigurable dual reflector antenna. In: Proceedings of the European CAP, Berlin (2009)

16. Datashvili, L.S., Baier, H., Wei, B.: Mechanical Investigations of in-space-Reconfigurable Reflecting Surfaces. https://www.researchgate.net/publication/228921635
17. Datashvili, L.S., Baier, H., Wei, B., et al.: Design of a morphing skin using flexible fiber composites for space-reconfigurable reflectors. In: AIAA/ASME/ASCE/AHS/ASC Structures, Structural Dynamics, & Materials Conference (2013)
18. Shubao, S., Siyang, S., Minglong, X., et al.: Mechanically reconfigurable reflector for future smart space antenna application. Smart Mater. Struct. **27**, 095014 (2018)
19. Kolken, H.M., Zadpoor, A.A.: Auxetic mechanical metamaterials. RSC. Adv. **7**, 5111–5129 (2017)
20. Bertoldi, K., Reis, P.M., Willshaw, S., Mullin, T.: Negative Poisson's ratio behavior induced by an elastic instability. Adv. Mater. **22**(3), 361–366 (2010)
21. Pozniak, A.A., Wojciechowski, K.W., Grima, J.N., et al.: Planar auxeticity from elliptic inclusions. Compos. Part B Eng. **94**(6), 379–388 (2016)

The Method of User Information Fusion Oriented to Manufacturing Service Value Net

Wenjia Wu, Lin Shi, Jiaojiao Xiao, and Changyou Zhang(✉)

Laboratory of Parallel Software and Computational Science, Institute of Software,
Chinese Academy of Sciences, Beijing, People's Republic of China
changyou@iscas.ac.cn

Abstract. Based on the business system users of manufacturing enterprises, building a value net platform including multi-enterprise and social service resources needs to integrate the user information and explore more value-added services. This paper proposed a method of user information fusion for manufacturing service value net. First, based on the request of the multi-party value net fusion, we researched the user information representation method and designed the data structure. And then, considering the protection of sensitive fields, the study of privacy fields protection methods based on homomorphic encryption methods should be carried out to ensure the utility of data. The last step was researching the multi-platform user information transmission based on the RESTful interface specification. The value net case system shows that the user information fusion method is realizable and can ensure the system users' security.

Keywords: Value net · User information · Fusion · Interface specification

1 Introduction

In the manufacturing service value net, user information is scattered in many different enterprise application systems, while there are many differences in the user information structure of each system [1], the heterogeneous user information is difficult to be used by the value net platform to develop value-added services for system users. There will be many difficulties in business innovation. So the work of integrating enterprise user information is particularly important. The manufacturing service value net platform integrates the user information of multiple enterprises so that it makes the management of users unified and skips the cold start phase. After logging in, platform users can use the value-added services provided by the platform to promote business innovation and improve the user experience.

This paper proposed a user information fusion method based on a homomorphic encryption algorithm. The value net platform defines a standard user information format, and then the enterprise systems preprocesses the user information, in which the systems mainly carries out homomorphic encryption on private data. Finally, the systems integrates the non-private data and encrypted private data into the value net platform through

© Springer Nature Switzerland AG 2021
Q. Zu et al. (Eds.): HCC 2020, LNCS 12634, pp. 68–74, 2021.
https://doi.org/10.1007/978-3-030-70626-5_7

a standardized RESTful API. Through this method, the platform only integrates meaningful system user information fields, and thus users can log in to obtain value-added services, which ensures the confidentiality of privacy data in system user information.

2 Related Work

2.1 Data Fusion Research Work

The small data fusion of research users includes four levels: basic database, association-level fusion, feature-level fusion, and demand-level fusion. The efficient promotion of small data fusion and innovation can be achieved through such process as data collection—data association—feature extraction—knowledge needs [2]. However, the characteristics of this fusion data are limited, and the fusion data can only be used for data mining analysis. Users add noise into data and encrypt it. After receiving the scrambled ciphertext, the cloud server integrates the data and sends the final result to the sensing platform, and then the platform decrypts the data [3]. However, the sensing platform integrates homogeneous data, whose structure is different from that of user information, and does not need to process sensor data. The data from different sources may involve different vocabulary and modeling granularity, which makes data fusion difficult [4]. A method based on a data processing pipeline is proposed, which takes a set of equivalent entity identifiers as input, and provide RDF triples and triple chain groupings of entities described in multiple sources as output. However, user information has multiple fields and is not a representation of RDF triples. The telecom network [5] uses UDC and adopts a hierarchical structure, which extracts user information from application logic and stores it in a logically unique data warehouse (UDR), and thus allows the access to the data warehouse for core network and business layer entities. The user information warehouse is a functional entity, which is unique in the operator's network. However, this data fusion method only supports the fusion of homogeneous data and requires the support of third-party storage devices.

2.2 Homomorphic Encryption Algorithm

The data is processed without exposing the original data by adopting homomorphic encryption technology, so that deep and unlimited analysis can be performed on the encrypted information without affecting its confidentiality [8]. NTRU [9] (Number Theory Research Unit) is a fully homomorphic encryption algorithm, which can perform various encrypted operations (addition, subtraction, multiplication and division, polynomial evaluation, exponent, logarithm, trigonometric function).

2.3 RESTful Interface Specification

REST (Representational State Transfer) is a software architecture style that provides a series of restriction guidelines for better creating web services. The web service conforming to the REST architectural style is called Restful web service [6, 7]. The core of RESTful is to focus on resources, not methods or messages. RESTful usually uses URL to locate resources, and uses HTTP verbs (GET, POST, DELETE, PUT, etc.) to describe operations on resources.

3   A User Information Fusion Method Base on Homomorphic Encryption Algorithm

According to its own demand, the value net platform defines the user information format. The user information in the platform has both private data and not-private data. The not private data can be preserved directly, however only if the private data have been encrypted by the system can they be preserved.

After the RESTful API [6, 7] specification were defined, the private data and the not private data will be transmitted to the value net platform and be preserved via RESTful API.

The not private data can be used directly, for example being used as login accounts; The value net platform applies the homomorphic computation to the private data, so the private data can be obtained a computed result without decryption.

3.1   The Data Structure of User Information

The different systems has different user information structure, so the value net platform should define its own data structure of user information.

Suppose that the user information of system in manufacturing service value net has m fields, represented by Attr.

$$\text{Attr} = (attr_0, attr_1, \ldots Attr_{j-1}, attr_j, attr_{j+1} \ldots . Attr_m)$$

Among them, $attr_0$, $attr_1$,... $Attr_{j-1}$ represent for j not private data items, attr J, $attr_j$, $attr_{j+1}$.... $Attr_m$ represent for m-j private data items in the value net platform.

Suppose there are n fields in the user information of the value net platform, and the format of the user information is expressed in tuples as follows:

$$U = (attr_0, attr_1, \ldots Attr_{i-1}, attr_i, attr_{i+1} \ldots . Attr_n, u_1)$$

Among them, $attr_0$, $attr_1$,... $attr_{i-1}$ represent for i not private data items, and attr $_i$, $attr_{i+1}$.... $attr_n$ represent for n-i private data items. ($attr_0$, $attr_1$,... $attr_{i-1}$) is a subset of ($attr_0$, $attr_1$,... $attr_{j-1}$),and ($attr_i$, $attr_{i+1}$.... $attr_n$) is a subset of ($attr_j$, $attr_{j+1}$.... attr $_m$). u1 represent for the name of system.

3.2   The Preprocessing of System User Information

Homomorphic encryption technology can be used for private data fields, so that the private data can be processed without exposing the original data. The encrypted data can be operated the same with raw data without decryption. The encrypted information can be analyzed deeply and infinitely without affecting its confidentiality.

1) Generating keys by the function of KeyGen(params)
2) The $Enc(pk, m)$ function is a process of mainly encrypting private data by keys. It is assumed that each user' s private data is a character string.
3) The system transmit the private encrypted data and the not private raw data together to the value net platform. The data format of transmission is as follows:

$$U = (attr0, attr_1, \ldots attr_{i-1}, Enc[attr_i], Enc[attr_{i+1}] \ldots Enc[attr_n], u_1)$$

3.3   The Transmission of User Information Based on RESTful

The value net platform should define the unified RESTful API specification for many systems, and deploy the service on the server of enterprises. The server can collect the user information of systems, then integrate the data on the platform.

Design URL. The CRUD operation of URL should be defined, including GET(Read), POST(Create), PUT(Update), PATCH(Partly Update), DELETE(Delete).

Regulate the Status Code. The server must response to each request of client. The response includes HTTP status and data.

The Response of Server. The return data format of API should not be pure text, but a JSON object, because only in this way, the standard structured data can be returned. The Content-Type in HTTP header of server response should be set to application/json.

The value net platform obtained the corresponding data format through the RESTful API in the enterprise server, which complete the fusion operation of system user information.

When the value net platform uses the fused user information, the methods of using private data and not private data are different.

The Use of the not Private Parts. When dealing with not private parts of user information, the value net platform obtain the original data directly, and the not private parts are be operated fast.

The Use of the Private Parts. The private parts of system user information are preserved as encrypted text. When dealing with the private parts, the value net platform dos not decrypt them, otherwise the platform can obtain the raw data of the private parts.

The platform can calculate the private encrypted data homomorphically without decryption. Any calculation operation will be converted into the corresponding bit operation at the computer bottom layer.

$$f(Enc[\text{attr } i], Enc[\text{attr } i+1] \dots Enc[\text{attr } n]) \longrightarrow Enc[F(\text{attr } i, \text{attr } i+1 \dots \text{attr } n)]$$

The platform can get the results by homomorphism calculation, then decrypt the results by $ec(f_{1,L}, \dots, f_{N,L}, c)$ function.

4   Case of Fusing User Information for Value Net

According to the user information fusion needs of the actual project, the goal is to simplify the research for the project. Assume the situation for user information fusing: in a value net composed three manufacturing enterprise systems, platform offers contract management, client management and other value-add services etc. In addition, the platform needs to analyze the data from user information. Therefore, this platform needs to fuse all three systems user information, however, each system has different structure of user information, furthermore some fields in the user information are private and cannot be directly transmitted to the value net platform.

Based on the above situation this chapter design a case of fusing of user information. This article purposed a method for user information fusion, and analyzed the fusion results.

4.1 Case Illustration

There are three systems from different companies, namely System1, System2, and System3. The user information data of the systems are different.

The user information data structure of System1 is: (user name, areas of expertise, title, hometown).

The user information data structure of System2 is: (user name, title).

The user information structure of System3 is: (username, areas of expertise, native place).

The user information structure of the platform is: (user name information, areas of expertise, system name). Among them, the not private data includes user name, system name, the private data is the areas of expertise.

4.2 Process of Fusing User Information

The steps of fusing user information are as follows (Fig. 1):

Step 1: preprocessing user information

The areas of expertise is the private data of the system, which needs to be preprocessed before transmission, that is, the areas of expertise is encrypted using the homomorphic encryption algorithm.

Step 2: Define the RESTful API specification

1) Design URL

Design a RESTful URL specification, which specifies the service request format of GET, POST, PUT, PATCH and DELET.

2) Specify the status code

200 EV_SUCCESS requested successfully, 201 EV_CREATED created successfully, 202 EV_ACCEPTED updated successfully, 401 EV_UNAUTHORIZED not authorized, 403 FORBIDDEN access deny, 404 NOTFOUND requested resource does not exist

3) Develop the data format returned by the server

The value net platform obtains user information in JSON format by utilizing the RESTful API of the system, which is directly stored in the designed database tables.

Step 3: System1, System2, and System3 transmit the non-private fields and encrypted private fields together to the value net platform.

Step 4: Value net platform integrate user information

The value net platform integrates user information transmitted by System1, System2, and System3.

Step 5: Fill in the missing fields of user information

Because System2's user information does not have field information in the areas of expertise, when System2 users log in the platform, they will be prompted to complete the field of expertise information.

4.3 The Use of Fusion Results

After logging on the platform, system users can access the services. In addition, the platform can calculate the sensitive data of system users homomorphically without

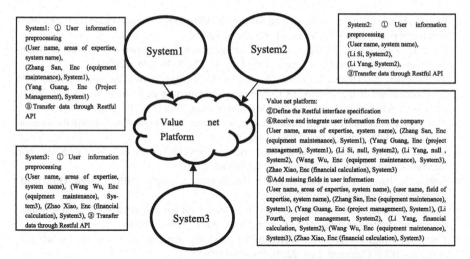

Fig. 1. The value net fusing user information of three systems

decryption. For example, it can count the number of users who are good at equipment maintenance. The calculation process is as follows:

f(Enc(equipment maintenance), Enc(project management), project management, Financial calculation, Enc(equipment maintenance), Enc(Financial calculation), Enc(English Translation)).

—> Enc[f(equipment maintenance, project management, project management, Financial calculation, equipment maintenance, Financial calculation, English Translation)] —> Enc(2).

Finally, we calculated the results of the value net platform through homomorphism through $Dec(f_{1,L}, f_{2,L}, f_{3,L}, n)$.

5 Summary

Our paper proposed a data format of standardized user information based on homomorphic encryption algorithm aiming at the problem of user information fusion in manufacturing value net. On the basis of ensuring the security and privacy of enterprise user information, it synchronizes the information to the platform. When the platform deals with the non-privacy parts of the user information, the interaction speed of data storage is fast. Finally, we proved the realization of this method through the value net case of user information fusion.

Acknowledgements. This paper is supported by the National Key Research and Development Program of China (2018YFB1701403), the Natural Science Foundation of China (61672508).

References

1. Gao, C., Sun, H.M.: Theoretical analysis of value creation of intelligent manufacturing enterprises based on value net. Modernization Manag. **40**(229(03)), 66–70 (2020)

2. Li, L.R., Deng, Z.H.: Research on small data integration of research users under the background of internet plus . Books Inf. Work **60**(06), 58–63 (2016)
3. Long, H., Zhang, S.K., Zhang, L.: Data fusion method based on privacy protection in swarm intelligence aware networks. Comput. Eng. Des. **41**(12), 3346–3352 (2020)
4. Thalhammer, A., Thoma, S., Harth, A., Studer, R.: Entity-centric data fusion on the web. In: Proceedings of the 28th ACM Conference on Hypertext and Social Media (HT 2017), pp. 25–34. Association for Computing Machinery, New York (2017)
5. Zhu, B., Fu, G., Zhu, A.H., Li, Y.B., Wu, Q.: Development strategy of user information fusion technology. Telecommun. Sci. **27**(S1), 302–305 (2011)
6. Ma, S.P., Chen, Y.J., Syu, Y., et al.: Test-oriented RESTful service discovery with semantic interface compatibility. IEEE Trans. Serv. Comput. 1 (2018)
7. Segura, S., Parejo, J.A., Troya, J., et al.: Metamorphic testing of RESTful web APIs. IEEE Trans. Softw. Eng. **44**(11), 1083–1099 (2018)
8. Abbas, A., Hidayet, A., Selcuk, U.A., et al.: A survey on homomorphic encryption schemes: theory and implementation. ACM Comput. Surv. **51**(4), 1–35 (2017)
9. Chen, R., Peng, D.: A novel NTRU-based handover authentication scheme for wireless networks. IEEE Commun. Lett. **22**(3), 586–589 (2018)

Image Fusion Method for Transformer Substation Based on NSCT and Visual Saliency

Fang Zhang[✉], Xin Dong, Minghui Liu, and Chengchang Liu

Dandong Power Supply of Liaoning Electric Power Co. Ltd. of State Grid, Dandong, China
ddjyuxin@yeah.net

Abstract. To solve the problems of the existing infrared and visible image fusion algorithms, such as the decrease of the contrast of the fusion image, the lack of the visual target, and the lack of the detail texture, an image fusion algorithm based on the visual saliency is proposed. Firstly, NSCT is used to decompose the two source images to obtain the corresponding low-frequency sub-band and a series of high-frequency sub-band; secondly, an improved FT algorithm is used to detect the visual saliency region of the low-frequency sub-band of different sources; Thirdly, according to the size of visual saliency, different weights are assigned to low-frequency sub-band of different sources, based on which fusion is carried out; fourthly, the high-frequency sub-band weight map is obtained by screening methods, and then the weighted image is used for fusion; finally, the final fusion image is obtained by inverse NSCT transform. The experimental results show that our method has a better visual effect and higher objective indicators than other classical image fusion methods.

Keywords: Image fusion · Visual saliency · Non-subsampled Contourlet transform · Infrared image · Visible image

1 Introduction

People's daily production and life are inseparable from the supply of electrical energy, so it is necessary to ensure reliable and stable operation of substation equipment. Among them, the substation equipment faults and abnormal monitoring is of great significance. According to statistics, temperature abnormality is one of the frequent faults of substation equipment [1], so the temperature detection of substation equipment is an important mission. At present, the common technique is to use an infrared (IR) imager to generate IR images to obtain the temperature information of the substation equipment, different temperatures show different colors, through the color information changes to determine the equipment's abnormal fever site. However, the IR image can only reflect the device's temperature change information, imaging fuzzy, gradient change slowly, and does not have the texture of the substation equipment information. When monitoring substation equipment, poses a huge obstacle to the identification of equipment types and subsequent processing. In contrast, visible (VIS) images can reflect the details and textures of a scene under certain illumination. Therefore, the fusion of IR images and VIS images can fuse

Q. Zu et al. (Eds.): HCC 2020, LNCS 12634, pp. 75–83, 2021.
https://doi.org/10.1007/978-3-030-70626-5_8

the information of the two images, i.e., it can retain the IR thermal characteristics of the substation equipment in the IR images, and also retain the rich texture detail information of the substation equipment to achieve complementary information, which is conducive to the subsequent better identification of the type of monitored substation equipment and the specific parts of the fault, which can improve the safety of the electrical equipment, and can improve the safety of the substation equipment. It is of great importance to operate reliably and reduce the cost of fault maintenance.

Currently, there are three main levels of image fusion methods, namely, pixel level, feature level, and decision level. Pixel-level fusion is widely used because it preserves the detailed texture information of the source image as much as possible and has a good fusion effect. The most important of the pixel-level fusion methods are the spatial domain decomposition-based and transform domain-based fusion methods [2]. Among them, the transform domain-based non-subsampled contour wave transform (NSCT) fusion method has developed rapidly in recent years. It has the properties of multi-resolution, multi-directional, anisotropy, translation invariance, and is an ultra-complete multiscale fusion method. In 2013, Song et al. [3] applied a variable-scale transient chaotic neural network-based image fusion method to the fusion of VIS and IR images of substation devices. Due to the insensitivity of the human eye to changes in grayscale information, the human eye cannot perceive changes in temperature information in the fused image, and cannot play a good role in image fusion in practical applications. Zhao et al. [4] proposed a VIS and IR image fusion algorithm based on the hidden Markov tree model of the contour wave transform domain. The fusion method is smoother and more detailed in subjective vision, and the edge information is more obvious. However, the information from the visible image is fused in the background region outside the substation equipment, which changes the overall temperature information of the fused image. In 2018, Shi et al. [5] proposed an improved VIS and IR image fusion method combining IHS transform and contour wave transform, which fuses the substation data, although the detailed texture information of the scene is better preserved, the IR information is poorly preserved Besides because it processes the IR color image into a grayscale image, it loses the color information that is more sensitive to human eye vision, making the fused image lose its application meaning in real life.

According to the shortcomings of the above algorithm, this paper proposes an image fusion method for substation equipment based on NSCT and visual saliency from the subjective visual perception of human beings. The visual saliency technique is applied to the image fusion of substation equipment, and the low-frequency sub-band are fused according to the visually salient regions of VIS and IR images, while the low-frequency information is given a certain weight to ensure that the detailed texture information is not lost. The high-frequency sub-band are fused using the maximum absolute value rule. The experimental results show that the algorithm is capable of preserving both the significant information and detailed information in the VIS and IR images, and has better subjective visual effects and higher objective evaluation indexes.

2 Basic Principle

2.1 YUV Color Space

YUV is divided into three components, where "Y" represents the brightness information, which is the grayscale value, while "U" and "V" are used to describe the color and saturation of the image. Information that is used to specify the color of a pixel. The color information of the IR image reflects the temperature changes of the equipment and scene, while the color information of the VIS light assists people to understand the objects in the real scene. When the two are fused, the color information in the VIS image is relatively less important, and if the color information in the VIS image is involved in the image fusion process, it may affect the IR thermal radiation information. Therefore, considering all aspects, this paper uses this coding algorithm to fuse IR and VIS images using only brightness information.

2.2 NSCT Transform

The NSCT transform is based on a non-subsampled pyramid (NSP) and a non-subsampled directional filter bank (NSDFB). As shown in Fig. 1, the NSP is first used to perform a tower-type multiscale decomposition of the source image to obtain low-frequency and high-frequency sub-band, and the NSDFB is used to filter the high-frequency sub-band in multiple directions, and each decomposition will yield sub-band in both horizontal and vertical directions. The low-frequency sub-band can be further decomposed in multiple scales, and finally, a low-frequency sub-band and a series of high-frequency sub-band with multiple directions can be obtained by NSCT transformation.

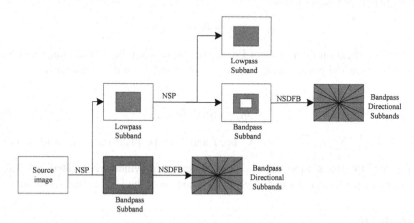

Fig. 1. NSCT decomposition diagram

2.3 FT Algorithm

Visual Saliency Detection (VSD) refers to the extraction of significant regions of an image by an intelligent algorithm that simulates human visual features. Radhakrishna Achantay et al. [6] proposed the FT visual saliency detection algorithm in 2009, which uses Gaussian differential filtering to calculate and therefore the saliency plot is given by Eq. 1:

$$S(x, y) = \|I_\mu - I_{\text{whc}}(x, y)\| \qquad (1)$$

where I_μ is the mean pixel value of the input I, I_{whc} is the blurred image by the Gaussian differential filter, $I_{\text{whc}}(x, y)$ is the pixel value at (x, y). Calculating the Euclidean distance between all pixel values and the mean pixel value of the input I will give a saliency map of the image.

2.4 Guided Filter

The FT algorithm uses Gaussian filter to blur the input image, and the kernel functions used in Gaussian filter are independent for the image to be processed, i.e. the same blurring operation is used for any input image. Therefore, Gaussian filter does not preserve the edge information of the image well, and there is no good visual saliency in the final obtained saliency map. Therefore, this paper improves on the FT algorithm by using a guide filter instead of Gaussian filter for blurring, which can preserve the edge information of the image while smoothing the image. The final output image is similar to the input image, but the edges of the image are similar to the guide image. The main principle of bootstrap filtering is to assume that there is a local linear relationship between the output of this bootstrap filter function and the bootstrap image in a pixel-centered window, which can be written as:

$$q_i = a_k p_i + b_k, \forall i \in \omega_k \qquad (2)$$

Taking the gradient on both sides of Eq. 2, we can see that the output image will have edges only when the guide image has edges. Deriving from the constraints, we know that:

$$q_i = \overline{a}_i p_i + \overline{b}_i \qquad (3)$$

where $\overline{a}_i = \frac{1}{|\omega|} \sum_{k \in \omega_i} a_k$, $\overline{b}_i = \frac{1}{|\omega|} \sum_{k \in \omega_i} b_k$, i and k is the pixel index, a_k and b_k is the coefficient of the linear function when the center of the window is located at k, $|\omega|$ is the number of pixels in the window, ω_i is the window of the $i-th$ kernel function.

3 Methods

First of all, the NSCT transformation is performed on the IR image I_{IR} and VIS image I_{VIS} respectively to obtain a low-frequency sub-band C_{j0}^{IR}, C_{j0}^{VIS} and a series of high-frequency sub-band C_{jk}^{IR}, C_{jk}^{VIS} ($jk > j0$), using the above-mentioned improved FT algorithm to detect the significance of the low-frequency sub-band of the IR and VIS images respectively, in which the source image is used as a guide map to retain as much significant information in the low-frequency sub-band, resulting in a low-frequency fusion sub-band C_{j0}^F; high-frequency sub-band coefficients using a large absolute value of the fusion strategy, resulting in high-frequency fusion sub-band $C_{jk}^F(jk > j0)$; finally through the inverse NSCT conversion can be obtained the final fusion image.

The low-frequency sub-band of an image is a global approximation of the image and contains most of the information of the image. Therefore, it is important to ensure the necessary texture detail in the low-frequency sub-band and to highlight the information of human interest. The brightness variation information of an IR image reflects the temperature variation of the device, while the low-frequency information of a VIS image retains most of the texture information of the scene and the device. It is, therefore, necessary to choose a fusion method that preserves both the luminance of the IR and the texture of the VIS information. Most of the traditional image fusion methods based on multi-scale decomposition use the rule of average fusion for low-frequency sub-band fusion, which reduces the contrast of the fused image and loses energy. To address the above problem, a strategy is proposed to determine the fusion weights of the low-frequency portion using a visually significant map. The significant regions of the low-frequency sub-band of the infrared and visible images are obtained using the improved FT algorithm described above, as shown in Fig. 2, where (a) represents the visible image, (b) represents the y component in the YUV color space of the VIS image, (c) represents the significant regions of the visible image obtained by the original FT algorithm, and (d) represents the significant regions of the VIS image obtained by the improved FT algorithm. (e) denotes the improved low-frequency sub-band fusion map, (f) denotes the IR image, (g) denotes the y component in the YUV color space of the infrared image, (h) denotes the infrared significant map obtained by the original FT algorithm, (i) denotes the infrared significant map obtained by the improved FT algorithm, and (j) denotes the low-frequency sub-band fusion map before the improvement; the resulting significant map is normalized to the [0,1] interval as the low-frequency sub-band fusion weights so that the salient information of the IR and VIS corresponding low-frequency sub-band can be preserved. At the same time, to preserve the global information of the low-frequency sub-band, only the IR information can be preserved in the non-significant region, i.e. the background region belonging to the image acquisition of the substation, because no texture detail needs to be preserved. Then the low-frequency fusion strategy for IR and VIS light is as Eq. 4.

$$C_{j0}^F(x, y) = \omega_{IR}(x, y) * C_{j0}^{IR}(x, y) + \omega_{VI}(x, y) * C_{j0}^{VI}(x, y)$$
$$+(1 - \max(\omega_{IR}(x, y), \omega_{VI}(x, y))) * C_{j0}^{IR}(x, y) \tag{4}$$

where $\omega_{IR}(x, y)$, $\omega_{VI}(x, y)$ are the significant plots of the low-frequency sub-band of the IR image and the VIS image, respectively. As can be seen from the (e) and (j) of Fig. 2, the improved low-frequency fusion map has more texture information while effectively preserving and enhancing the significant information of the infrared image in the substation, consistent with the visual perception of the human eye.

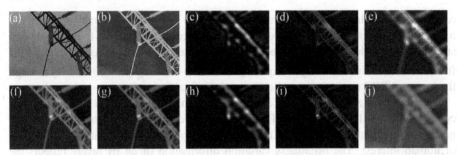

Fig. 2. Visual salience. (a) VIS image; (b) y component in YUV color space of VIS image; (c) VIS saliency image obtained by original FT algorithm; (d) VIS saliency image obtained by improved FT algorithm; (e) fusion image of low frequency sub-band after improvement; (f) IR image; (g) y component in YUV color space of IR image; (h) IR saliency image obtained by original FT algorithm; (i) modified IR saliency map obtained by FT algorithm; (j) low frequency sub-band fusion before improvement.

4 Experiment and Results

To evaluate our fusion method, four sets of IR and VIS images are used for simulation experiments. In order to illustrate the effectiveness of this algorithm, several classical image fusion methods are used for comparison, as shown in Fig. 3, which gives the fusion images generated by different image fusion methods, and the fusion methods from top to bottom are CVT [7], GTF [8], MSVD [9], WaveLet [10], NSCT [11], and the proposed method in this paper. The NSCT algorithm is based on the original FT visual saliency detection algorithm, and the fusion strategy for the low-frequency sub-band uses a weighted average strategy, while the maximum fusion strategy is chosen for the high-frequency sub-band fusion.

As can be seen from the Fig. 3, the WaveLet, the MSVD and the CVT have poor target saliency, and there are virtual edges for the fusion of the substation image, the background contains unnecessary VIS light information, making the final fusion image visual effect is poor, and the IR color information receives interference. GTF, NSCT and OURS algorithms have better fusion effect, which can preserve the information of the two source images and retain the rich detail texture on the substation equipment, and have better subjective visual feeling. However, the NSCT algorithm reduces the contrast in the background region and provides poor fusion because it considers both IR and VIS information in the insignificant region. Both the method in this paper and the GTF method effectively retain the color information of the IR image in the background region, which makes the overall color information of the generated fusion image similar to the IR image, and can better use the IR color changes to perceive temperature information changes in the detection of substation equipment, which is more in line with the visual perception of the human eye.

In order to evaluate the fusion results more objectively, the fusion images are evaluated using entropy (EN), standard deviation (STD), structural similarity (SSIM), correlation (CC), spatial frequency (SF), and visual information fidelity (VIFF) evaluation metrics. EN is an index to evaluate the amount of image information, the larger the EN,

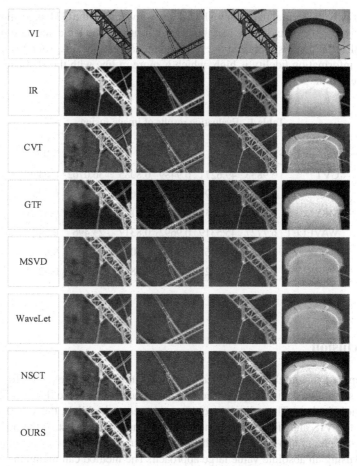

Fig. 3. Fusion results of different algorithms. From top to bottom, it represents (a) visible image; (b) infrared image; (c) CVT; (d) GTF; (e) MSVD; (f) WaveLet; (g) NSCT; (h) OURS.

the richer the image information; STD reflects the change of image contrast, the larger the STD, the better the visual effect; weighted SSIM reflects the degree of similarity between the fused image and the source image, the larger the value indicates that the structural information of the image is more similar; CC represents the similarity with the content of the source image; SF can measure the gradient distribution of the image, the larger the spatial frequency, the better the image, VIFF is evaluated from the perspective of visual perception of the human eye. The calculation results of each index are shown in Table 1, from which it can be seen that among the six evaluation indexes, the EN, STD and VIFF indexes of our method rank first, SSIM and SF rank fourth and third, respectively, but they all have a small gap with the best index values. The reason for the lower CC indexes may be due to the enhancement of the visually significant areas in the fusion image of this paper's method. A comparison of the objective evaluation metrics between this paper's method and the GTF method, which has better subjective results, shows that only the spatial frequency SF is slightly lower than the GTF algorithm, while the other five metrics are higher than the GTF algorithm. The large standard deviation indicates that the contrast between the fused image target and background is high, and the visual information fidelity is high, which is more in line with the characteristics of human eye vision.

Table 1. Evaluation results of image fusion

Evalutions	EN	SD	SSIM	CC	SF	VIFF
CVT	6.9853	49.6271	0.9489	0.8085	36.0762	0.5479
GTF	0.9801	60.5792	0.7742	0.7674	34.0220	0.5058
MSVD	6.8635	46.2259	0.9355	0.8172	31.8594	0.5133
WAVELET	6.7727	45.5319	0.9692	0.8317	23.496	0.5421
NSCT	7.0243	57.8965	0.9676	0.80785	5.1371	0.6238
OURS	7.0276	64.532	0.9378	0.7837	33.6321	0.6247

5 Conclusion

In this paper, a fusion algorithm based on NSCT and visual saliency of IR and VIS images of substation devices is proposed for the characteristics of the IR and VIS images of substation devices. The saliency technique is used to highlight the visually salient targets of the VIS and IR images in the substation and retain most of the information in the low-frequency sub-band, while the high-frequency self-contained detail texture is preserved using an absolute value large approach. The method can maintain consistent color information with the IR image in the background region, which is better suited for the detection of temperature anomalies in substation equipment. The algorithm is simple and easy to implement, and at the same time, it can guarantee better evaluation indexes than other classical algorithms in the fusion of IR and VIS images of substation equipment. It is a simple and effective algorithm for the fusion of IR and VIS images of substation devices.

References

1. Zhao, Z.: Research on Registration Method of Infrared/Visible Image of Power Equipment. North China Electric Power University, Beijing (2009)
2. Stathaki, T.: Image Fusion: Algorithms and Applications. Elsevier, Amsterdam (2011)
3. Song, X.: Key Technology Research of Intelligent Iterative Inspection Robot in Substation. Changsha University of Science and Technology, Changsha (2013)
4. Zhao, Z., Guang, Z., Gao, Q., Wang, K.: Infrared and visible images fusion of electricity transmission equipment using CT-domain hidden Markov tree model. High Voltage Eng. **39**(11), 2642–2649 (2013)

5. Shi, Y.: Infrared Image and Visible Light Image Fusion Method and its Application in Electric Power Equipment Monitoring. Xi'an University of Technology, Xi'an (2018)
6. Achanta, R., Hemami, S.S., Estrada, F.J., et al.: Frequency-tuned salient region detection. In: Computer Vision and Pattern Recognition, pp. 1597–1604 (2009)
7. Choi, M., Kim, R.Y., Nam, M., et al.: Fusion of multispectral and panchromatic Satellite images using the curvelet transform. IEEE Geosci. Remote Sens. Lett. **2**(2), 136–140 (2005)
8. Ma, J., Chen, C., Li, C., et al.: Infrared and visible image fusion via gradient transfer and total variation minimization. Inf. Fusion **31**, 100–109 (2016)
9. Naidu, V.: Image fusion technique using multi-resolution singular value decomposition. Defence Sci. J. **61**(5), 479–484 (2011)
10. Qu, G., Zhang, D., Yan, P., et al.: Medical image fusion by wavelet transform modulus maxima. Opt. Express **9**(4), 184–190 (2001)
11. Adu, J., Gan, J., Wang, Y., et al.: Image fusion based on nonsubsampled contourlet transform for infrared and visible light image. Infrared Phys. Technol. **61**, 94–100 (2013)

Permission Dispatching Mechanism Inside and Outside of the Warranty Period for Equipment Maintenance Service System

Yujie Liu[1,2], Wenjia Wu[1], Wen Bo[1], Chen Han[1], Quanxin Zhang[2], and Changyou Zhang[1(✉)]

[1] Laboratory of Parallel Software and Computational Science, Institute of Software, Chinese Academy of Sciences, Beijing, People's Republic of China
changyou@iscas.ac.cn

[2] School of Computer Science and Technology, Beijing Institute of Technology, Beijing, People's Republic of China

Abstract. It is a vital service process for equipment maintenance to ensure regular operation. The equipment transitions from inside of the warranty period to outside of the warranty period and the staff involved in the maintenance service will transform from manufacturer personnel to a mixed team of user teams and third-party companies. The permissions will change with their responsibilities in the equipment maintenance service system. In this paper, we take the maintenance service system of shield tunneling machine equipment as a case, which design the personnel structure of the mixed maintenance team and establish an authorization mechanism to support the multi-team work process of equipment maintenance. We design a user group-based hierarchical control strategy that combines the equipment maintenance service system's resource structure. At the end of this paper, though the simulation-based on Colored Petri Net CPN Tools, we verify the correctness and effectiveness of the permission dispatching mechanism.

Keywords: Equipment maintenance · Service system · Inside and outside of the warranty period · Permission dispatching mechanism · User group

1 Introduction

With the rapid development of the Chinese manufacturing industry and the modernization process is gradually accelerating. Large-scale equipment is widely used in construction in various fields. The importance of the equipment maintenance system is noticeable. This article focuses on the access control mechanism of the maintenance system. David Ferraiolo and Rick Kuhn first proposed role-based access control RBAC in the 1990s and created the first RBAC model [1].

The users of the equipment maintenance system and the aircraft maintenance support system are related to the maintenance process. For the complex permission dispatching requirements of aircraft maintenance support information systems, literature [2] Xun Zhang introduced "organization" based on the RBAC model and proposed an improved

Q. Zu et al. (Eds.): HCC 2020, LNCS 12634, pp. 84–90, 2021.
https://doi.org/10.1007/978-3-030-70626-5_9

access control model to achieve fine-grained access control. To solve the problem of a large number of users and scattered physical locations in the rail transition integrated monitoring system, based on the analysis of the integrated monitoring system' user profile, literature [3] Min Deng proposed a user access control model based on "user-group-permission". However, the model does not relate to the issue of organizational structure. Literature [4] Ke-jun Sheng proposed an access control model based on the organizational structure characteristics of enterprises and government departments. Although the minimum authorization of the role is addressed, it does not solve the problem of changes in each department' permissions in different periods. Literature [5] Elisa Bertino introduced TRBAC to support periodic enabling and disabling permissions. Role-triggered actions can be executed immediately or delayed till a specified time. Although this access control mechanism solves the problem of permission changes, the unified management of permissions is still lacking. Literature [6, 7] Muhammad Umar Aftab and others improved the role-based access control model to increase the system's security, but they did not consider the changes in permissions over different periods.

The contributions that we describe in this paper may be summarized thus. This paper proposes a permission dispatching mechanism for the shield tunneling machine maintenance service system. We add an "organizational structure" module to the RBAC model, which is more convenient to manage permissions for users in different positions. In addition, combining with the resource structure of the equipment maintenance service system, a hierarchical control mechanism based on user groups is designed, which is conducive to the change of permission transition from inside of the warranty period to outside of the warranty period.

2 The Resource of Equipment Maintenance System

2.1 System Resource Structure

The resources of the equipment maintenance system are divided into three categories: operation, menu, and page. The number of menu resources is small and fixed, and each menu resource corresponds to a functional module, which covers a wide range. A page contains multiple operations, multiple pages form a functional module, and there is a progressive relationship between resources. For example, if you only have the "view user information" operation but no "user information" page resource, you still cannot view users' information.

According to the characteristics of the functions, we classify the resources of the equipment maintenance system. Menu resources refer to the content of the menu bar in the maintenance system. Each menu resource corresponds to a service. Page resources refer to the website pages in the maintenance system. Controlling page resources can prevent external personnel from accessing corresponding functions through the page address. Operation resources refer to the methods of operating various resources in the maintenance system, which are the more granular division of system resources.

2.2 System Resource Structure

The resources in the equipment maintenance service system are divided into menu resources, page resources, and operations. The number of menu resources is small and fixed, and each menu resource corresponds to a functional module, which covers a wide range. A page contains multiple operations; multiple pages form a functional module. There is a progressive relationship between resources. For example, if you only have the "view user information" operation but no "user information" page resource, you still cannot view user information.

To achieve a specific function, users need to combine three types of resources. The user needs to have the "maintenance items" menu resource, the "user information" page resource, and the "view user information" operation to obtain user information in the maintenance system. In the maintenance system, these combined resources are managed uniformly by the "permission". In the organizational structure, the responsibilities of the project manager, deputy manager, and other positions remain unchanged. The resource permissions that need to be bound are fixed, and the "roles" manage these permissions lists.

Due to the timeliness of the after-sales service of the manufacturer, the equipment maintaining during the warranty period is the responsibility of the manufacturer. However, outside the warranty period, the constructor has more options in equipment maintenance, which also raises the issue of permission changes during the transition from inside to outside of the warranty period. Manage users with the same responsibilities can simplify the problem of changing permissions and avoid repeated operations through the "user group" and "organizational structure", which can also solve the problems caused by the transition from inside of the warranty period to outside of the warranty period.

3   Permission Dispatching Mechanism

3.1   Permission Dispatching Model

The ministers and members of each department have a clear division of labor. Due to the distinct responsibilities of users, the resources and permissions they have are also different. It is necessary to categorize the permissions and bind different roles in volume authorization to simplify the user management process. When transitioning from inside the warranty period to outside of the warranty period, different departments' permission will change.

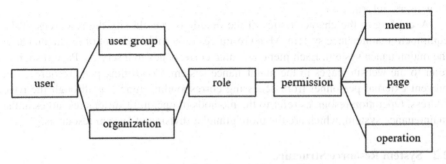

Fig. 1. Permission dispatching model.

In this paper, we introduce the concepts of "user group" and "organizational structure", which make the management and distribution of system permissions more convenient, Fig. 1. A user group can be understood as a collection of organizations with the same privileges, such as an electrician class or a maintenance class, which has the permissions to perform maintenance tasks and submit maintenance results during the maintenance of the equipment. Binding roles with these permissions to user groups can simplify the authorization process. The organizational structure is used to represent the logical structure of the personnel in each department. As the system administrator, the project manager has the permission to grant matched roles to subordinates.

3.2   Hierarchical Control Strategy Based on User Group

In the maintenance system, the user group includes members of the same "class" in the physical world. Establishing user groups can improve the limitations of only official companies to provide services and realize multi-enterprise collaborative services. Besides, based on user groups' hierarchical control strategy, it can fulfill reasonable resource allocation and improve service efficiency.

Dividing the user level of the equipment maintenance system and exploring the main influencing factors of level division are the core tasks of hierarchical control. According to management purposes, we improve the management methods required by each level's functions to achieve optimized management and promote management effectiveness. First, user groups will add users based on their positions and departments during the warranty period. All users in the user group have corresponding permissions together. Secondly, outside the warranty period, the maintenance team formed by the manufacturer's personnel, will log off in the maintenance system. After exiting user groups, users do not have the corresponding permissions any longer. According to the third-party maintenance company's scale, service price, customer reviews, etc., the construction party selects a new maintenance team to enter the maintenance system. Finally, the newly authorized maintenance team continues to complete maintenance tasks. They have the permissions to download maintenance manuals, view historical work orders, and so on, which can complete the work handover efficiently.

The establishment of user groups realizes batch management of various departments and rationally allocates resources and improves the limitation of only providing services by official enterprises, making it possible for multi-enterprise collaborative services and improving service efficiency.

3.3 Permission Dispatching Mechanism Verification

To study the rationality and functionality of the internal and external permission dispatching mechanism, this paper establishes a colored Petri net model based on the user permission dispatching inside of the warranty period and outside of the warranty period. The corresponding relations of the elements of the Colored Petri Net are shown in Table 1.

For the permission dispatching mechanism's network model, the organizations need to consider the transition of each role' permissions from inside the warranty period to outside of the warranty period. For example, outside the warranty period, office staff have the permissions to view, hire, and dismiss maintenance personnel from third-party companies. The finance department needs to add the permissions to pay third-party company maintenance fees and parts replacement costs. The maintenance engineers no longer only delegates tasks to the maintenance team workers but also has the permission to issue maintenance tasks to the third-party enterprise's maintenance team. Simultaneously, the maintenance team of a third-party enterprise enters the maintenance system and needs to add functions such as changing user information, receiving maintenance tasks, submitting work reports, and viewing work evaluations.

Table 1. Element correspondence

System elements	Petri Net elements
Permission dispatching node	Transition
Permission management node	Place
Permission	Token
Permission control	Connection
Permission change	Token color
User	Token value

We use CPN-tools to establish a user permission dispatching model, Fig. 2. In the user permission dispatching model, the engineer loses the factory personnel's original management permission during the warranty period but adds the third-party maintenance company's permissions to issue tasks. According to the demand analysis and simulation of resource scenarios, the change of permissions inside the warranty period outside the warranty period makes labor division within the organization clearer.

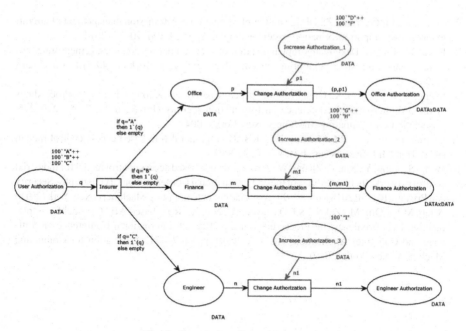

Fig. 2. Permission dispatching model.

4 Conclusions

This paper takes the maintenance service system of shield tunneling machine equipment as a case to analyze the organizational structure of the shield project and the composition of the personnel responsible for maintenance tasks. We design the organizational structure permissions to simplify the personnel change process and the permission dispatching mechanism to support the multi-team work process of equipment maintenance. By analyzing the equipment maintenance service system's resource structure and control requirements, a hierarchical control strategy based on user groups is designed to help third-party companies enter the equipment maintenance system outside the warranty period. At the end of the article, the simulation based on the Colored Petri Net CPN Tools verifies the correctness and effectiveness of the permission dispatching mechanism during and outside the warranty period of the equipment service system.

Acknowledgments. This paper is supported by the National Key Research and Development Program of China (2018YFB1701400), the Natural Science Foundation of China (U1636213, 61672508).

References

1. Sandhu, R., Coyne, E.J., Feinstein, H.L., et al.: Role-based access control models. IEEE Comput. **29**(2), 38–47 (1996)

2. Zhang, X., Li, Q., Yang, Z.: Design and implementation of permission management of aircraft maintenance support information system. Comput. Appl. **34**(S1), 70–73 (2014)
3. Deng, M., Liu, T.: Design and implementation of user and permission management functions of rail transit integrated monitoring system. Mechatron. Eng. Technol. **49**(5),165–167, 213 (2020)
4. Sheng, K., Liu, J., Liu, X.: Organizational structure based access control model. In: Proceedings of the 3rd international conference on Information security (InfoSecu 2004), pp. 210–215. Association for Computing Machinery, New York (2004)
5. Bertino, E., Bonatti, P.A., Ferrari, E.: TRBAC: a temporal role-based access control model. ACM Trans. Inf. Syst. Secur. **4**(3), 191–233 (2001)
6. Wang, S., Yang, Y., Xia, T., Zhang, W.: A role and node based access control model for industrial control network. In: Proceedings of the 2nd International Conference on Cryptography, Security and Privacy (ICCSP 2018), pp. 89–94. Association for Computing Machinery, New York (2018)
7. Aftab, M.U., Qin, M., Zakria, S.F.Q., Javed, A., Nie, X.: Role-based ABAC model for implementing least privileges. In: Proceedings of the 2019 8th International Conference on Software and Computer Applications (ICSCA 2019), pp. 467–471. Association for Computing Machinery, New York (2019)

New Technology Development and Application Trends Based on Monitoring of Patent Data

Ce Peng$^{(\boxtimes)}$, Tie Bao$^{(\boxtimes)}$, and Ming Sun

Jilin University, Changchun 130012, China
pengce18@mails.jlu.edu.cn, baotie@jlu.edu.cn

Abstract. As an important manifestation of science and technology, patents contain many aspects of information. The combination of multiple patents can dig out the development trend of a technology. Therefore, it is very important to understand them for mastering the trend of a technology and the decision-making and deployment of future development of the enterprise. Existing methods cannot provide enough analysis and forecast information. Therefore, a domain technology analysis model based on patent data is proposed in this paper. The visualization diagram constructed by this model can vividly display the multi-dimensional characteristics of the technology in the patent data, including the provincial and municipal distribution of technology domains in the patent, the development trend of sub-domains, the main technology composition of the patent, etc. which helps the enterprises quickly understand the status quo of technology development behind patent data. At the same time, the comprehensive analysis diagram generated can help us understand the relationship between the main team composition and technology in the technology domain, and make better decisions in future research. Finally, patents in the deep learning domain are researched in this paper.

Keywords: Analysis model · Technology characteristics · Data analysis and development trend

1 Introduction

With the advent of the knowledge age in the 21st century, our county has entered a critical period of economic and social development, which is also a period of strategic opportunities for our county to move from a major intellectual property country to a strong intellectual property country. As important manifestation of technology, patent is an important measure and means for innovation and development [1]. The core of innovation is technological innovation. Patent is important expression of innovation value [2]. At present, use of patent data for research at home and abroad can be divided into two mainstreams. One of them is to perceive the industry situation through analysis of patent data [3], and a method of technology trend analysis is proposed. The author analyzed the frequency and time changes of patented technical terms, so as to identify the existing technologies that led to emergence of new technologies, and discovered the most critical technologies [4]. Through research on patent application that uses face recognition as

Q. Zu et al. (Eds.): HCC 2020, LNCS 12634, pp. 91–102, 2021.
https://doi.org/10.1007/978-3-030-70626-5_10

the data source, the author mainly used statistical analysis methods and social network analysis methods, and conducted advanced patent analysis in many aspects, such as main distribution, main institutions, research focus, and key facial recognition patents [5]. Three quantitative models are proposed to analyze Apple's patents. The technology map is built by using time series regression and multiple linear regression, and then the technology behind the patent is analyzed by using clustering [6]. A model is proposed to use the recommended existing patents to evaluate the value of newly granted patents. By analyzing the time pattern trends of related patents and proving that these trends are highly correlated with the patent value [7]. A patent value evaluation method based on the potential citation network is proposed, the corresponding algorithm is designed to evaluate the value of each patent, and a patent value improvement algorithm is proposed to deal with the problem of low algorithm efficiency [8]. A patent prediction scheme is proposed, the implementation process of the scheme is discussed, and example verification is performed according to the railway transportation patent.

Considering that the existing patent website cannot provide sufficient technical analysis and monitoring of technology development trends from massive patent data, this paper proposes a method based on the development trend of emerging technologies in the field of patent data monitoring. The paper makes the following contributions:

- Proposing a domain technology analysis model based on the patent data.
- Introducing a method to build a domain technology analysis model based on the patent data.
- Using the patent data analysis model proposed by us, realizing monitoring of technology development trends in a certain field.

2 Technology Analysis Model

Issuance of a patent contains many aspects of information. The combination of multiple patents can dig out the development trend of a domain technology. For emergence of emerging technologies, in order to help scholars or enterprises quickly understand the characteristics and development trends of technologies, we suggest dividing a technology-related patent into sub-domains according to the research direction of scholars, and use a multi-dimensional model to represent the characteristics of each direction. Therefore, two dimensions were adopted in our technology analysis model based on the patent data. The domain technology analysis model based on the patent data proposed in this paper was shown in Fig. 1. It was composed of two dimensions of patent-related character information and patent domain technology information, and visualization was used for analysis. The attribute information contained in each dimension will be introduced in details next.

2.1 Technology Information

In order to introduce and understand a certain technology in an orderly manner, we further divided them into technology composition, technology trend, national distribution of

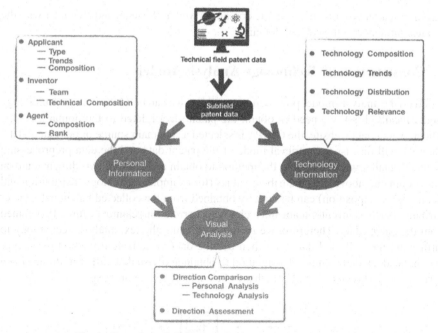

Fig. 1. Domain technical analysis model based on patent data

technology and technology correlation. Technology composition shall be based on sub-classification of the international patent classification number, and sub-classification can effectively reflect the technology composition behind the patent; national distribution of technology can effectively show the status quo of technology development in different provinces; the last type, technology correlation, is usually ignored by many people. We can find that a patent contains multiple technologies. We often only pay attention to the main technology of a patent, but ignore the secondary technology involved in the patent. Therefore, it is very necessary to deeply understand the implicit relationship between the main technology and the secondary technology of a certain domain patent, so as to effectively dig the correlation between the technologies.

2.2 Personal Information

The current dimension shows the basic information of the personnel related to the specific direction of the patent, including applicant, agent and inventor. Each category contains valuable and different attribute information that we need to understand. Most of the applicants are enterprises or universities. Through this type of information, we can further dig the applicant's technology composition and technology development trend; the agent part can provide the number of different categories of patents represented by the agency. The inventor's information is the most valuable among personnel information. We can accurately locate a certain inventor according to the patent IPC and patent article keyword extraction technology, so as to further dig the inventor team and its technology composition. In addition, other inventors can use this attribute to judge the current research progress of a specific domain as well as the team and character with the greatest influence in the current domain.

3 Construction of Technology Analysis Model

Based on the analysis model proposed by us above, we can build the domain technology model locally. First, we need to obtain the patent data related to the domain. As the official patent data website, the CNIPA is selected as the main source of our patent data. Then we will discuss the retrieval mode of the patent data and the data preprocessing method. Finally, we will discuss the method to obtain and analyze two-dimension data we recommend above. Some attribute values (for example, technology distribution and technology composition) can be directly obtained from the collected patent data. Other attributes (such as inventor team and technology correlation) cannot be directly obtained from the patent data. Therefore, we recommend using the text analysis technology to further dig the collected data, and then use the data visualization method to analyze the obtained attribute values. The method to obtain and visualize different dimensional attributes was shown in Fig. 2, which will be explained in detail next.

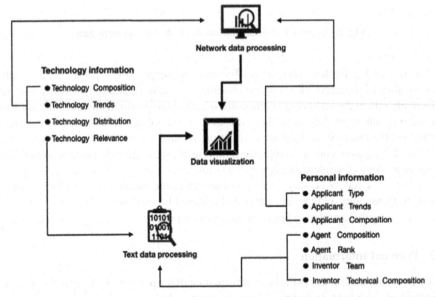

Fig. 2. Methods for obtaining and visualizing properties

3.1 Data Acquisition and Pre-processing

There are many categories of patents in different countries. For example, patents in China and Japan are divided into invention patents, utility model patents and appearance design patents, while the patents in the U.S. are divided into plant patents, invention patents and design patents [9]. The patents cover a wide range of domains. Therefore, it is the most critical step in the patent data collection process to accurately obtain the useful data from the massive patent data. This paper mainly takes Chinese invention patents as the data source, searches for the patent data according to the International Patent Classification (IPC) Index [10], and sorts the retrieved data into local files. For the data preprocessing part, data cleaning and unified processing of data format shall be completed. Data cleaning includes de-duplication and error correction of patent applicant, inventor, agency, classification number and other fields, and ultimately ensures that all patent data collected is maintained consistent.

3.2 Network Data Processing

As shown in Fig. 2, After actual research on the patent data, we believe that applicants are roughly divided into the following four categories, namely enterprises, universities, research institutes and others. Considering that some patents include the above multiple categories at the same time, such as "collaborative innovation" patents, in-depth research on the collaborative innovation patents was not conducted in this paper. If necessary, it is possible to further dig the applicant information in the patent data by building a university-enterprise patent cooperation network analysis graph, so as to explore the mode, depth and breadth of cooperation between universities and enterprises. Therefore, the types of applicants in this paper are simply divided into four categories above.

3.3 Text Data Processing

Inventor Team Analysis. The analysis on inventors is of great value in the process of patent data analysis. Taking the inventors as the research portal, it can carry out detailed analysis on individual inventors, clarify the main research technology domain that they are good at, and understand the composition of the main R&D personnel or R&D team of the target enterprise in various domains, so as to obtain effective technical information. Therefore, the method of correlation analysis is applied in this paper. According to the actual situation of the data, filtering is performed by setting the lower limit of lift and support.

$$S = P(A\&B) \tag{1}$$

$$C = P(A\&B)/P(A) \tag{2}$$

$$L = \left(P(A\&B)/P(A)\right)/P(B) = P(A\&B)/P(A)/P(B) \tag{3}$$

Construction of Technical Relevance. We can find that a patent contains multiple technologies. We often only pay attention to the main technology of a patent, but ignore the secondary technology involved in the patent. Therefore, it is very necessary to deeply understand the implicit relationship between the main technology and the secondary technology of a certain domain patent, so as to effectively dig the correlation between the technologies. Considering that there are many categories of technologies, and the technologies in each patent data may have certain specific associations, it is considered to use a dynamic relation diagram for display. In order to make the relationship between skills more general, we recommend setting a threshold to control the connecting line between skills. The threshold calculation formula is shown in Formula (4).

$$J_{ab} = \frac{p(a \cap b)}{p(a)} + \frac{p(a \cap b)}{p(b)} \tag{4}$$

$$\begin{cases} 1 \ J_{ab} \geq 0.5 \\ 0 \ J_{ab} < 0.5 \end{cases}$$

3.4 Data Visualization Analysis

The data visualization part of this chapter mainly conducted detailed visualization analysis on the geographic location of patents in the deep learning domain, sub-domain development trend and main technology components, and provided a comprehensive analysis diagram.

Patent Location Analysis. The pre-processed patent data was used to make statistics on the application location of each piece of patent information. It was the most intuitive to display the application location by using 3D map and 3D histogram, and 3D histogram was used in 3D map to indicate the number of jobs. In order to improve the display effects, it was suggested to set the basic unit of the histogram to compress the number of patents.

Subdomain Trend Analysis. After obtaining the pre-processed patent keywords, we divided the domains into sub-domains according to the keywords. It was recommended that applicants be divided into four categories, namely universities, research institutes, enterprises and others. The number of applicant compositions in the above four categories and the number of sub-domains were counted respectively, and then the number of applicant compositions counted within ten years were saved into an excel form (Fig. 3). Considering that data had three-dimensional characteristics, we recommended using a 3D bar chart for display. The x-axis represents the time dimension, the y-axis represents the category dimension, and the z-axis represents the quantity. Different colors were used to divide the quantity.

Main Technical Composition Analysis. The IPC number was intercepted from the patent data. First, the patent data needed to be processed hierarchically according to the IPC number, and the processed data was stored in excel (Fig. 4). The dynamic sunburst chart was used for the patent data processed hierarchically to show the quantitative relationship between the patent technology categories. In the sunburst chart, the first ring was a part, the second ring was the category in each part, and the outermost ring was the sub-category in each category. The width of the used ring corresponded to the data traffic, which was more suitable for visualization analysis on traffic and other data. Different colors were used to distinguish technology categories, which could effectively show the quantitative relationship between patent data categories in the deep learning domain. Finally, the data was analyzed in depth according to the International Patent Classification.

	2011	2012	2013	2014	2015	2016	2017	2018	2019	2020
CNN	12	15	24	36	128	236	789	3287	7634	2873
RNN	1	1	3	24	98	117	342	1427	4237	987
GAN	1	4	6	8	36	45	114	356	2012	356
School	1	12	26	34	42	68	253	568	2835	1023
Scientific	3	6	18	19	54	64	54	213	1256	574
Enterprise	1	7	18	26	78	81	91	175	2753	998
Other	0	8	4	16	17	21	37	203	635	241

Fig. 3. Excel table of field development trends

	source	target	value		source	target	value
1	Total	CNN	33187	15	G06	G06F	5243
2	Total	RNN	8549	16	G06	G06Q	1963
3	Total	GAN	3673	17	Total	GAN	3673
6	CNN	G	31494	18	H	H04	1936
7	CNN	H	2099	19	A	A61	1038
8	CNN	A	1294	20	B	B60	214
9	G	G06	29476	21	B	B25	170
10	G	G01	1912	22	A	A01	106
11	G	G10	995	23	A	A63	83
12	G06	G06N	18643	24	A	A47	44
13	G06	G06K	17714	25	H04	H04N	1035
14	G06	G06T	7709	26	A61	A61B	951

Fig. 4. Main technologies constitute excel table

4   Empirical Analysis

We collected 49,274 patents related to deep learning through IPC retrieval. A total of 45,409 valid patent data in the deep learning domain were obtained after data preprocessing, including 33,187 patents related to convolutional neural network, 8,549 patents related to recurrent neural network, and 3,673 patents related to generative adversarial network. Python and related libraries were adopted for the experiment display part.

4.1   Provincial Distribution Map

We set the basic unit as 80, and produced a provincial and municipal distribution map of deep learning-related patents (Fig. 5). We could see that among the patents in the

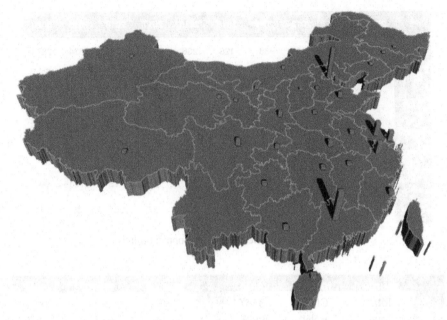

Fig. 5. Provincial and municipal distribution map

deep learning domain, the top three provinces and cities are Beijing (9,315), Guangdong (6,933) and Jiangsu (3,536) in terms of the number of applications. The total number of patents published in the deep learning domain in the three provinces and cities accounted for 43.6% of the total number of patents in this domain.

4.2 Subdomain Trend Chart

We produced a trend map of deep learning sub-domains in the past ten years (Fig. 6). As a whole, the number of patents in various sub-domains of the deep learning domain had a rising trend in the past 10 years. The popular research direction in the deep learning domain was CNN, and the number of patent applications has increased significantly since 2014. The number of patent applications in 2019 reached the current peak in this domain, and exceeded 10,000. In terms of applicant composition, the patent applicants in the deep learning domain were mainly enterprises and universities, accounting for 48.3% and 47.9% of the total number respectively. In terms of the growth trend, RNN was the popular research direction in the deep learning domain from 2011 to 2013.

4.3 Domain Technology Composition Chart

Through the IPC number, we produced a technology composition diagram of the deep learning domain (Fig. 7). In terms of major classification, G had the largest proportion in the deep learning domain, followed by A. In terms of sub-classification, G06 had the largest proportion. Among G06, G06K, G06F and G06T were the current research hotspots in the domain, followed by A61B and H04L. Based on the above information,

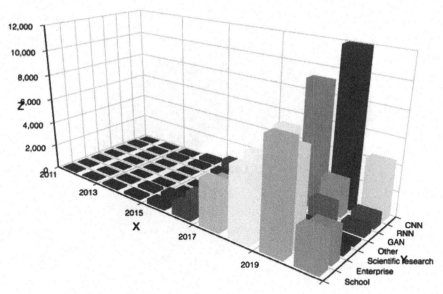

Fig. 6. Development trend chart of deep learning subfields

we can conclude that the current research in the deep learning domain mainly focuses on image data processing and calculation method. Through the conclusion drawn in 4.2, it is not difficult to find that this is because CNN has gradually become a popular research direction in the deep learning domain since 2014. Class A technology domain also has a large proportion, because the deep learning domain mainly involves medicine, game entertainment and household equipment. In medicine, doctors can identify the physical health of a person by photographing or scanning the facial features of the patient. In entertainment, some games can fit the player's facial expressions to the role of the virtual game. On home equipment, we can use the face recognition technology to verify the personal identity.

4.4 Comprehensive Analysis Diagram

We produced a comprehensive analysis diagram of the deep learning domain (Fig. 8), based on the technology correlation, main team composition and other information obtained above, a descriptive text was automatically generated based on actual data, and monitoring effects of technology development in the deep learning domain with the patent data was realized.

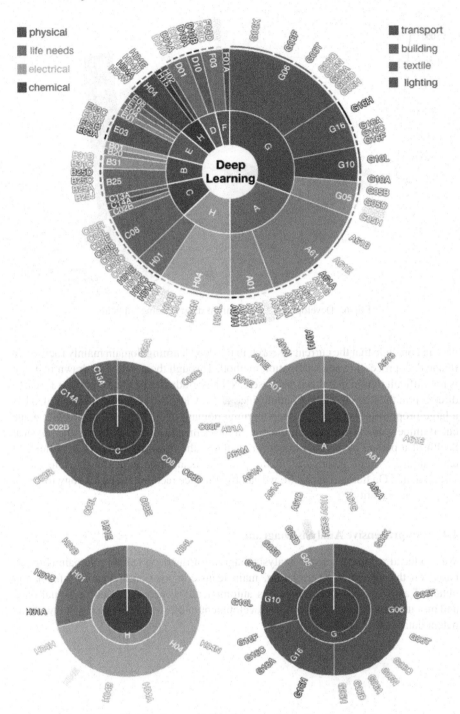

Fig. 7. Technical composition diagram of deep learning domain

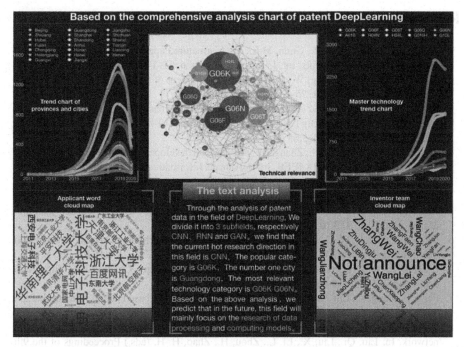

Fig. 8. Comprehensive analysis diagram of deep learning fieldss

5   Conclusion and Discussion

This paper proposes a method for monitoring development of emerging technologies based on the patent data. First, we propose a domain technology analysis model based on the patent data, and then explain a method to build the analysis model we proposed. Finally, we conduct detailed visualization analysis on provincial and municipal technology distribution of the deep learning domain, sub-domain development trend and main technology components of patents, and gives a comprehensive analysis diagram. It is verified that the method proposed in this paper has achieved the expected effects. At the same time, our method also has shortcomings. The data source of this paper has certain limitations. The selected domain patents do not cover all types of patent data on the patent websites, resulting in limited representativeness of patent data. The research objects selected are limited to the data collected from domestic patent websites. However, the current development momentum of the technology industry is rapid, and a variety of patents related to emerging technologies have emerged abroad. Our subsequent research shall cover the patent websites in multiple countries, and explore the development status of a certain domain technology at home and abroad through comparative analysis. Only in this way can we conduct in-depth research and analysis on the technology behind the patent data in a wider range and at a deeper level.

References

1. Ashton, W.B., Kinzey, B.R., Gunn, M.E., Jr.: A structured approach for monitoring science and technology developments. Int. J. Technol. Manag. **6**, 91–111 (2011)
2. Chen, L., Zhang, Z., Shang, W.: Reviews on development of patent citation research. New Technol. Libr. Inf. Serv. **Z1**, 75–81 (2013)
3. Segev, A., Jung, C., Jung, S.: Analysis of technology trends based on big data. In: 2013 IEEE International Congress on Big Data, Santa Clara, CA, pp. 419–420 (2013). https://doi.org/10.1109/BigData.Congress.2013.65
4. Yu, J., Zhu, S., Xu, B., Li, S., Zhang, M.: Research on development of face recognition based on patent analysis. In: 2018 IEEE 3rd Advanced Information Technology, Electronic and Automation Control Conference (IAEAC), Chongqing, pp. 1835–1839 (2018). https://doi.org/10.1109/IAEAC.2018.8577919
5. Jun, S., Park, S.S.: Examining technological innovation of Apple using patent analysis. Ind. Manag. Data Syst. **113**(6), 890–907 (2013)
6. Oh, S., Lei, Z., Lee, W.-C., Yen, J.: Patent evaluation based on technological trajectory revealed in relevant prior patents. In: Tseng, V.S., Ho, T.B., Zhou, Z.-H., Chen, A.L.P., Kao, H.-Y. (eds.) PAKDD 2014. LNCS (LNAI), vol. 8443, pp. 545–556. Springer, Cham (2014). https://doi.org/10.1007/978-3-319-06608-0_45
7. Feng, L., Peng, Z., Liu, B., Che, D.: A latent-citation-network based patent value evaluation method. J. Comput. Res. Dev. **52**(3), 649–660 (2015)
8. Zhang, Y., Wang, Q.: Patent prediction based on long short-term memory recurrent neural network. In: Liu, Qi., Liu, X., Li, L., Zhou, H., Zhao, H.-H. (eds.) Proceedings of the 9th International Conference on Computer Engineering and Networks. AISC, vol. 1143, pp. 291–299. Springer, Singapore (2021). https://doi.org/10.1007/978-981-15-3753-0_28
9. Liu, K.: Introduction of patent retrieval on U.S. patent and trademark office website. China Invent. Pat. **9**, 38–39 (2014)
10. Bo, H., Makarov, M.: The sixth edition of the IPC. World Pat. Inf. **17**(1), 5 (1995). https://doi.org/10.1016/0172-2190(94)00063-R

A K-means Clustering Optimization Algorithm for Spatiotemporal Trajectory Data

Yanling Lu[1,2], Jingshan Wei[1], Shunyan Li[1], Junfen Zhou[1], Jingwen Li[1,2(✉)], Jianwu Jiang[2], and Zhipeng Su[1]

[1] Guilin University of Technology, Guilin 541004, China
Luyl2014@glut.edu.cn, 358498163@qq.com
[2] Guangxi Key Laboratory of Spatial Information and Geomatics,
Guilin 541004, Guangxi, China

Abstract. It is a hotspot problem to quickly extract valuable information and knowledge hidden in the complex, different types, fuzzy and huge amount of space-time trajectory data. In the space-time trajectory data clustering method, according to the existing deficiencies of the classical K-means algorithm, the mathematical distance method and effective iteration method are used to select the initial clustering center to optimize the K-means algorithm, which improves the accuracy and efficiency of the algorithm. Based on MATLAB experimental simulation platform, the comparison experiments between the classical algorithm and the optimized algorithm, the applicability test of the performance test, and the comparison test with the classical algorithm were designed. The experimental results show that the optimized K-means randomly selected initial clustering center is more accurate, which can avoid the drawbacks caused by randomly selected initial clustering center to a certain extent and has better clustering effect on sample data, and at the same time avoid the K-means clustering algorithm falling into the dilemma of local optimal solution in the clustering process.

Keywords: K-means · Spatio-temporal trajectory data · Clustering algorithm · Data mining

How to effectively mine potential valuable information from complex data structures, diverse data formats, and spatio-temporal trajectory data with unclear boundaries has become the most important problem in spatial data mining [1]. The K-means algorithm is a typical algorithm in data mining analysis and was proposed by Macqueen in 1967 to improve the performance of the K-means clustering algorithm in terms of the number of class clusters K-values, successively fusing intelligent algorithms such as immune algorithm, genetic algorithm, and ant colony algorithm for cluster mining [2–4]. 2010 Wang Qiang et al. [5] proposed an optimization algorithm for the traditional K-means algorithm aggregation result that is prone to change with the beginning of the set midpoint, and demonstrated that its optimization algorithm has higher data accuracy and algorithm reliability. In 2013, Yongjing Zhang [6] proposed a new IU-MK-means Clustering Algorithm (K-means Clustering Algorithm based on Improved UPGMA and Max-min Distance Algorithm) that can be used to optimize the choice of initial cluster centers

© Springer Nature Switzerland AG 2021
Q. Zu et al. (Eds.): HCC 2020, LNCS 12634, pp. 103–113, 2021.
https://doi.org/10.1007/978-3-030-70626-5_11

for the drawbacks of the dependence of traditional K-means clustering algorithms. 2019 Jianren Wang et al. [7] studied the K-value uncertainty problem in the experimental process of the traditional K-means clustering algorithm based on the elbow method, and the improved ET-SSE algorithm can more accurately determine the choice of clustering K-values and improve the accuracy and stability of its algorithm. The K-means-based algorithm's adjustment parameter K-value is the number of clusters, the initial cluster point of the algorithm is arbitrarily set, the center is randomly chosen, and the cluster result changes with the change of the arbitrarily set aggregation point, thus requiring manual setting of the number of clusters to be carried out. In this paper, a combination of mathematical geometric distance and effective iteration is used to set the initial aggregation point of the experimental data set, which reduces the probability of the method falling into the dilemma of local optimal solution and also reduces the number of iterations of the cluster mining algorithm in the experimental process, thus improving the traditional K-means mining method based on the selection of the optimal cluster center point to compensate for the shortcomings of the classical K-means algorithm that randomly selects the initial cluster center.

1   K-means Clustering Algorithm

1.1   The Principle of Algorithms

Clustering algorithm is one of the important techniques in the field of data mining, in which K-means clustering algorithm classifies the track data into clusters and selects the clustering center reasonably for experiments, with obvious clustering effect and simple operation, which has the advantages of easy implementation and high efficiency. In the K-means clustering algorithm, initial clustering centers that are farther away from the actual best clustering centers tend to produce more noisy data or outliers to be clustered into the results during the clustering process, and if the initial clustering centers of different class clusters are too close, there is a high probability that the data sample points that should have been input into a particular cluster will be divided into different class clusters. For example, suppose there is a spatiotemporal trajectory data set x:x = [x1, x2...xn], the data set is aggregated into K class clusters C, C = [C1, C2...Cn], the minimization loss function is shown in Eq. (1), and the clustering center is shown in Eq. (2).

$$E = \sum_{i=1}^{k} \sum_{x \in C_i} ||x - \mu_i|| \tag{1}$$

$$\mu_i = \frac{1}{|C_i|} \sum_{x \in C_i} x \tag{2}$$

K is the number of clusters in a cluster, if the number of clusters is small, there will be too few clustering centers, which will lead to some clusters with small sample size to be discarded, if K is too large, a cluster will be misinterpreted into many different clusters, the more distant clusters will be rejected by the algorithm, which greatly affects the performance of the clustering algorithm. The practical performance of clustering algorithms. The K-means clustering algorithm groups the data sets according to the clustering results of the experiments, and different application scenarios use different similarity measures to cluster the set of track data samples.

1.2 Similarity Measures

The complexity of the temporal and spatial dimensions, combined with the constraints of environmental and technological conditions, makes the calculation of the similarity of space-time trajectories more difficult.The degree of similarity between trajectory data is generally measured using inter-trajectory similarity, thus requiring a uniform measurement of trajectories in the road network space combining space and time.The Eulidean distance is a more common space-time trajectory distance metric scheme in research, which first samples equal time intervals according to temporal characteristics, simultaneously performs Eulidean distance calculations, and sums the results superimposed on each other.The distance between trajectories is calculated as shown in Eqs. (3) and (4).

$$Euclidean(tr_i, tr_j) = \sum_{m=1}^{n} dis(D_m^i, D_m^j) \qquad (3)$$

$$dis(D_m^i, D_m^j) = \sqrt{(D_{m,x}^i - D_{m,x}^j)^2 + (D_{m,y}^i - D_{m,y}^j)^2} \qquad (4)$$

Where $Euclidean(tr_i, tr_j)$ denotes the Euclidean distance between spacetime trajectories, D_m^i denotes the mth point on the tr_i trajectory, $D_{m,x}^i$ and $D_{m,y}^i$ are the x and y dimensions of D_m^i, respectively, and $dis\left(D_m^i, D_m^j\right)$ embodies the trajectories tr_i and tr_j. $dis\left(D_m^i, D_m^j\right)$ embodies the distance values of the trajectories tr_i and tr_j at the point m position.

For applications where data accuracy is required or where precision is required, a mix of external perturbations in the data collection process will have a significant impact on the final result. The actual sampling process needs:

- Due to systematic errors and other uncertainties, the sampling data is prone to certain time deviations, Eulidean distance in the calculation of different space-time trajectories on the distance between the various sampling points, the need to sample data points at the same time, otherwise, you need to take the data interpolation continuous function value method.
- The trajectory is spatially limited in the network, and the trajectory lines in spacetime exhibit stable and high overlap characteristics.The data clustering algorithm can effectively analyze the important data in trajectories, cluster the trajectory segments into clusters, effectively analyze the reasons for the formation of hotspot regions, and at the same time, organically combine the important dimension of trajectory similarity, i.e., time dimension, which can more efficiently and rapidly excavate the features of geospatial data, so as to cluster the spatial and temporal trajectories, and use the clustering results of spatial and temporal trajectories to excavate related hotspot regions.

2 Optimization of K-means Clustering Algorithm for Spatiotemporal Trajectory Data

2.1 Space-Time Trajectory Data Pre-processing

Spatiotemporal trajectory data is the most important metric in the data mining process, and raw data noise point processing can effectively remove the disturbing signals mixed in the data samples, thus improving the overall performance [8]. In practice, different data processing schemes need to be designed for different experimental research environments as well as the practical needs of users. The spatio-temporal trajectory data studied in this paper are mainly offline trajectory data, and the data pre-processing process mainly includes: offline trajectory compression, trajectory filtering, feature extraction, and trajectory data cleaning.

Off-line compression of trajectory data is one of the common methods of preprocessing, mainly using the TDTR algorithm for time and trajectory distance, which reduces the amount of data without compromising accuracy and can better ensure the quality of data samples; Trajectory filtering is the key link in the pre-processing stage, using Kalman filtering to process the original data, which can effectively remove the noise interference in the data and make the speed and position information data in the trajectory more accurate [9]. For the feature extraction of spatiotemporal trajectory data, principal component analysis is mainly used to most disperse the data along the direction of maximum variance of high-latitude data. The feature extraction scheme can determine the features of the data samples initially, which has important reference value for the selection of the cluster number of K-means clustering algorithm [10]; Space-time trajectory data cleaning refers to the effective elimination of redundant and noisy points from the original data of space-time trajectory to obtain more accurate and as undisturbed experimental data as possible, so as to ensure that the experiment can dig out the information required by users.

Through trajectory feature extraction in trajectory data pre-processing, the K-means algorithm avoids the uncertainty of relying on manual experience for K-value selection when the number of clusters is initially determined.

2.2 Improved K-means Clustering Algorithm for Spatiotemporal Trajectory Data

In the K-means optimization algorithm, the unreasonable location of the randomly selected center point can lead to deviations in the collection of clustering samples and result in reduced accuracy and stability in the clustering process, so the algorithm is optimized as follows. The initial clustering center point Ki will be randomly selected in the set of trajectory data samples, and the subset Qi will be randomly selected in the data samples for clustering to obtain the clustering center Li. Si in the state of Li and Ki squared and maximum is selected as the initial clustering center, which can make the generated center point diffuse, and at the same time, the collection data gradually concentrate to the clustering center and update the clustering center, the optimization process of this algorithm can to some extent improve the problem that the initial clustering center point randomly selected resulting in the poor stability of the algorithm (Fig. 1).

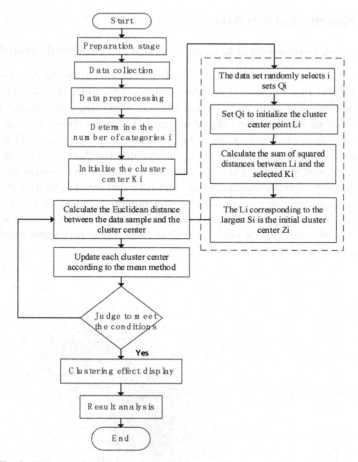

Fig. 1. K-means algorithm optimization process for space-time trajectory data.

It can be seen that the classical K-means algorithm is more likely to reduce the clustering efficiency and deteriorate the clustering effect if the clustering center points are generated completely randomly when dealing with a large amount of trajectory data. And after the algorithm is optimized, it mainly randomly selects a part of the trajectory data sample to form a small collection, and calls the algorithm in the small collection to regularly select the initial center, which can reduce the dispersion of points. At the same time, a portion of the data in the application sample is clustered first, which effectively improves the running rate and clustering efficiency of the algorithm. And a small part of the cluster collection is computed to update the cluster center, and then the objects of the dataset are assigned to the center point, and as the algorithm iterates and the change of the cluster center gradually decreases to the center stability, so that the clustering process ends. Based on the above optimization process of K-means clustering algorithm, a simulation platform is built by MATLAB to further analyze and evaluate the clustering effect of the classical K-means algorithm and the improved optimized K-means algorithm on the data.

3 Experiments and Analysis

3.1 Simulation Comparison of K-means Classical and Optimized Algorithms

Three sets of Gaussian-distributed data with statistical characteristics totaling 5000 were selected for simulating the preprocessed data samples of the trajectory data, while overlaying the random data generated by random seeds in the range [0,1] respectively as a clustered data sample design, as shown in Fig. 2. The figure shows the relative concentration of data in categories A, B, and C, indicating sample points with trajectory characteristics and conforming to a Gaussian distribution.But there are still some messy points that represent randomly added data points.The points with Gaussian distribution characteristics represent the simulated trajectory data, and the random data are used to simulate the uncertain disturbances in the spatiotemporal trajectory data and the noise perturbations in the data acquisition process, etc. The improved Gaussian distribution characteristics plus the random points make the clustering experimental results more convincing.

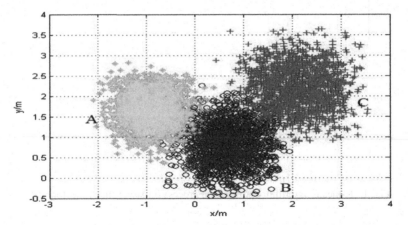

Fig. 2. Distribution of data samples for 5000 groups (Gaussian distribution + random points)

The classical K-means and optimized K-means clustering algorithms were used to cluster 5000 sets of experimental data consisting of Gaussian eigen-distributions and random numbers, respectively, and the experimental results of the algorithms are shown in Figs. 3 and 4.

Figure 3 shows that class A and B sample data are concentrated in the corresponding center, the effect is better, class C has some data discrete center, distributed out of the center is not aggregated by the class C cluster, resulting in a bias in the clustering effect. The experiments of the optimization algorithm are shown in Fig. 4, and the three types of experimental data are aggregated according to their proximity to the optimal center point and the appropriate distance between the clustering centers, showing good clustering results. Not only are the A, B, and C data clustered more centrally, but the three types of data are clearly spaced so that each category can be clearly identified. Compared with the classical algorithm, the optimized K-means algorithm can more accurately determine

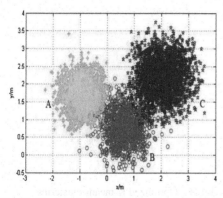

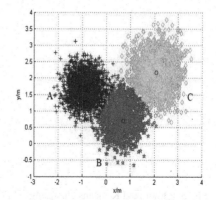

Fig. 3. Plot of experimental results of classical K-means clustering algorithm

Fig. 4. Experimental results for optimizing the K-means clustering algorithm

the clustering center of the sample data and make the sample data better identify the clustering center with different clustering conditions, and effectively converge to the nearest clustering center by numerical calculation and position update during iteration, thus efficiently completing the clustering experiments on big data samples (Gaussian feature distribution + random points).

3.2 Testing the Applicability of Optimal K-means Algorithm Under Data Sample Changes

In order to further verify the applicability of the optimization algorithm to the data sample detection, on the basis of the optimization K-means algorithm, the experimental data sample size is adjusted and the clustering analysis of the optimization algorithm is carried out to further verify whether the applicability of the optimization algorithm to the data sample capacity is more inclusive. According to the experimental conditions and basis, on the basis of the clustering classification cluster number K value of 3, the sample capacity of data of 3000, 6000, 8000 and 12000 were selected into the validation, and the results are shown in Fig. 5, Fig. 6, Fig. 7 and Fig. 8 respectively.

As shown in Fig. 8, the clustering experiment was conducted using 3000 data samples (Gaussian features + random). Three classes of data, A, B and C, have good concentration effect, and the data of classes B and C have intersection at the edges of the data, but the classification can still be discerned. All three classes of data have a part of data points discrete, and class A is the most concentrated at the center of clustering.

Using 6000 data samples (Gaussian features + random) for clustering experiments, the data sample capacity is twice as much as 3000 groups, the clustering effect of the three classes of data is obvious and the algorithm is more stable. Increasing the data sample capacity, the data of class A is more concentrated at the center point, but the clustering effect of class C data changes most obviously. Compared with class B data, not only the data objects are more concentrated and discrete centroids are relatively less; than class B data, the data capacity is more, which indicates that in the data samples

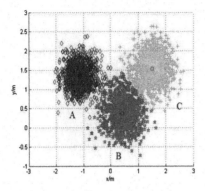

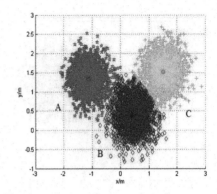

Fig. 5. Optimized K-means clustering algorithm 3000 groups of data clustering

Fig. 6. Optimized K-means clustering algorithm for 6000 groups of data clustering

conforming to the distribution of trajectory characteristics and randomly generated data, the data conforming to the clustering principle are closer to the centroids of class C data.

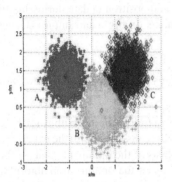

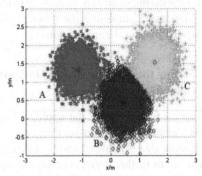

Fig. 7. Optimized K-means clustering algorithm for clustering 8000 groups of data

Fig. 8. Optimized K-means clustering algorithm for clustering 12000 groups of data

Using 8000 groups of data samples (Gaussian features + random) for clustering, the degree of data dispersion, the three classes of data are produced with edge data at a distance from the center point, although in line with the corresponding classification principles, but not fully aggregated, Class C images have more discrete data and the most obvious dispersion effect; Class A data have the least discrete data, the highest degree of aggregation and better results. Comparing 3000 groups and 8000 groups of data images, it is found that class A centroids in the case of less or more data, the clustering effect is more stable and produce the least discrete data; class C centroid position is more suitable for the middle 6000 groups of data sample size, before and after comparison in this group of data, the most obvious change in clustering expression good clustering effect and less noisy data.

And using 12,000 sets of data samples (Gaussian features + random) for clustering, it is found that the clustering indicates that the maximum effect is achieved, and all three classes of data are better done with the produced data. The clustering boundaries of class B data and class C data are more obvious, and the range of center selection under the principle of clustering can be explored; the larger the data sample capacity, the more samples meet the clustering criterion, and the law of data at a greater distance shows up, accommodating more data in the clustering, the less noisy data from the center point.

It can be seen that in different data sample cases (such as 3000, 6000, 8000, 12000), the data discrete point transformation, class B data discrete points from more reduce to tend to stabilize and find the clustering law; class A data in the sample data at least or most, the noise points are less. In the clustering effect, the result of class A data is more stable without obvious changes, and the effect is good; the clustering effect of class B and C data becomes more and more obvious as the amount of data increases. The algorithm is more obvious in the clustering effect, and the data samples gathered in the center of the clusters are more obvious and concentrated as the data samples increase.

3.3 Optimization of K-means Algorithm Performance Detection Experiments Under K-value Variation

Usually, the selection of K-values relies on empirical judgments, and considering that the location of K-values has an important impact on the clustering results, the design adjusts K-values to detect and verify the performance of the optimized algorithm.The experimental data consisted of 6000 sets of sample points composed of Gaussian feature distribution and random data, and the optimization algorithm was used to perform the experiments under adjusting the cluster number K values K = 2, K = 3, K = 4, and K = 5, respectively, and the results are shown in Fig. 9, Fig. 10, Fig. 11, and Fig. 12.

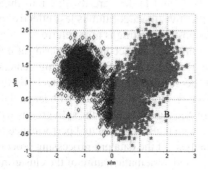 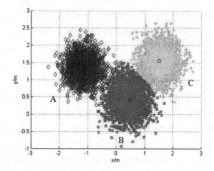

Fig. 9. Optimization K-means algorithm 6000 sets of data K = 2 clustering results.

Fig. 10. Optimization K-means algorithm 6000 sets of data K = 3 clustering results.

It can be seen that the optimal setting value for K is 3.Using the same data sample, experiments were conducted with K set to 2, 3, 4, and 5, respectively. When K = 2, although there are two clusters of circle points and only class A and B data, but there are still three clusters can be seen position, so the K value set unreasonable. When K = 3, the experimental data are classified into three circle centers, the data of class A, class B and class C have a certain range interval, the boundaries of each data category

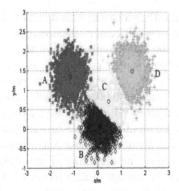
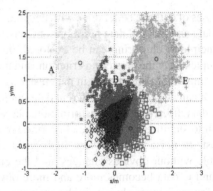

Fig. 11. Results of K = 4 clustering experiments for optimizing the K-means algorithm for 6000 sets of data.

Fig. 12. Results of K = 5 clustering experiments for optimizing the K-means algorithm for 6000 sets of data.

are obvious and the distribution of noise data is more balanced without a large number of mutually cross-identified positions, and the clustering effect shows good. When K = 4, there are four clustering circles and the data are also distributed according to the clustering rules to form four types of data with certain patterns. Category A and D data are more reasonably located, with data allocated to the vicinity of the corresponding centroid based on distance; Category B and C data are located closer together, with close intersection and unclear boundaries between the data. When K = 5, the figure shows five clustering circles, where class E data are best clustered but the noisy data are more scattered; class A data have the next best effect, with boundary lines clearly classified accurately and slightly noisy.Although class B, C and D data are divided into three categories, the graphical classification is less effective, the algorithm is less effective in classifying them closer together, and the center position setting is not preferred.

Therefore, the setting of the K-value has an important impact on the clustering effect. If the K-value of the algorithm is set too small, there will be few clustering centers, and clusters with small sample size will be difficult to be identified effectively by the algorithm. If the number of classifications K is set too large, some groups with larger data sample points in the original data will be split by the algorithm, and the split groups will also interfere with other well-classified groups, which affects the actual performance of the algorithm. Thus, conducting an effective analysis of the data pre-processing process and K-value selection can improve the scientific and systematic nature of the clustering algorithm.

4   Conclusion

As a popular research field of big data mining in recent years, the improvement and optimization of space-time trajectory data clustering algorithm in pursuit of more accurate and faster data mining efficiency is a hot issue in research.The main drawback of the classical K-means algorithm, which uses random selection of initial centroids, is the randomness of the selected centroids, which is prone to low accuracy and falling into local optimal false clustering.In this paper, an optimization study of the K-means cluster mining algorithm for spatiotemporal trajectory data is carried out based on the inadequacy of the classical algorithm, and clustering experiments are conducted using 5000 sets of spatiotemporal trajectory sample data (Gaussian feature distribution + random data sample points). Compared with the classical K-means, the optimized K-means clustering algorithm can, to a certain extent, accurately determine the clustering center of the sample data clusters, so that the sample data can be better identified to the clustering center of different clustering conditions, and the sample data can be automatically aggregated to the cluster center that meets the corresponding clustering conditions according to its own characteristics, thus improving the accuracy and efficiency of the optimization algorithm.

References

1. Song, J.: Research on Improvement of K-means Clustering Algorithm. Anhui University, Hefei (2016)
2. Chudova, D., Gaffney, S., Mjolsnes, E., et al.: Translation invariant mixture models for curve clustering. In: Proceedings of the 9h ACM SIG KDD Int Conference on Knowledge Discovery and Data Mining, pp. 79–88. ACM, New York (2003)
3. Alon, J., Sclaroff, S., Kollios, G., et al.: Discovering clusters in motion time-series data. In: Proceedings of the 2003 IEEE Conference on Computer Vision and Pattern Recognition, pp. Los Alamitos. IEEE Computer Society, Los Alamitos (2003)
4. Nanni, M., Pedreschi, D.: Tim e-focused clustering of trajectories of moving objects. J. Intell. Inf. Syst. 27(3), 267–289 (2006)
5. Wang, Q., Jiang, Z.: Improved K-Means Initial Clustering Center Selection Algorithm. Guangxi Normal University, Guilin (2010)
6. Zhang, Y.: Improved K-Means Algorithm Based on Optimizing Initial Cluster Centers. Northeast Normal University, Changchun (2013)
7. Wang, J., Ma, X., Duan, G.: Improved K-means clustering K-value selection algorithm. Comput. Eng. Appl. 55(08), 27–33 (2019)
8. Niu, Q., Chen, Y.: Hybrid clustering algorithm based on KNN and MCL. J. Guilin Univ. Technol. 35(01), 181–186 (2015)
9. Wang, D.: Inertial Navigation Technology Based on Kalman Filtering. North China University of Technology, Changchun (2018)
10. Feng, Q., Jiaxin, H., Huang, N., Wang, J.: Improved PTAS for the constrained k -means problem. J. Comb. Optim. 37(4), 1091–1110 (2019)
11. Deng, X., Chaomurilige, W., Guo, J.: A survey of cluster senter initialization algorithms. J. China Acad. Electron. Inf. Technol. 14(04), 354–359+372 (2019)
12. Li, Y., Shi, H., Jiao, L., et al.: Quantum-inspired evolutionary clustering algorithm based on manifold distance. Acta Electronica Sinica 39(10), 2343–2347 (2011)
13. Zhang, L., Chen, Y., Ji, Y., et al.: Research on K-means algorithm based on density. Appl. Res. Comput. 28(11), 4071–4073+4085 (2011)
14. Jiang, T., Yin, X., Ma, R., et al.: Bus load situation awareness based on the k-means clustering and fuzzy neural networks. J. Electric Power Sci. Technol. 35(03), 46–54 (2020)
15. Zhou, S., Xu, Z., Tang, X.: Method for determining optimal number of clusters in k-means clustering algorithm. J. Comput. Appl. 30(08), 1995–1998 (2010)

Temporal and Spatial Changes of Ecosystem Health in Guangdong Province in Recent 11 Years

Nan Zhang, Yong Xu$^{(\boxtimes)}$, Shiqing Dou, Juanli Jing, and Hanbo Zhang

College of Surveying, Mapping and Geographic Information, Guilin University of Technology, Guilin 541000, China

1415439699@qq.com, xuyongjiangsu@163.com

Abstract. As a coastal province in southern China, the economy of Guangdong has been rapidly developed after the implementation of the reform and opening up. In the past decades, the authority of Guangdong has emphasized ecological health construction while paying attention to economic development. Meanwhile, the construction of "the Belt and Road Initiatives" has posed great challenges to the health of the local ecosystem. Based on the Vigour -Organization-Resilience Model (VOR), the ecological health assessment system was established to evaluate and analyze the ecosystem health from 2009 to 2019 of Guangdong. Results showed that the ecosystem health of Guangdong showed a slight improvement for the study period, The areas with greater than level 2 had decreased by 8.42%. The overall ecological system health varied from general to good. The ecosystem health of the study area exhibited obvious spatial heterogeneity. The level of the ecosystem health in the northern part was better than the southern part. The ecosystem health of the study area decreased slightly at the beginning and then increased significantly, and the ecosystem health showed a stable and positive trend.

Keywords: Guangdong region · Ecosystem · Health assessment · VOR model

1 Preface

Regional ecosystem health is the core of comprehensive evaluation of ecosystem health, and it is also an important foundation for environmental protection and sustainable development [1]. It has become a hot spot and trend in ecological environment research and social development. Therefore, a reasonable evaluation of the regional ecological health status and timely discovery of the regional ecosystem health status play a very important role in the sustainable development of the regional economy [2].

VOR (vigour-Organization-resilience model, vitality-organization-resilience model) as an objective, large-scale, scientific and reasonable ecosystem health evaluation model has a wide range of applications in regional ecosystem health research. Many scholars use the VOR model to analyze the regional ecosystem, Such as Dianchi Lake Basin [3], the central coast of Jiangsu [4], Wenchuan earthquake area [5], etc. to conduct

© Springer Nature Switzerland AG 2021
Q. Zu et al. (Eds.): HCC 2020, LNCS 12634, pp. 114–125, 2021.
https://doi.org/10.1007/978-3-030-70626-5_12

health evaluation research, objectively reflect the health of the regional ecosystem. Due to its special geographical location and climatic conditions, Guangdong Province [6] has very rich natural resources, but with the development of social economy and the impact of human activities, the pressure on the ecological environment and sustainable development of the region is facing huge challenges. Therefore, this article attempts to use the VOR model, remote sensing and GIS and other means to build an ecological health evaluation system for Guangdong, using the analytic hierarchy process to calculate the weight of each system index and calculate the ecological health index, and discuss the spatial distribution and temporal changes of ecosystem health in Guangdong Features, in order to provide theoretical guidance and scientific basis for the rational development and utilization of local natural resources and the protection of regional ecosystem health.

2 Overview of the Study Area

Guangdong Province [7] is located in the southernmost part of mainland China, between 109°39′–117°19′E east longitude and 20°13′–25°31′N north latitude, bordering Fujian in the east, Jiangxi and Hunan in the north, Guangxi in the west, South China Sea in the south, and the Pearl River Estuary borders the administrative regions of Hong Kong and Macau respectively. The whole terrain is mostly northeast mountain range. Zhujiang City, located in the south-central part of the autonomous region, has the largest plain in Guangdong, the "Pearl River Delta Plain". Guangdong is located at a low latitude, with the Tropic of Cancer traversing the middle of the region. It belongs to the Asian-East Asian monsoon climate zone with an average annual temperature of 19–24 °C. Most of the region has a warm climate, abundant heat, abundant rain, clear dry and wet, insignificant seasonal changes, moderate sunshine, less winter and more summer [8]. Guangdong has a long coastline and more excellent port resources. It is an important channel for domestic and foreign traffic and trade in our country, and it is the most convenient access to the sea and a window for opening to the outside world in southern China (Fig. 1).

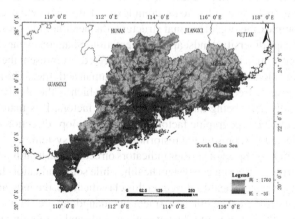

Fig. 1. Location of study area

3 Data Collection and Preprocessing

Collected MODIS satellite remote sensing images covering the entire study area in 2009, 2014 and 2019, including MCD12Q1, MOD17A3HGF, MOD13A3, MOD11A2. Used for the extraction of ground feature types in the study area and calculation of ecological health indicators The digital elevation model data comes from (https://srtm.csi.cgiar.org/srtmdata/), The rest of the data comes from (https://ladsweb.modaps.eosdis.nasa.gov/).

The image processing first uses MODIS reprojection tool (MODIS reprojection tool, MRT) to mosaic, project, resampling and data format conversion processing of the image, and combine the administrative vector boundary map of Guangdong area to crop the processed image. MCD12Q1 data [9] is the data classified by the supervisory decision tree, including 5 types of land cover, of which there are 17 types of ground features in Type 1. Supported by ArcGIS 10.6, the study area landscape is divided into 7 types: cultivated land, woodland, grassland, water body, shrubland, construction land and bare land by reclassification method. The interpreted images are stitched, mosaic and cropped. Obtain the landscape type map of the study area. Use Fragstats4.2 to extract the landscape pattern index, and use ArcGIS 10.6 for spatial analysis and mapping [10].

4 Research Methods

4.1 Construction of Evaluation Index System

This paper selects vitality, organizational power and resilience from three perspectives of resource environment, landscape ecology, and human activities to form the ecosystem health evaluation subsystem in Guangdong. Vitality refers to the energy accumulation and nutrient cycling of an ecosystem. It is a major indicator of the metabolism and productivity of the system and is determined by the coverage rate of green plants and light cooperation. Photosynthesis interacts with temperature and vegetation growth. Therefore, normalized difference vegetation index NDVI [11], net vegetation primary productivity NPP [12] and surface temperature value [13] were selected to represent the system vitality. Organizational power refers to the complexity of an ecosystem. Generally speaking, the more complex the organizational structure, the healthier the ecosystem will be. Therefore, landscape diversity, landscape fragmentation, Shannon evenness index, average patch area and human disturbance index were selected to represent the complexity of ecosystem organization structure and the impact of human activities. The restoring force refers to the relative degree of the system elasticity, which is the ability of the system to recover gradually after being disturbed by external factors. It is mainly affected by landscape elements and topographic factors, so the slope, slope direction and ecological elasticity are selected to represent the restoring force of the system.

Due to the complex impact of various indicators on ecosystem health, some indicators have a positive driving effect on ecosystem health, while some indicators have a negative driving effect on ecosystem health. Based on this classification, the indicator system and indicator types of ecological health assessment in Guangdong are shown in Table 1.

By referring to the research of relevant scholars [14] and combining with the actual situation of the research area, the specific scores of ecological elasticity of different landscape types are shown in Table 2.

Table 1. Ecosystem health evaluation index system and index types in Guangdong

Evaluation model	Evaluation subsystem	Evaluation index	Index type
Ecosystem health assessment	Vitality	NDVI	Positive
		NPP	Positive
		LST	Negative
	Organization	Landscape diversity	Positive
		Landscape fragmentation	Negative
		Shannon uniformity index	Positive
		Average patch area	Positive
		Human interference index	Negative
	Resilience	Slope	Negative
		Aspect	Negative
		Ecological resilience	Positive

Note: Ecological resilience is the ability to self-regulate and recover when the ecosystem is under pressure or damage [15].

Table 2. Ecological elasticity score table of different landscape types

Landscape type	Arable land	Woodland	Grass	Water	Bush	Construction land	Bare land
Ecological resilience score	0.5	0.9	0.7	0.9	0.8	0.2	0

4.2 Standardization of Evaluation Indicators

Ecosystems are complex and changeable. Therefore, the criteria for evaluating ecosystem health are also multi-indicators. As the data sources and types of different indicators are not the same in various aspects, in order to make the indicators comparable to each other, this article adopts the range method [16]. The indicators in Table 2 are standardized, and the calculation formula is as follows:

1) Positive health significance index:

$$Y = (X - X_{min})/(X_{max} - X_{min}) \tag{1}$$

2) Negative health significance index:

$$Y = (X_{max} - X)/(X_{max} - X_{min}) \tag{2}$$

Where: Y is the standardized value of the indicator, X is the actual value of the indicator, Xmax is the maximum value in the indicator observation data, and Xmin is the minimum value in the indicator observation data.

4.3 Determination of Evaluation Index Weight

Ecosystem health evaluation is multi-objective, multi-criteria and complex. Analytic hierarchy process (AHP) [17] is adopted to decompose the complex evaluation system, and the decision-making problem that is difficult to be fully quantified is reduced to a multi-level single-objective problem, and the weight value of each index is scientifically determined. The weight value between the calculated evaluation index and the subsystem and the ecological health evaluation system is shown in Table 3.

Table 3. Ecological health evaluation index system and its weights in Guangdong

Evaluation model	Subsystems and weights	Evaluation index	Relative to subsystem weights	Relative to the weight of ecological health assessment system
VOR ecosystem health evaluation	Vitality (0.5483)	NDVI	0.2299	0.0497
		NPP	0.6479	0.1400
		LST	0.1222	0.0264
	Organization (0.1383)	Landscape diversity	0.4062	0.2768
		Landscape fragmentation	0.0675	0.0460
		Shannon uniformity index	0.2182	0.1487
		Average patch area	0.2182	0.1487
		Human interference index	0.0898	0.0612
	Resilience (0.3134)	Slope	0.6333	0.0650
		Aspect	0.1062	0.0109
		Ecological resilience	0.2605	0.0267

4.4 Ecosystem Health Assessment Model

This paper uses vitality, organization and resilience to reflect the health of the ecosystem. The comprehensive index method is used to calculate the evaluation indicators and subsystems and the ecological health indexes of the subsystems and ecosystems. A method often used in ecosystem health assessment research, so the following aspects should be considered when establishing an assessment model [18]:

The subsystem evaluation model is:

$$HI_i = \sum_{i=1}^{n} x_{ij} y_{ij} \tag{3}$$

Where Hi is the ecosystem health score of the i-th subsystem; is the value obtained by standardizing the j-th index of the i-th subsystem of Xij; Yj is the weight of the j-th index relative to the target layer; n is the number of evaluation indexes number.

The VOR model is:

$$HI = HI_V \times HI_O \times HI_R \qquad (4)$$

Where: HI is the ecosystem health index; HIV is the ecosystem health vitality index; HIO is the ecosystem health organization index; HIR is the ecosystem health resilience index.

4.5 Ecosystem Health Classification

According to the calculated ecosystem health index based on the existing research, the ecosystem health level of the study area is divided into 5 levels according to the equal discontinuity method [19], as shown in Table 4.

Table 4. Classification of ecosystem health

Health level	Health score	State of health
Level 1	0.8–1.0	Healthy
Level 2	0.6–0.8	Good
Level 3	0.4–0.6	General
Level 4	0.2–0.4	Poor
Level 5	0.0–0.2	Bad

5 Result Analysis

5.1 Health Status of Ecosystem Vitality

The spatial distribution of ecosystem vitality levels and the area occupied are shown in Fig. 2 and Fig. 3:

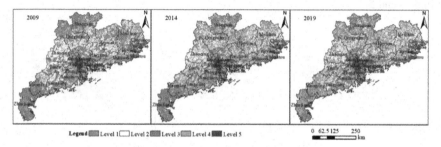

Fig. 2. Spatial distribution of ecosystem vitality levels

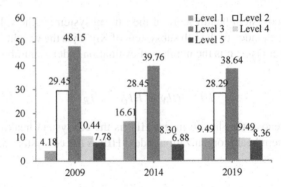

Fig. 3. Proportion of area of each level of ecosystem vitality

From the perspective of spatial distribution, the areas with high levels of vitality and health above Grade 3 in Guangdong account for about four-fifths of the entire study area, and most of them are concentrated in the northeastern cities bordered by Guangzhou and Huizhou, as well as Zhaoqing and Jiangmen cities. In the southwestern city of the boundary, this part of the relatively high terrain corresponds to the features of woodland, shrubs and grasslands. Areas with lower vitality levels are mainly concentrated in Foshan, Dongguan, Guangzhou, Shenzhen and other cities. The vitality level of the study area has changed to different degrees in the past 11 years, and the overall vitality level has shown a downward trend. It can be seen from the figure above that the proportion of the area of the second-level vitality area first increased to the peak in 2014 and then greatly decreased. By 2014, the proportions of the third-level and second-level areas were respectively higher Regional transformation, the overall ecological system vitality is on the rise. In 2019, the proportion of the area of the first-level healthy area plummeted, and the proportion of the fourth-level and fifth-level areas increased accordingly. The main reason is that with the development of the economy and the increase in urbanization, Although the local government has taken certain measures to protect the ecological environment, the intensity of human utilization of natural resources has increased, resulting in increased pressure on the ecosystem and reduced overall vitality.

5.2 Health Status of Ecosystem Organization

The spatial distribution and area share of ecosystem organization power levels are shown in Figs. 4 and 5:

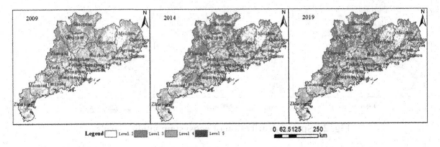

Fig. 4. Spatial distribution of ecosystem organization power level

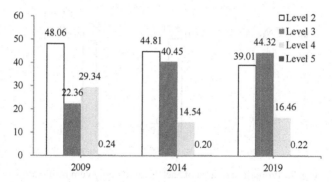

Fig. 5. Proportion of the area of each level of ecosystem organization

As illustrated, nearly 11 years, the organization in the study area for the secondary area ratio showed a trend of slow decline and level 3 area accounted for a significant increase in, four research area proportion was significantly decreased first and then slowly rising trend. The area of the second-level area of the study area has decreased by 9.05%, the area of the third-level area has increased by 21.95%, and the proportions of the area of the fourth and fifth-level areas have first decreased and then increased. Although after 2014, the government actively responded to the "Implementation Opinions" jointly issued by the Central Government and the State Council, advocated the concept of green development, actively promoted the construction of ecological civilization, and protected the ecological environment. Due to increased human interference, a large number of grasslands and bushes were destroyed, and the average plate area decreased. The area of woodland has increased significantly, and the diversity and fragmentation of landscapes have increased slightly. The organization of the ecosystem in the study area has slightly deteriorated in 2019 compared with 2014, but has improved significantly compared with 2009.

5.3 Health Status of Ecosystem Resilience

The spatial distribution of ecosystem resilience levels and the area share are shown in Fig. 6 and Fig. 7:

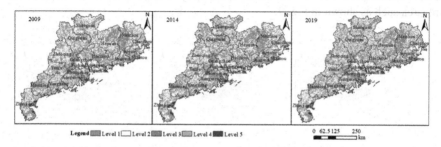

Fig. 6. Spatial distribution of ecosystem resilience levels

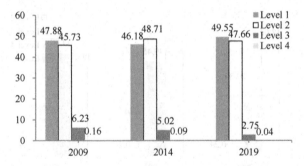

Fig. 7. Proportion of area of each level of ecosystem resilience

As illustrated, the resilience grades of the entire study area are mixed. The resilience grades have been relatively stable in 11 years. The northern areas such as Qingyuan and Shaoguan are significantly inferior to the surrounding cities. The entire ecosystem health of restoring force in the study area is in good condition. The proportion of areas with first-level resilience has increased by 1.67%, and the proportion of first-level areas has approached 50%. The proportion of the second-level area showed a downward trend and then increased, while the proportion of the area with a third-level resilience decreased in a stepwise manner, with an average annual decrease of 0.32%. The proportion of the area with a second-level resilience and above in the study area increased by 3.6%. Both rose nearly 0.33%. The overall resilience level of the study area tends to be stable. It can be seen from this that with the development of the economy, people's damage to the natural environment has further increased. In the past 11 years, with the implementation of the government's ecological civilization construction policy, the proportion of forest land in the study area has increased, and people have become more concerned about nature. Certain actions were taken to protect the environment to compensate for the damage caused by human activities to the natural environment, and the overall ecological resilience of the study area tended to be stable.

5.4 Analysis of Ecosystem Health

Integrated ecosystem vitality, organization and resilience from three aspects, we mapped the distribution level of ecosystem health in the study area space as shown in Fig. 8, each rank area such as shown in Fig. 9.

Fig. 8. Distribution of ecosystem health levels

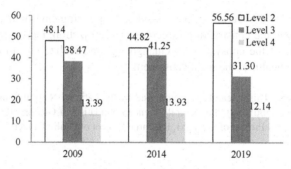

Fig. 9. Area ratio of each grade

From the perspective of spatial distribution, the ecological health status of the three periods in the research area is generally better than that in the northern region with Yunfu city, Foshan City, Huizhou City and Meizhou City as the boundary. The overall ecological health status of the study area has been in a stable state during the 11 years. The areas with good ecological conditions have an upward trend. The area of areas with normal ecological health status first increased slightly and then decreased significantly. The area occupied has decreased by 0.65% annually, and the area of areas with poor ecological health tends to be more stable. As shown in Fig. 8 and Fig. 9, the ecosystem health has stabilized from 2009 to 2014. The area ratios of the secondary and tertiary regions are transformed into each other. However, from 2010 to 2019, the areas with the second-level ecosystem health status increased significantly, the areas with poor ecosystem health status gradually decreased, while the areas with better ecological health status increased significantly. The reason is that with the construction of the "Belt and Road" after 2009, with the development of social economy and frequent human activities, people's interference index to the natural environment has increased. Human activities have damaged the natural environment, the fragmentation of landscape has increased, and the elasticity of ecosystems has decreased. As a result, the ecological and environmental health status of the study area has been transformed from the second-level to the third-level area, but the area of the second-level area in 2019 has greatly increased compared with the previous year. This shows that the local government has actively promoted ecological civilization in response to the national call With the development of construction, people are aware of ecological health problems and take actions to remedy them, make up for the damage to the environment caused by the economic development stage, and make the health of the ecosystem show a trend of improvement.

6 Conclusion

The ecological health of Guangdong has shown a slight improvement in the past 11 years, and areas with better ecosystem health have increased slightly, Although the proportion of areas with general health conditions or more in the three periods is generally stable, compared with the previous two periods, the proportion of areas with good health in the study area has increased slightly, indicating that the ecosystem health has a slight upward trend. Although the rapid development of social economy has caused great damage to the environment, the local government has issued corresponding environmental protection policies to promote the construction of ecological civilization, increase the protection and management of the ecosystem in Guangdong, and realize the improvement and restoration of the health of the local ecosystem.

Acknowledgements. This research was supported by the national Natural Science Foundation of China (41801071;42061059); the Natural Science Foundation of Guangxi (2020JJB150025); Bagui Scholars Special Project of Guangxi; Research Project of Guilin University of Technology (GUIQDJJ2019046).

References

1. Zhang, F., Zhang, J., Rina, W., et al.: Ecosystem health assessment based on DPSIRM framework and health distance model in Nansi Lake, China. Stoch. Env. Res. Risk Assess. **30**(4), 1235–1247 (2016)
2. Wang, Z., Tang, L., Qiu, Q., et al.: Assessment of regional ecosystem health-a case study of the golden triangle of southern Fujian Province, China. Int. J. Environ. Res. Public Health **15**(4), 802 (2018)
3. Zhang, Y.: VOR model-based Multi-scale Evaluation of ecosystem health in Dianchi Basin. Yunnan University of Finance and Economics (2020)
4. Hongjie, S., Dong, Z., Shunjie, S., et al.: Ecosystem health assessment in the central coastal zone of Jiangsu Province based on coupling model and remote sensing. J. Ecol. **38**(19), 7102–7112 (2008)
5. Zhu Jie-yuan, L., Hui-ting, W.-F., et al.: Ecosystem health evaluation in the recovery period of Wenchuan earthquake disaster areas. Acta Ecol. Sin. **38**(24), 9001–9011 (2008)
6. Ping, X.: Remote sensing monitoring and comprehensive assessment method for ecological environment of land resources – a case study of Guangdong Province. Anhui Agric. Sci. **48**(11), 77–81+84 (2020)
7. Yuliang, L.: Study on sedimentary characteristics and paleoclimate of changba formation in Danxia Basin, Shaoguan city, Guangdong Province. Donghua University of Technology (2018)
8. Jiang, C., Zhifeng, W., Cheng, J., et al.: Relative effects of climate fluctuation and land cover change on net primary productivity of vegetation in Guangdong province. J. Trop. Subtrop. Flora **24**(04), 397–405 (2016)
9. Yabo, H., Shunbao, L.: Automatic extraction of land cover samples from multi-source data. J. Remote Sens. **21**(5), 757–766 (2017)
10. Lixin, N., Lan, M., Yunkai, Z., et al.: Spatio-temporal changes of coastal ecosystem health in Jiangsu province based on PSR model. China Environ. Sci. **36**(02), 534–543 (2016)
11. Junjun, H.: Spatio-temporal variation characteristics of NDVI and its response to meteorological factors in the Pearl River Delta region based on MODIS data. J. Ecol. Environ. **28**(09), 1722–1730 (2019)
12. Shuang, L., Honghua, R.: Analysis of forest biomass and NPP spatial pattern in Guangdong and Guangxi based on geostatistics. J. Ecol. **32**(09), 2502–2509 (2013)
13. Jianbo, X., Zhifeng, X., Linyi, Z., et al.: Hj-1b thermal infrared LST inversion and its error accuracy analysis using partial differential. J. Wuhan Univ. (Inf. Sci. Ed.) **41**(11), 1505–1511 (2016)
14. Min-hua, L., Gui-hua, D.: Ecosystem health assessment in Qinhuangdao region supported by RS and GIS. Geogr. Res. **05**, 930–938 (2006)

15. Qian, N., Zhou, X., Ji, Z., et al.: Analysis of ecosystem elastic changes in Karst Mountain cities—taking Guiyang City as an example. Res. Environ. Yangtze Basin **28**(03), 722–730 (2019)
16. Caihe, T., Wei, Z., Qiang, C., et al.: Ecosystem health assessment based on PSR model: taking Chongqing section of the three gorges reservoir area as an example. J. Guilin Univ. Technol. **39**(03), 713–720 (2019)
17. Ye, X., Fan, Y., Changzhou, Y.: Xiongan City wetland ecological health evaluation based on analysis of landscape pattern. Acta Ecol. Sin. **40**(20), 7132–7142 (2020)
18. Yunkai, Z., Jiahu, B.X.J.: Study on the temporal and spatial changes of ecosystem health in Poyang Lake area in recent 17 years. Acta Sci. Circum. **32**(04), 1008–1017 (2012)
19. Peng, C.: A landscape-scale regional ecological health assessment based on remote sensing and GIS: a case study of the new bay city area. Acta Sci. Circum. **10**, 1744–1752 (2007)

Salient Attention Model and Classes Imbalance Remission for Video Anomaly Analysis with Weak Label

Hang Zhou, Huifen Xia, Yongzhao Zhan$^{(\boxtimes)}$, and Qirong Mao

School of Computer Science and Telecommunication Engineering, Jiangsu University, Zhenjiang 212013, Jiangsu, China
henrryzh@qq.com, xhf_ishere@qq.com, {yzzhan,mao_qr}@ujs.edu.cn

Abstract. Recently, weakly supervised anomaly detection has got more and more attention. In several security fields, realizing what kind of anomaly happened may be beneficial for security person who have preparation to deal with. However, lots of studies use global features aggregation or topK mean, and it exists feature dilution for anomaly. An attention model is proposed to generate the segment scores, i.e. we propose a salient selection way based on attention model to efficiently detect and classify the anomaly event. With these selected highlighted features, graphs are constructed. Graph convolutional network (GCN) is powerful to learn the embedding features, anomaly event can be expressed more strongly to classify with GCN. Because normal events are common and easy to collect, there is a problem that the normal and abnormal data are imbalance. An abnormal-focal loss is adapted to reduce influence of large normal data, and augment the margin of normal and different anomaly events. The experiments on UCF-Crime show that proposed methods can achieve the best performance. The AUC score is 81.54%, and 0.46% higher than state-of-the-art method. We obtain 58.26% accuracy for classification, and the normal and anomalies are separated better.

Keywords: Salient attention model · Classes imbalance · Graph convolutional network · Abnormal-focal loss

1 Introduction

With the development of the information technology, anomaly detection in surveillance video has been attracted more and more attention in security field. It takes too long to process the videos and find out the anomaly manually. Intelligent anomaly detection [1, 2] becomes a key technology to alert people or relive the manual viewing pressure, but there are so many semantic information in videos. With the limit of the dataset collection, it's hard to just focus on abnormal events.

Along with the advancements of deep learning and video analyze technology, the complex video anomaly detection (CVAD) is divided into two genres. Because abnormal events deviate from normal, the first one is unary classification solved by unsupervised methods [3–6], which only apply the normal video to train in the training phase. Without

© Springer Nature Switzerland AG 2021
Q. Zu et al. (Eds.): HCC 2020, LNCS 12634, pp. 126–135, 2021.
https://doi.org/10.1007/978-3-030-70626-5_13

the anomaly information induction, this method may be not sensitive for some abnormal events. The second method is binary classification [7–12] which adopts the normal and anomaly message with video labels. Review of previous studies, multiple instance learning (MIL) has been mostly exploited for CVAD with weak label.

In this paper, we also treat the CVAD as MIL task. Previous researches have only considered whether anomalies occurred, thus, MIL can suit well. The model performance degrades when classification tasks are involved. In many scenes, knowing what kinds of abnormal events happening could be a preparatory role for those police. We combine the anomaly detection with anomaly classification like the weakly supervised action localization. Firstly, attention model is adopted to dig the abnormal clips, and then we aggregate the salient departments to classify the anomaly events. The attention model is employed to achieve the detection part. As for the classification, we use topK selection to generate a highlight feature to improve the performance, and adapt abnormal-focal loss to classify abnormal events. The main contributions of our work are summarized below:

- A topK selection method based on attention model is proposed which can aggregate the video feature to get more powerful embeddings.
- An abnormal-focal loss is proposed to relieve the imbalance influence, where exists imbalance between normal and anomaly events. The abnormal events can be classified better with the loss.
- The experiments are conducted on UCF-Crime. The results demonstrate our framework can get excellent performance compared with current methods.

2 Related Work

Video Anomaly Detection: Video anomaly detection has been researched for many years as a challenging problem. A statistical method is applied to estimate the probability density and determining whether events follow a normal or abnormal distribution in [1, 2]. [3] proposes Gaussian mixture model (GMM) with Markov chains for detection with anomalous features, and the sparse learning [4] of normal patterns is a better way to make inferences. These methods are not universal which depend on the experience of suitable features.

Video analysis with deep learning has receive more and more success in recent years. Researchers define an open-set problem for this detection task. The anomaly is boundless, they want to dig the abnormal events different from normal. Several researchers apply generative model [5, 6] to reconstruct or predict the normal frames in the training phase with only normal videos. In order to improve the detection performance, a few anomaly information is needed to be utilized. [7] proposes a complex scenarios abnormal event detection method as a baseline, the MIL measure is adopted to address the problem. [10–12] considers that the supervised way could improve the training effects, transforming the weakly label to supervised label is essential so that they proposal a noise cleaner to get fine grained labels. Review the recent years' studies, anomaly classification is not involved. In this paper, we incorporate the classification with detection problem. Attention model is used to assign the scores to all the features, and topK indices embeddings are selected to aggregate more expressive features.

Weakly Supervised Action Localization (WSAD): WSAD [13–16] is one type of temporal action localization. There are two main methods to solve the problem. One is based on class-agnostic attention model with class activation sequence (CAS), the other relieves influences of background frames which are dynamic and difficult to split from the context of action frames. Our work is almost similar to the first one. From the salient anomaly event clips to classify them, the difference is that the background is also important in our framework.

3 Proposed Method

Suppose that there are N training video $\{V_i\}_{n=1}^N$ with video labels $\{Y_n\}_{n=1}^N$, where Y_i is C-dimensional one-hot label. Especially, the normal label is zero vector. We split a video $V_i = \{v_{ij}\}_{j=1}^T$ into T segments, each segment is composed of 16 consecutive frames. The goal of our work is to detection which segment is abnormal and what kind of anomaly has happened.

In order to relieve the influence of features dilution and imbalance [17] of normal class and anomaly classes, salient selection and abnormal-focal loss are proposed to solve these problems. The overall architecture is illustrated in Fig. 1. We use the I3D [18] model to extract the local spatial-temporal features. The features are applied to generate the probability of segments by attention model. After sorting of each segment score to select the topK scores, graph is considered to represent the relations between features similarity and segment scores. Two views of graph [19–21] are composed of nodes and edges to learning robust embeddings. In classification phase, an abnormal-focal loss is applied to solve classes imbalance problem, so that we can classify anomaly events better.

3.1 Anomaly Detection Model

Feature Extraction. Each video V_i has been divided into 16-frame non-overlapping segments. In training phase, we sample a fixed number of T segments from each video because of some long and variable video lengths. Then we feed segments into pre-trained I3D model to get embedding features $X \in \mathbb{R}^{T \times F}$. The extraction operation can be defined as Eq. (1):

$$X = Extra_{i3d}(V_i) \tag{1}$$

Anomaly Detection. Anomaly detection is to locate where the event has happened in a video. Utilizing regression method to solve it is suitable. The same as [12, 13], 1D convolution neural network is applied to capture the time dependence. This can be formalized as follows:

$$S = \sigma(f_{conv}(X; \phi)) \tag{2}$$

where σ is sigmoid function. $S \in \mathbb{R}^T$ is class-agnostic attention scores for T segments. ϕ denotes training parameters in CNN layer.

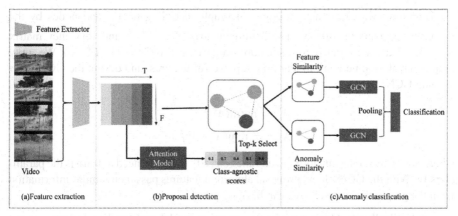

Fig. 1. Overview of our proposed method. Three parts have been exhibited: (a) feature extraction, (b) anomaly detection, and (c) anomaly event classification

3.2 Anomaly Features Selection

In [12], topK mean-pooling is applied to aggregate features, but the features are easy to be diluted in large scale anomaly video with more background class. So as to highlight these features, we consider a new topK approach which is changed to select features according to segments score. It can be defined as following:

$$
\begin{cases}
index = \text{topK}(\text{rank}(S_i), k) \\
\tilde{X} = X(index, :) \\
\tilde{S} = S(index, :)
\end{cases}
\tag{3}
$$

where rank() is the sort function, topK(, k) is to get indexes of the first k values. The fine-grained candidate abnormal features $\tilde{X}$ and scores $\tilde{S}$ are generated by the indexes slices of X and S.

3.3 Anomaly Classification

In order to explore better representation of anomaly event, graph convolution network is adopted to learn the relationship of each event. The same as [10], two views of affiliation have been considered to construct these graphs.

Feature Similarity Graph. Feature similarity is the first way used to measure them. We consider the cosine function as the measurement. It is defined as follows:

$$
f\left(\tilde{X}_i, \tilde{X}_j\right) = \text{cosine}\left(\tilde{X}_i, \tilde{X}_j\right) = \frac{\tilde{X}_i^{\mathsf{T}} \tilde{X}_j}{\left\|\tilde{X}_i\right\|_2^2 \cdot \left\|\tilde{X}_j\right\|_2^2}
\tag{4}
$$

where $\tilde{X}_i, \tilde{X}_j \in \mathbb{R}^F$ represent the $i^{\text{th}}, j^{\text{th}}$ segment. f measures the similarity of every two segments.

$$
A_{ij}^F = \begin{cases}
f(\tilde{X}_i, \tilde{X}_j) & f(\tilde{X}_i, \tilde{X}_j) > \alpha \\
0 & else
\end{cases}
\tag{5}
$$

Therefore, we can construct edges of graph according to A_{ij}^F, and nodes by the proposal segments features $\tilde{X}$. The features matrix $\tilde{X} \in \mathbb{R}^{T \times F}$ and adjacency matrix $A_F \in \mathbb{R}^{T \times T}$ are the sign to express feature similarity graph $G^A = (V, E^F, X)$. We apply graph convolution network to learn more powerful features, and design the GCN layer as plain GCN:

$$\hat{A}^F = \text{norm}(A^F)$$
$$\tilde{X}_{l+1}^F = \text{relu}(\hat{A}^F \tilde{X}_l W_l^F) \tag{6}$$

where norm() is Laplacian normalized function [10]. What we need to learn is the parameters W. With the GCN block, proposal abnormal features pass their similar information from the graph, the same classes can be aggregated.

Anomaly Similarity Graph. From the attention model and top-K selection, we get candidate segments and scores. Different scores represent different level of anomaly, capturing the same level can also help us classify different anomaly. As same as above feature similarity, anomaly similarity is defined as follows:

$$A_{ij}^A = e^{-\left|\tilde{S}_i - \tilde{S}_j\right|} \tag{7}$$

where $\tilde{S}_i$, $\tilde{S}_j$ are the ith, j^{th} candidate scores. A^A is the affiliation matrix of abnormal level. The anomaly similarity graph $G^A = (V, E^A, X)$ is contributed by features matrix X and edge matrix A^A. Then, GCN block is utilized to learn same levels message delivering, same as above:

$$\hat{A}^A = \text{norm}(A^A)$$
$$\tilde{X}_{l+1}^A = \text{relu}(\hat{A}^A \tilde{X}_l W_l^A) \tag{8}$$

Finally we perform an operation of weighted fusion between $\tilde{X}_{l+1}^A$ and $\tilde{X}_{l+1}^F$, and project the fusion feature to $Z \in \mathbb{R}^D$, softmax is used for classification. The operation can be defined as follows:

$$Z = \tilde{S}^{\text{T}}(\theta \cdot \tilde{X}_{l+1}^A + (1 - \theta)\tilde{X}_{l+1}^F)$$
$$P = \text{softmax}(Z) \tag{9}$$

3.4 Loss Function

Localization and classification are our work. We use MSE to measure localization:

$$L_{loc} = \frac{1}{n} \sum_{i=1}^{n} (max(S_i) - L_{V_i})^2 \tag{10}$$

where L_{V_i} is the label of whether the ith video is anomalous. *Max* acquires the maximal abnormal probability of a video. According to the loss function, the proposal segment can be detected more precisely, and it is beneficial to classify.

In the classification phase, the dataset is imbalance [14] when considering the classes. There are lots of normal data and several different anomaly data. In order to mitigate the imbalance influence and avoid model to learn normal pattern mostly, the abnormal-focal loss (AFL) function is proposed and defined as follows:

$$L_{cls} = \begin{cases} -\beta^a(1 - P_{a;i})^\gamma \log(P_{a;i}) & (y_{a;i} = 1) \\ -\beta^n(max(P_n) - t)^\gamma \log(1 - max(P_n) + t) & (y_n = 0) \end{cases} \quad (11)$$

where β^a, β^n is the abnormal rate and normal rate in a batch data, $P_{a;i}$ is the score of ith anomaly class, P_n is the score of all the predict classes with normal label. t is used to adjust the normal score float and avoid gradient explosion and disappearance.

Finally, a total loss is weighted by above two loss, the loss is defined as follows.

$$L = \lambda L_{loc} + L_{cls} \quad (12)$$

4 Experiments

4.1 Experimental Setting and Metric

Dataset. UCF-Crime [7] is the weakly labeled anomaly dataset frequently-used for CVAD. It has 13 types of anomalous events of 1900 long unnumbered videos with video-level labels. There are 1610 training videos which consists of 800 normal videos and 810 abnormal videos. 290 videos are used for test with temporal annotations.

Evaluation Metric. Following standard evaluation metrics as [7], in order to measure detection performance, we computer the area under the ROC curve (AUC) as the metric. Meanwhile, false alert is important for people to care. False alarm rate is applied for another metric. The accuracy is also applied to measure the classifier.

Implementation Details. In feature extraction phase, we use the pre-trained I3D model on Kinetics to extract every 16 frames which are resized to 224×224, and fix the number of input segments T to 400 with random perturbation. All the best hyperparameters are defined as: $k = 0.125$ in Eq. (3), $\theta = 0.5$ in Eq. (9), $\gamma = 2$ in Eq. (11), $\lambda = 0.5$ in Eq. (12).

4.2 Results and Analysis

In order to select best value of k in topK selection phase, the comparison experiment is needed and shown in Table 1. It shows the influence of different k value, the best performance can be obtained when the value of k is 0.125.

Table 2 summarizes the results of AUC comparisons. We compared most of current popular methods. Our method achieves best AUC in current techniques. It advances about 0.46% from SOTA performance [10]. Compared with the weakly supervised action location algorithm STPN, our method obtains higher performance. The false alarm of ours is also the lowest in test videos. The classification experiments are almost self-comparisons.

Table 1. Comparison on UCF-Crime under different value of k in TopK

k value	AUC (%)	Accuracy (%)
0.5000	80.81	52.67
0.2500	81.07	56.55
0.1250	81.54	58.26
0.0625	80.85	55.74

Table 2. Frame level AUC (%) comparisons.

Methods	AUC (%)	False alarm (%)
Binary	50.00	–
Hasan et al. [5]	50.60	27.2
Lu et al. [6]	65.51	3.1
Sultani et al. [7]	75.54	1.9
Zhang et al. [9]	78.66	–
Zhu et al. [8]	79.0	–
STPN [13]	78.88	2.1
Zhong et al. [10]	81.08	2.8
Ours	**81.54**	**1.9**

Table 3 demonstrates the results under two kinds of supervision, the train set and test set setting are different for them. As same as [10], the supervised methods also need to be contrasted. We can see that our network still performs best. In order to explore the salient features selection, our proposal method and STPN are applied to aggregate all the features. STPN only get 53.22% accuracy, and our method's accuracy is 58.26% with 5.04% increase against STPN. With the attention model to choose which features are most abnormal, anomaly classes can be recognized more exactly. It also can be improved when considering of the imbalance of normal and abnormal events. With anomaly focal loss, our method brings the accuracy from 57.24% to 58.26%.

In order to better show our detection performances, we visualize the result in Fig. 2. Figure 2 shows the detection results in two test videos. The orange lines in the graph represent the ground truth label under frame level, and the blue wavy lines describe the prediction probability of frames. Compared with curve of STPN in explosion and shooting, our method can obtain more smooth curve. The area of ground truth can be overlapped exactly by our prediction curve. Though there are several areas shaking, the areas can be processed by seconds level division. If the continue frames level is short, the anomaly event may be filtered.

Table 3. Accuracy of anomaly recognition experiments, cross-entropy (CE), abnormal-focal loss (AFL)

	Methods	Accuracy (%)
Supervised	C3D + MA [8]	26.10
	TCNN + MA [8]	31.00
	Ours + AFL	39.30
Weakly supervised	STPN + CE [13]	53.22
	STPN + AFL [13]	56.21
	Ours + CE	57.24
	Ours + AFL	58.26

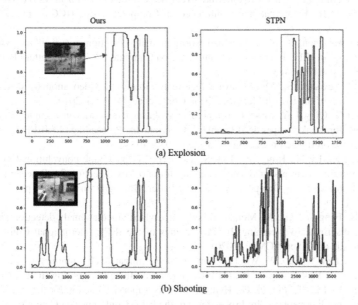

(a) Explosion

(b) Shooting

Fig. 2. Visualizations of different classes and detection

5 Conclusion

In this paper, due to lack of fine-grain labels of segments, an attention model is proposed to select the salient features and enhance the anomaly embedding features. With the graph network, anomaly can be efficiently learnt. It exists data imbalance between abnormal and normal data. We propose the anomaly focal loss to relieve the imbalance effect. The experiments in UCF-Crime demonstrate our network can significantly improve the anomaly detection performance compared to existing SOTA methods. In the future work, anomaly videos have different anomaly events. We want to learn the multi label anomaly events with label learning. Meanwhile, some anomaly events are ambiguous to distinguish. We want to margin them to achieve higher performance.

References

1. Pawar, K., Attar, V.: Deep learning approaches for video-based anomalous activity detection. World Wide Web **22**(2), 571–601 (2018). https://doi.org/10.1007/s11280-018-0582-1
2. Li, W.X., Mahadevan, V., Vasconcelos, N.: Anomaly detection and localization in crowded scenes. IEEE Trans. Pattern Anal. Mach. Intell. **36**(1), 18–32 (2014)
3. Leyva, R., Sanchez, V., Li, C.: Video anomaly detection with compact feature sets for online performance. IEEE Trans. on Image Process. **26**(7), 3463–3478 (2017)
4. Luo, W.X., Liu, W., Gao, S.H.: A revisit of sparse coding based anomaly detection in stacked RNN framework. In: Proceedings of IEEE International Conference on Computer Vision (ICCV), pp. 341–349 (2017)
5. Hasan, M., Choi, J., Neumann, J., Roy-Chowdhury, K.A., Davis, L.S.: Learning temporal regularity in video sequences. In: Proceedings of IEEE International Conference on Computer Vision and Pattern Recognition (CVPR), pp. 733–742 (2016)
6. Lu, C.W., Shi, J.P., Jia, J.Y.: Abnormal event detection at 150 FPS in MATLAB. In: Proceedings of IEEE International Conference on Computer Vision (ICCV), pp. 2720–2727 (2013)
7. Sultani, W., Chen, C., Shah, M.: Real-world anomaly detection in surveillance videos. In: Proceedings of IEEE International Conference on Computer Vision and Pattern Recognition (CVPR), pp. 6479–6488 (2018)
8. Zhu, Y., Newsam, S.: Motion-aware feature for improved video anomaly detection. In: Proceedings of the British Machine Vision Conference (BMVC) (2019)
9. Zhang, J.G., Qing, L.Y., Miao, J.: Temporal convolutional network with complementary inner bag loss for weakly supervised anomaly detection. In: Proceedings of IEEE International Conference on Image Processing (ICIP), pp. 4030–4034 (2019)
10. Zhong, J.X., Li, N., Kong, W., Liu, S., Li, T.H., Li, G.: Graph convolutional label noise cleaner: train a plug-and-play action classier for anomaly detection. In: Proceedings of the IEEE Conference on Computer Vision and Pattern Recognition (CVPR), pp. 1237–1246 (2019)
11. Wan, B., Fang, Y., Xia, X., Mei, J.: Weakly supervised video anomaly detection via center-guided discriminative learning. In: Proceedings of the IEEE International Conference on Multimedia and Expo (ICME) (2020)
12. Sun, L., Chen, Y., Luo, W., Wu, H., Zhang, C.: Discriminative clip mining for video anomaly detection. In: Proceedings of the IEEE Conference on Image Processing (ICIP) (2020)
13. Nguyen, P., Liu, T., Prasad, G., Han, B.: Weakly supervised action localization by sparse temporal pooling network. In: Proceedings of the IEEE Conference on Computer Vision and Pattern Recognition (CVPR), pp. 6752–6761 (2018)
14. Nguyen, P., Ramanan, D., Fowlkes, C.: Weakly-supervised action localization with background modeling. In: Proceedings of the IEEE Conference on Computer Vision (ICCV), pp. 5501–5510 (2019)
15. Lee, P., Uh, Y., Byun, H.: Background suppression network for weakly-supervised temporal action localization. In: AAAI Conference on Artificial Intelligence (AAAI) (2019)
16. Liu, D., Jiang, T., Wang, Y.: Completeness modeling and context separation for weakly supervised temporal action localization. In: Proceedings of the IEEE Conference on Computer Vision and Pattern Recognition (CVPR), pp. 1298–1307 (2019)
17. Lin, T., Goyal, P., Girshick, R., He, K.M., Dollar, P.: Focal loss for dense object detection. IEEE Trans. Pattern Anal. Mach. Intell. **42**(2), 318–327 (2020)
18. Carreira, J., Zisserman, A.: Quo vadis, action recognition? a new model and the kinetics dataset. In: Proceedings of the IEEE Conference on Computer Vision and Pattern Recognition (CVPR), pp. 6299–6308 (2017)

19. Kipf, T.N., Welling, M.: Semi-supervised classification with graph convolutional networks. In: Proceedings of the IEEE Conference on Learning Representations (ICLR) (2017)
20. Wang, X., Gupta, A.: Videos as space-time region graphs. In: Ferrari, V., Hebert, M., Sminchisescu, C., Weiss, Y. (eds.) ECCV 2018. LNCS, vol. 11209, pp. 413–431. Springer, Cham (2018). https://doi.org/10.1007/978-3-030-01228-1_25
21. Wu, P., et al.: Not only look, but also listen: learning multimodal violence detection under weak supervision. In: Vedaldi, A., Bischof, H., Brox, T., Frahm, J.-M. (eds.) ECCV 2020. LNCS, vol. 12375, pp. 322–339. Springer, Cham (2020). https://doi.org/10.1007/978-3-030-58577-8_20

A Data Fusion Model Based on Multi-source Non-real-Time Intelligence

Yuanyuan Qin and Songyuan Gu[✉]

China Academy of Electronics and Information Technology, Beijing, China
qinyuan98@163.com, gusongyuan614@163.com

Abstract. In order to solve the problem of target fusion in complex battle environment, a data fusion model based on multi-source intelligence is established. Weight quantification analysis and fusion of non-real-time intelligence source are realized based on this model, the evidence reasoning method is used to solve the uncertainty in decision-making, and the accuracy of data fusion is improved by integrating non-real-time and real-time intelligence source. In general, this model considers more sufficient and comprehensive influencing factors, and the computational process of this model is intelligent.

Keywords: Multi-source intelligence · Data fusion · Decision support · Non-real-time intelligence

With continuous advance of military technology and upgrade of equipment, all kinds of camouflage and jamming technologies are widely applied, and thus target analysis relies on single intelligence source could not satisfy the demands of modern warfare and it's necessary to develop fast and valid multi-source intelligence fusion model to improve analysis capability for target threats.

A target threat assessment method is proposed and a non-real-time intelligence weight quantification model and a corresponding fusion algorithm are given in this paper for target threat quantification and fusion of prewar non-real-time intelligence and wartime intelligence. The final target threat is also given integrating characteristics of non-real-time and real-time intelligence to provide basis for subsequent decisions.

1 Background Analysis

Multi-Source data fusion is to combine data and information from multiple sensors to obtain more detailed and accurate reasoning results than from single sensor. In terms of target threat assessment, the current research mainly focuses on the analysis, judgment and situation generation of real-time intelligence of which the major drawbacks lie in 2 aspects [1]. On the one hand there are no solutions to generate threat automatically according to prewar information obtained from non-real-time intelligence sources such as technical reconnaissance, forces reconnaissance, aerospace reconnaissance and open reconnaissance, etc. [2]. On the other hand, completeness, accuracy, real-time and reliability of prewar non-real-time intelligence and fusion of non-real-time and real-time intelligence are not considered [3].

© Springer Nature Switzerland AG 2021
Q. Zu et al. (Eds.): HCC 2020, LNCS 12634, pp. 136–142, 2021.
https://doi.org/10.1007/978-3-030-70626-5_14

2 Data Fusion Model Design

2.1 Non-real-Time Intelligence Source

The non-real-time intelligence source consists of technical reconnaissance, forces reconnaissance, aerospace reconnaissance and open reconnaissance. Technical reconnaissance is used for detecting telemetry of enemy missile launch, aircraft takeoff, etc., acquiring current location of enemy forces and equipment and identifying enemy equipment types [4]. Forces reconnaissance is used for acquiring deployment location and attempt of enemy forces [5]. Aerospace reconnaissance means, e.g., imaging reconnaissance satellites, aircrafts are used for acquiring enemy forces deployment, weapon types and movements [6]. Open reconnaissance is used for acquiring enemy attempts and weapon types [7]. All kinds of fusion results of prewar reconnaissance and wartime intelligence are gotten considering state, movement and type information of targets which all the 4 kinds of non-real-time intelligence source provide.

Definition of Intelligence Source Attributes. 4 kinds of non-real-time intelligence sources play different roles in synthetic decision. A weight quantification model would be established according to the critical attributes of intelligence sources to analyze weights of all kinds of intelligence participating in target threat decision.

Completeness. Completeness is the integrity of intelligence which non-real-time intelligence sources could provide. For target information, there are the following critical elements:

The greater the number n of effective elements is obtained, the more complete the intelligence would be (Table 1).

Table 1. Completeness

Completeness	Very complete	Complete	Fair	Incomplete	Very incomplete
Criteria	$n_1 \leq n$	$n_2 \leq n < n_1$	$n_3 \leq n < n_2$	$n_4 \leq n < n_3$	$n < n_4$
Value B_1	1	0.8	0.6	0.4	0.2

Accuracy. Accuracy refers to the accuracy degree of intelligence, which is divided into 5 grades. The lower the relative error e of obtained critical elements of the target is, the higher the accuracy would be (Table 2).

Table 2. Accuracy

Accuracy	Very accurate	Accurate	Average	Inaccurate	Very inaccurate
Criteria	$e \leq e_1$	$e_1 \leq e < e_2$	$e_2 \leq e < e_3$	$e_3 \leq e < e_4$	$e_4 \leq e$
Value B_2	1	0.8	0.6	0.4	0.2

Real-time. Real-time represents the real time degree of intelligence acquisition which is also divided into 5 grades, and it can be characterized as the difference between the time intelligence was generated and the current time. The smaller time difference is, the better the real-time of the intelligence would be (Table 3).

Table 3. Real-time

Real-time	Very real-time	Real-time	Average	Non-real-time	Very non-real-time
Criteria	$f \leq f_1$	$f_1 \leq f < f_2$	$f_2 \leq f < f_3$	$f_3 \leq f < f_4$	$f_4 \leq f$
Value B_3	1	0.8	0.6	0.4	0.2

Reliability. Reliability represents the credibility of intelligence which is divided into 5 grades as well and can be characterized as the anti-interference capability or the capability in identifying false targets (Table 4).

Table 4. Reliability

Reliability	Very reliable	Reliable	Average	Unreliable	Very unreliable
Value B_4	1	0.8	0.6	0.4	0.2

Quantification Model. The model consists of 4 layers, namely the objective layer, the criterion layer and the object layer. Weights of the 4 kinds of non-real-time intelligence could be obtained in the objective layer, the criterion layer includes the 4 critical attributes of intelligence sources and the object layer consists of the 4 kinds of non-real-time intelligence introduced above. The hierarchical model structure is shown as Fig. 1.

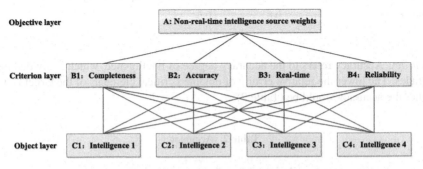

Fig. 1. Hierarchical model structure

First of all, the judgment matrix *A-B* would be constructed which represents the relative importance between each criterion in terms of overall objective (Table 5).

Table 5. Matrix A-B

A	B_1	B_2	B_3	B_4
B_1	1	$B12$	B_{13}	B_{14}
B_2	B_{21}	1	B_{23}	B_{24}
B_3	B_{31}	B_{32}	1	B_{34}
B_4	B_{41}	B_{42}	B_{43}	1

The matrix element B_{mn} represents the importance degree of the n-th criterion in relation to the m-th criterion which is determined by experts. The maximum eigenvalue λ_{max} and the eigenvector $\varpi = [\varpi_1, \varpi_2, \varpi_3, \varpi_4]$ would be calculated based on the matrix value. The element ϖ_n in the eigenvector is the weight of the n-th criterion in the criterion layer. Then the judgment matrix $B_i - C$ would be constructed which represents the relative importance between each object in terms of the criterion layer B_i (Table 6).

Table 6. Matrix B_i-C

Bn	C_1	C_2	C_3	C_4
C_1	1	$C12$	C_{13}	C_{14}
C_2	C_{21}	1	C_{23}	C_{24}
C_3	C_{31}	C_{32}	1	C_{34}
C_4	C_{41}	C_{42}	C_{43}	1

The C_{ij} in the judgment matrix above represents the importance of the i-th reconnaissance source in relation to the j-th reconnaissance source which could be calculated by the following formula $B_n : C_{ij} = B_{ni}/B_{nj}$, where B_{ni} represents the quantized value corresponding to the assessment level of the critical attribute B_n of the i-th reconnaissance source. On this basis the maximum eigenvalue λ_{max} and the eigenvector $\varpi_n = [\varpi_1^n, \varpi_2^n, \varpi_3^n, \varpi_4^n]$ of the matrix B_i-C could be calculated, and the element ϖ_i^n in the eigenvector above represents the weight of the i-th reconnaissance source in relation to the n-th criterion.

Finally, the weight matrix could be obtained as follow (Table 7):

Then the weight coefficient of the i-th reconnaissance source could be calculated through the following formula:

$$\varpi_i = \sum_{n=1}^{4} \varpi_n \varpi_i^n \tag{1}$$

For convenience of calculations, target threat H here is divided into 5 grades, namely $H = \{H_1, H_2, H_3, H_4, H_5\}$, each of which represents the threat is extremely low, low, average, high or extremely high.

Table 7. Weight coefficient matrix

	B_1	B_2	B_3	B_4	Weight coefficient
	ϖ_1	ϖ_2	ϖ_3	ϖ_4	
C_1	ϖ_1^1	ϖ_1^2	ϖ_1^3	ϖ_1^4	ϖ_1
C_2	ϖ_2^1	ϖ_2^2	ϖ_2^3	ϖ_2^4	ϖ_2
C_3	ϖ_3^1	ϖ_3^2	ϖ_3^3	ϖ_3^4	ϖ_3
C_4	ϖ_4^1	ϖ_4^2	ϖ_4^3	ϖ_4^4	ϖ_4

Given a situation, intelligence analysts could assign confidence on each level of threats respectively for the obtained 4 kinds of intelligence. Confidence represents the uncertainty of the assessment which is denoted as $m_j(H_i)$, where $j = 1, 2, 3, 4$, $i = 1, 2, 3, 4, 5$, namely the confidence level of threat level H_i for the situation that the j-th intelligence source provides [8].

2.2 Fusion Algorithm

The evidence reasoning method [9, 10] is used to fuse the threats from the 4 non-real-time intelligence sources on the basis of the weights of the intelligence sources and it could solve the uncertainty in decision-making well and gives a more reasonable result.

Considering the quantized weight of each intelligence source, the weighted confidence assignment could be obtained as follows:

$$m_j^{\varpi}(H_i) = \varpi_i m_j(H_i)$$

$$m_j^{\varpi}(H) = 1 - \sum_{i=1}^{N} m_i^{\varpi}(H_i), j = 1, 2, ..., L \tag{2}$$

The confidence assignment of each level of threats after fusion and the residual confidence assignment could be calculated iteratively through the following formula:

$$M_{j+1}(H_i) = K_{j+1}(M_j(H_i)m_{j+1}^{\varpi}(H_i) + M_j(H)m_{j+1}^{\varpi}(H) + M_j(H)m_{j+1}^{\varpi}(H_i)) \tag{3}$$

$$M_{j+1}(H) = K_{j+1}(M_j(H)m_{j+1}^{\varpi}(H_i)) \tag{4}$$

Where $M_{i+1}(H_i)$ represents the confidence of threat level i assessed by the fusion result of the first $j + 1$ intelligence sources, $M_{i+1}(H)$ represents the residual confidence assignment after the fusion of the first $j + 1$ intelligence sources, and K_{i+1} represents normalizing factor which could be calculated by the following formula:

$$K_{j+1} = [1 - \sum_{t=1}^{N} \sum_{i=1;i \neq t}^{N} M_j(H_t)m_{j+1}^{\varpi}(H_i)]^{-1} \tag{5}$$

Through iterative calculations, the confidence assignment $M_L(H)$ of each level of threats after the fusion of L intelligence sources could be obtained.

2.3 Fusion Calculation

The confidence assignment of each level of threats after fusion of multi-source intelligence and the residual confidence assignment are calculated in the method described above. The calculation of threat range is given below for visual representation of fusion results. Let the 5-grade threat be $u = \{u(H_1), \ldots, u(H_N)\}$, $N = 5$ and take its value as $u = \{0.2, 0.4, 0.6, 0.8, 1\}$, then the threat range could be calculated as the following formula:

$$
u_{max} = \sum_{i=1}^{N} (u(H_i)M_L(H_i)) + u(H_N)M_L(H)
$$

$$
u_{min} = \sum_{i=1}^{N} (u(H_i)M_L(H_i)) + u(H_1)M_L(H)
$$

$$
u_{mean} = (u_{max} + u_{min})/2
$$

(6)

Where the threat range is denoted as $[u_{min}, u_{max}]$ and the threat mean is denoted as u_{mean}.

2.4 Threat Integrating of Non-real-Time and Real-Time Intelligence Sources

Let the target threat from real-time intelligence source be u_p and the target threat from non-real-time intelligence source obtained from the calculation above be u_{mean}. The final target threat u_c could be obtained through integrating the 2 threat values aforementioned as the following formula:

$$
u_c = w_1 u_p + w_2 u_{mean}
$$

(7)

Where w_1 and w_2 represent the weights of real-time and non-real-time intelligence sources assigned by decision makers and $w_1 + w_2 = 1$.

3 Conclusion

The data fusion model in this paper introduces various kinds of non-real-time intelligence into calculation and obtains fusion results at different times and by which situation assessment and operational decision could be automatically accomplished. This model considers completeness, accuracy, real-time and reliability of prewar non-real-time intelligence and integrates the influences of prewar non-real-time intelligence and wartime real-time intelligence for threats.

References

1. Shi, X., Hong, G., Weimin, S., Dong, T., Chen, X.: Study of target threat assessment for ground surveillance radar. Acta Armamentarii **36**(219(06)), 1128–1135 (2015)
2. Li, Z., Li, Z.: Research on the method of threat assessment in intelligent decision support system based on CGF. Command Control Simul. **32**(22705), 20–23 (2010)

3. Xunqiang, H., Xie, X., Yang, Y.: Study on modeling of perception behavior of CGF entity in battlefield. Electron. Opt. Control **17**(150(12)), 44–48 (2010)
4. Guo, Z.: Study of technology of satellite reconnaissance and anti-reconnaissance for ground missile launching equipment. Beijing Institute of Technology (2016)
5. Xiaoliang, Z., Qun, C., Huazhen, Z.: The evaluation of reconnaissance capabilities of ECM reconnaissance troops based on multilevel grey theory. Radar ECM **032**(004), 15–17 (2012)
6. Wei, F.: Reconnaisance and anti-reconnaissance of aerospace navigation in high technology war. Mod. Def. Technol. **28**(004), 22–26 (2000)
7. Huang, Q., Li, L.: An analysis of network open source military intelligence. Natl. Def. Sci. Technol. **38**(306(05)), 79–83 (2017)
8. Wang, Y., Liu, S., Zhang, W., Wang, Y.: Treat assessment method with uncertain attribute weight based on intuitionistic fuzzy multi-attribute decision. Acta Electronica Sinica **42**(382(12)), 2509–2514 (2014)
9. Zhijie, Z., Taoyuan, L., Fangzhi, L., et al.: An evaluation method of equipment support resources based on evidential reasoning. Control and Decis. **33**(06), 83–89 (2018)
10. Xiong, N., Wang, Y.: Multiple attribute decision method based on improved fuzzy entropy and evidential reasoning. J. Comput. Appl. **38**(10), 55–60 (2018)

Ancient Chinese Lexicon Construction Based on Unsupervised Algorithm of Minimum Entropy and CBDB Optimization

Yuyao Li[1,2(✉)], Jinhao Liang[2], and Xiujuan Huang[3]

[1] Guangzhou Key Laboratory of Multilingual Intelligent Processing, Guangdong University of Foreign Studies, Guangzhou, China
everybit@163.com

[2] School of Information Science and Technology, Guangdong University of Foreign Studies, Guangzhou, China

[3] Faculty of Chinese Language and Culture, Guangdong University of Foreign Studies, Guangzhou, China

Abstract. Ancient Chinese text segmentation is the basic work of the intelligentization of ancient books. In this paper, an unsupervised lexicon construction algorithm based on the minimum entropy model is applied to a large-scale ancient text corpus, and a dictionary composed of high-frequency co-occurring neighbor characters is extracted. Two experiments were performed on this lexicon. Firstly, the experimental results of ancient text segmentation are compared before and after the lexicon is imported into the word segmentation tool. Secondly, the words such as person's name, place name, official name and person relationship in CDBD are added to the lexicon, and then the experimental results of ancient text segmentation before and after the optimized lexicon is imported into the word segmentation tool are compared. The above two experimental results show that the lexicon has different enhancement effects on the segmentation effect of ancient texts in different periods, and the optimization effect of CDBD data is not obvious. This article is one of the few works that applies monolingual word segmentation to ancient Chinese word segmentation. The work of this paper enriches the research in related fields.

Keywords: Ancient Chinese word segmentation · Minimum entropy model · Unsupervised algorithm · CDBD

1 Introduction

Chinese Word Segmentation (CWS) is a necessary preprocessing technique in many Chinese NLP tasks. In this paper, by adopted a unsupervised lexicon construction method based on minimum entropy model [1], we directly create a lexicon calculated from a large-scale ancient text corpus spanning multiple periods without additional labeled corpus or other reference corpus. On this basis, we try to combine some existing word segmentation tools, such as Jieba, an well-known CWS tool, to perform word segmentation experiments on ancient texts. Additionally, for the purpose of optimizing the lexicon, we

Q. Zu et al. (Eds.): HCC 2020, LNCS 12634, pp. 143–149, 2021.
https://doi.org/10.1007/978-3-030-70626-5_15

also tested the effect of adding the CDBD [2] vocabulary to the above lexicon and importing it into Jieba for the word segmentation experiment. The China Biographical Database (CBDB) is a freely accessible relational database with biographical information about approximately 471,000 individuals, primarily from the 7th through 19th centuries.

2 Related Work

Generally, CWS methods are divided into two types: bilingual word segmentation methods that use parallel corpus for word segmentation and monolingual word segmentation methods that only apply to one language.

2.1 Bilingual Word Segmentation

In bilingual word segmentation, languages without obvious separators but with high-performance segmentation tools can also be regarded as anchor languages. For example, Chu et al. [3] improved CWS based on the Sino-Japanese parallel corpus. Although there are no spaces between Japanese words, the F1 score of the Japanese segmenter toolkit can reach up to 99%, so they used Japanese as the anchor language in CWS. In addition, they also used common Chinese characters shared between Chinese and Japanese to optimize CWS. Chao Che et al. [4] used segmented modern Chinese as the anchor, and used the shared characters between ancient and modern Chinese to extract common words to achieve ACWS.

2.2 Monolingual Word Segmentation

The research field of CWS has evolved from dictionary or rule-based, tag sequence-based to neural network-based. Zheng et al. [5] use character embedding in a local window as input to predict the position label of a single character. Liu et al. [6] proposed a neural segmentation model that combines neural network and semi-CRF. Despite using different structures, the performance of neural segmentation models highly depends on the number of labeled corpora.

Due to the lack of a large-scale tagged corpus, most of the above methods cannot be applied to ACWS. The minimum entropy model used in this paper is a typical monolingual word segmentation method that does not require annotated corpus to segment large-scale ancient texts.

3 Models and Algorithms

3.1 Minimum Entropy Model

It could be assumed that the information volume of each word equals $(-logp_w)$, then the time required to memorize this batch of corpus is $(-\sum_{w \in corpus} logp_w)$ when performing word by word summation in the corpus. When considering repeated words in corpus, the time needed equals $(-\sum_{w \in corpus} N_w logp_w)$, where N_w stands for number for word of w in the corpus. In this way, the average amount of information for each word when memorizing words can be calculated as formula (1), where T_w and T_c stand for total number of words and characters respectively; $p_w = N_w/T_w$ is the frequency for word of w; l_w is the character numbers for word of w, then $N_w * l_w$ equals total number of characters.

$$L = -\sum_{w \in corpus} \frac{N_w}{T_w} log p_w = \frac{-\sum_{w \in word\_list} \frac{N_w}{T_w} log p_w}{\frac{T_c}{T_w}} = \frac{-\sum_{w \in word\_list} p_w log p_w}{\frac{-\sum_{w \in word\_list} N_w l_w}{T_w}}$$

$$= \frac{-\sum_{w \in word\_list} \frac{N_w}{T_w} log p_w}{\sum_{w \in word\_list} \frac{N_w}{T_w} l_w} = \frac{-\sum_{w \in word\_list} p_w log p_w}{-\sum_{w \in word\_list} p_w l_w}$$

$$(1)$$

Therefore, minimizing L becomes the main goal for constructing unsupervised lexicon. The optimization direction of the minimization of L is mainly to examine the point mutual information, PMI $(a, b) = ln \frac{p_{ab}}{p_a p_b}$, when merging two adjacent elements a and b into one item, where a and b can be character or word. Under the necessary condition of PMI (a, b) > 1, the greater PMI(a, b), the more conducive to the reduction of uncertainty when the two elements are merged, then the two should be merged. In contrast, when the PMI value is less than a certain threshold, these two elements should be divided.

3.2 Unsupervised Lexicon Construction Algorithm

The algorithm for constructing lexicon based on the model is as follows:

Statistics: Counting the frequency of each word (p_a, p_b) from the large-scale ancient Chinese corpus, and the co-occurrence frequency of two adjacent words (p_{ab}).

Segmentation: Set the threshold *min_prob* of the co-occurrence frequency and the threshold *min_pmi* of PMI respectively, and then separate the neighboring items of (p_a, p_b) when $p_{ab} < min\_prob$ or $ln \frac{p_{ab}}{p_a p_b} < min\_pmi$.

Truncation: counting the frequency p_w of each "quasi-word", and only retaining the one that $p_w > min\_prob$; these retained words form a preliminary wordlist.

Elimination: Arrange the candidate words in the wordlist from most to least, and then delete each candidate word in the lexicon in turn, use the remaining words and word frequency to cut the candidate word, and calculate the original word and sub The mutual information between words; if the mutual information is greater than 1, the word is restored; otherwise, the word frequency of the segmented sub-words is kept deleted and updated.

Removing Redundant Words: Sort the candidate words by word length of the lexicon from longest to shortest, and then perform the following loop operation:

- Delete each candidate word in the lexicon in turn,
- Use the remaining words and their frequency information to cut the candidate word to multiple sub-words,

– Compute the PMI value between the original word and those sub-words;
– If the PMI value is greater than 1, the word is restored; otherwise, the candidate word is kept deleted and each sub-word is inserted into the wordlist and their frequency is updated.

4 Word Segmentation Experiment Based on Large-Scale Ancient Chinese Corpus

The large-scale ancient Chinese corpus used in this experiment contains texts from many dynasties, mainly in the Tang, Song, Yuan, Ming, and Qing dynasties. The content covers literature, medicine, historical records, engineering technology and other disciplines, as well as various stylistic such as poetry, fiction, opera, and political theory, etc., with a capacity of 2.53 G and a total of 8182 texts.

This experiment consists of two stages: the unsupervised lexicon construction experiment based on the minimum entropy model and the word segmentation experiment.

4.1 Unsupervised Lexicon Construction Experiment

The lexicon construction experiment employs a python implementation of the unsupervised algorithm based on the minimum entropy model [7]. A word-list containing 1,992,531 words with different number of characters was calculated after the python code run on this ancient Chinese corpus. In order to observe segmentation results more clearly, these words of different lengths are divide into 6 types, including 1c-word, 2c-word, 3c-word, 4c-word, 5c-word, here c stands for character. The TOP10 of 1–5 character words are shown in Table 1.

Table 1. List of TOP15 words of each length the word-list

SN	1c-word	2c-word	3c-word	4c-word	5c-word
1	Zhi	YiWei	TianXiaZhi	BingBuShangShu	AnWanXingTongPu
2	Yi	BuDe	BuKeYi	ZhongShuSheRen	RuWuTongZiDa
3	Wei	BuKe	DaJiangJun	LiBuShangShu	JiJiRuLüLing
4	Ye	TianXia	GeYiLiang	JianChaYuShi	DuZhiHuiQianShi
5	Er	SuoYi	DaXueShi	LiBuShangShu	DianZhongShiYuShi
6	Ren	BuZhi	WeiXiMo	YiQianWuFen	ChuShengJiZongLu
7	Zhe	XianSheng	JiShiZhong	XiaoBianBuLi	MeiFuSanShiWan
8	You	BuNeng	ShiErYue	YuShiDaFu	PaoLieQuPiQi
9	Bu	RuCi	YuNeiGe	CanZhiZhengShi	BuZhongYiQiTang
10	Yi	ErBu	BuKeBu	DuanZhuShuoWen	DaQingGaoZongFa

In general, the word-list has distinct characteristics of ancient Chinese. And it can be seen that for words with 3–5-characters, there are various proprietary words commonly used in ancient times, including official titles, such as "DaXueShi", "JianChaYuShi", "DianZhongShiYuShi"; or traditional Chinese medicine terms, such as "WeiXiMo", "XiaoBianBuLi", "ChuShengHuiFang", "DaoXiLuoWeiSan", "BuZhongYiQiTang" and Taoist terms, such as "JiJiRuLüLing". It can be seen that the unsupervised lexicon construction algorithm based on the minimum entropy model can capture the common collocations of neighboring characters in the text; the calculated result provides a high-quality basic word-list for the study of ancient Chinese word segmentation.

In order to compare in subsequent word segmentation experiments, this paper extracts the words with the value >=100 as a high-value word-list which we name as Lexicon1.

4.2 Lexicon Optimized Based on CDBD

CDBD provides more detailed data from multiple angles such as official titles, place names, personal relationships, and social organizations. For example, it contains about 624370 ancient names (in table of ZZZ_NAMES), about 32170 detailed official titles (in table of OFFICE_CODES), about 30630 ancient book titles (in table of TEXT_CODES), and about 2500 detailed social organization names (in table of SOCIAL_INSTITUTION_NAME_CODES) including various academies and monasteries. The above words are added to the Lexicon1 and named as Lexicon2.

4.3 Comparative Experiment of Ancient Chinese Word Segmentation

The comparison experiment of ancient Chinese word segmentation is designed as follows:

Ancient Texts Selection: Select fragments of ancient Chinese texts from multiple periods, and perform artificial word segmentation on each text separately, and use artificial word segmentation as the basis for the judgment of experimental results; the selected ancient text includes the following four fragments:

Fragment 1: Mencius (Warring States Period) Sect. 2 Articles and Sentences of Liang Huiwang (the second volume).

Fragment 2: Fan ZhongYan (Song) Fan Wenzheng set·Si Min Poetry(Shi Nong Gong Shang) Nong.

Fragment 3: An YuShi (Ming) Bao Zheng's cases (Sect. 2 Lady Ding suffered humiliation to avenge her revenge, A bad monk hides others' wife and kill her husband Paragraph 1).

Fragment 4: Guo Xiaoting (Qing) The Complete Biography of Monk (Excerpts).

Word Segmentation: Use *Jieba (A python library for Chinese word segmentation)* directly to segment the ancient text fragments, and the segmentation results obtained are referred to as the *original results*;

Adding Custom Dictionary: Use Lexicon1 and Lexicon2 as custom dictionary of *Jieba separately*, and then segment the ancient text fragments to obtain the word segmentation result, and referred to as the *Le1_res* and *Le2_res*, which correspond to lexicon1 and lexicon2 respectively;

Result Compare: Compare the *Le1_res* and *Le2_res* with the *original results.* Take the artificial word segmentation result as a positive example, and the Lexicons as a negative example, and calculate precision and recall Rate (Recall), F1 value respectively.

5 Results

The results of the word segmentation experiment are shown in Table 2.

Table 2. Experimental results of segmentation of ancient Chinese text

Fragment 1				Fragment 2			
Results	Precision	Recall	F1	results	precision	Recall	F1
Original	0.67	0.91	0.77	original	0.69	0.89	0.78
Le1_res	0.79	0.95	0.86	Le1_res	0.72	0.95	0.82
Le2_res	0.76	0.94	0.84	Le2_res	0.75	0.94	0.83
Fragment 3				Fragment 4			
Results	Precision	Recall	F1	results	precision	Recall	F1
Original	0.77	0.93	0.84	original	0.77	0.91	0.83
Le1_res	0.77	0.94	0.85	Le1_res	0.73	0.91	0.81
Le2_res	0.79	0.94	0.86	Le2_res	0.74	0.94	0.83

From Table 2, the difference in the segmentation effect of ancient texts in different periods is related to the differences in the morphology and syntax of the ancient texts in different periods. When comparing Fragment 1 and 2 with Fragment 3 and 4. The ancient grammar features of Fragment 1 and Fragment 2 are more obvious, while Fragment 3 and Fragment 4 are closer to modern Chinese. This may be the fundamental reason why the segmentation effect of Fragment 1 and Fragment 2 has improved significantly. Furthermore the difference in the segmentation effect of ancient texts in different periods is related to the degree of optimization of the word segmentation tool itself too; Inferred from the experimental results, Jieba has been well optimized for modern texts, such as novels, romances, and scripts etc.

6 Summary

Ancient Chinese text segmentation is one of the important breakthroughs in the intelligentization of ancient books. In this paper, an unsupervised lexicon construction algorithm based on the minimum entropy model is used to extract a word-list composed of high-frequency co-occurring neighbor characters in a large-scale ancient text corpus; and combining the word-list with existing word segmentation tools to perform ancient text segmentation experiment. The experimental results show that the method in this paper has different enhancement effects on the word segmentation effect of ancient texts in different periods, which shows that the word-list has the validity within a certain range. Therefore, the in-depth optimization of the word-list is an important direction in the follow-up work. In addition, this article is one of the few works that apply monolingual word segmentation to ACWS. The work of this article has enriched the research in related fields.

Acknowledgement. This work is funded by Characteristic Innovation Project (No. 19TS15) of Guangdong University of Foreign Studies.

References

1. Su, J.: Principle of minimum entropy (2): Thesaurus construction of decisively [J/OL] (2018). https://kexue.fm/archives/5476.
2. Harvard University, Academia Sinica, and Peking University, China Biographical Database Project (Cbdb) [M/OL]. https://projects.iq.harvard.edu/chinesecbdb. Accessed 24 Apr 2018
3. Chu, C., Nakazawa, T., Kawahara, D., et al.: Chinese-Japanese machine translation exploiting Chinese characters. ACM Trans. Asian Lang. Inf. Process. **12**(4), 1–25 (2013)
4. Che, C., Zhao, H., Wu, X., Zhou, D., Zhang, Q.: A word segmentation method of ancient Chinese based on word alignment. In: Tang, J., Kan, M.-Y., Zhao, D., Li, S., Zan, H. (eds.) NLPCC 2019. LNCS (LNAI), vol. 11838, pp. 761–772. Springer, Cham (2019). https://doi.org/10.1007/978-3-030-32233-5_59
5. Zheng, X., Chen, H., Xu, T.: Deep learning for Chinese word segmentation and POS tagging. In: Proceedings of the Proceedings of the 2013 Conference on Empirical Methods in Natural Language Processing, Seattle, Washington, USA, F October 2013. Association for Computational Linguistics
6. Liu, Y., Che, W., Guo, J., Qin, B., Lium T.: Exploring segment representations for neural segmentation models. In: Proceedings of the Twenty-Fifth International Joint Conference on Artificial Intelligence. Aaai Press, New York, USA, pp. 2880–2886 (2016)
7. Su, J.: NLP library based on the principle of minimum entropy: nlp zero [J/OL] (2018). https://kexue.fm/archives/5597

Availability Analysis of GNSS RAIM Based on Maximum Undetectable Position Bias

Shitai Wang[1,2] and Min Yin[1,2(✉)]

[1] College of Geamatics and Geoinformation, Guilin
University of Technology, Guilin 541004, China
416721782@qq.com
[2] Guangxi Key Laboratory of Spatial Information and Geomatics,
Guilin University of Technology, Guilin 541004, China

Abstract. The combined use of multiple constellations of the Global Navigation Satellite System (GNSS) provides a great possibility to improve the integrity of the satellite navigation; however, with the increase of the number of satellites, the probability of multiple outliers increases. Although a number of studies analyzed the performance of the combined GNSS Receiver Autonomous Integrity Monitoring (RAIM), the analysis of the case of multiple outliers is still lacking. In this paper, a Maximum Undetectable Position Bias (MUPB) scheme, which is based on Bias Integrity Threat (BIT), is used to analyze the horizontal availability of RAIM with multiple outliers. We have analyzed the theoretical background for BIT and MUPB. Next, detailed simulations and analyses with single or combined GPS/BDS (BeiDou Satellite Navigation System)/GLONASS constellations were conducted to evaluate the performance of RAIM. In the simulated scenarios with three constellations (GPS + BDS + GLONASS) and 2 or 3 outliers occurring simultaneously, we compared the MUPB and the Horizontal Alert Level (HAL) of different aviation approach phases and analyzed the results.

Keywords: BIT · MUPB · Multiple outliers · RAIM · Horizontal integrity

1 Introduction

As GNSS has been increasingly used in a wide range of applications, including safety-of-life and liability critical operation, it is essential to guarantee the reliability of GNSS navigation solutions. Integrity, which refers to the function of fault detection of the GNSS system, can provide a timely alarm when the fault is detected and can ensure that the GNSS system works well in a specified range. Integrity is defined by the Radio Technical Commission for Aeronautics (RTCA) as follows: it is the ability of the system to provide timely warning to users when the system cannot be used for navigation.

The Receiver Autonomous Integrity Monitoring (RAIM) technique monitors the integrity of the navigation signals independently of any external monitors via measurement consistency test operations. RAIM includes a Faults Detection (FD) function and Faults Exclusion (FE) function under the certain false alarm probability and missed

© Springer Nature Switzerland AG 2021
Q. Zu et al. (Eds.): HCC 2020, LNCS 12634, pp. 150–160, 2021.
https://doi.org/10.1007/978-3-030-70626-5_16

detection probability [1]. To guarantee that the false alarm probability and missed detection probability meet the demand of RAIM, the RAIM availability based on the satellite's geometry and measurement noises must have a certain quality assurance. The common methods to determine the satellite geometry configuration that is required by RAIM probability are the maximum Horizontal Dilution of Precision (HDOP) method, Approximate Radial Protection (ARP) method and Horizontal Protection Level (HPL) method [1]. They calculate the protection limit according to the distribution of the test statistics, which are based on the complete geometric distribution; because these statistics do not consider the measurement noise, the actual positioning error may exceed the limit value [2].

Bias Integrity Threat (BIT) was proposed by P.B. Ober in 1996 [3]; the key point of BIT is that the integrity of the system depends not only on the geometric configuration of the satellites but also on the measurement noise. BIT reflects the linear relationship between positioning error square and the non-central Chi-square distribution parameters, the Maximum Undetectable Position Bias (MUPB) based on BIT can be compared with the Horizontal Alert Level (HAL) to determine the availability. The MUPB based on the least square method is suitable for the application in multi-constellation and multi-fault environment because it considers the geometric distribution and the measure noise [3].

The simultaneous operation of multiple navigation systems provides a great possibility for improving the integrity of the satellite navigation system; thus, it is necessary to analyze the performance of different integrated systems. Hewitson et al. (2006) used reliability and separability measures of the global simulation results to assess the integrity performance levels of standalone GPS and integrated GPS/GLONASS, GPS/Galileo and GPS/GLONASS/Galileo systems [4]. Xu and Yang (2013) analyzed the RAIM availability of standalone GPS and integrated GPS/BDS, GPS/Galileo and GPS/BDS/Galileo systems based on HPL and VPL [5]. Their simulations are all based on the assumption that there is one fault in the standalone or integrated systems. With the development of GNSS and the increase of the number of satellites, the probability of multiple outliers increases. As a result, in this study, the scenarios of standalone GPS or GLONASS and integrated GPS/BDS, GPS/GLONASS, BDS/GLONASS and GPS/BDS/GLONASS systems with 2 or 3 faults are simulated and the MUPB for these situations are calculated. In addition, we compared the MUPB and the HAL of different aviation approach phases and then analyzed the results.

2 Methods

2.1 The Estimation of Positions and Bias

Assuming there are N satellites that can be measured, the functional model is

$$y = Ax + \varepsilon \qquad (1)$$

Where $y_{n \times 1}$ is a vector of the differences of measurement pseudoranges and the approximation; X is the unknown vector, including three user position correction parameters and receiver clock error correction parameter; A is the design matrix; and ε is the vector of measurement noises, which follows a Gaussian distribution with mean 0 and variance σ_0^2. P is the observation weight matrix. The least square solution is

$$\hat{x} = (A^T PA)^{-1} A^T Py = Ny \tag{2}$$

$$N = (A^T PA)^{-1} A^T P \tag{3}$$

Thus, the residual vector is

$$v = Q_v Py = Dy \tag{4}$$

$$Q_v = P^{-1} - A \cdot (A^T PA)^{-1} \cdot A^T \tag{5}$$

$$D = I - (A^T PA)^{-1} A^T P = I - AN \tag{6}$$

where Q_v is the covariance matrix of residual vector.

v is residual vector of measurement noises and faults and follows a Gaussian distribution $N(\mu_v, R_v)$.

2.2 Calculation of $\delta HDOP_{max}$

Assuming that the ith observation has a fault and the remaining satellites are in a poor geometry, it is difficult to provide precise location information and to determine whether the ith observation is correct. If that observation with a fault directly affects the formation of the optimal observation geometry, then the fault will result in a much larger positioning error than the positioning accuracy [1]. Thus, the change of Horizontal Dilution of Precision (HDOP) when the ith observation is removed is an important value for the positioning accuracy and the detection of fault. $\delta HDOP$ is defined as

$$HDOP_i^2 - HDOP^2 = \delta HDOP_i^2 \tag{7}$$

where $HDOP_i$ is the HDOP for which the ith observation is removed.

2.3 Bias Integrity Threat (BIT)

Assuming that the measurement noises have same covariance matrix ($R_v = \sigma^2 I$), the covariance matrix of unknown parameters is $R_{\hat{x}} = \sigma^2 (A^T A)^{-1}$, and the variance sum of unknown parameters is $\sigma_x^2 + \sigma_y^2 + \sigma_z^2 + \sigma_b^2 = \text{trace}(R_{\hat{x}}) = \sigma^2 \cdot DOP^2$, where DOP is Dilution of Precision. The satellite geometry and the accuracy of measurement should be considered when searching for the satellite group, the exiting of which will result in the fastest decline in accuracy. BIT is defined as [3]:

$$BIT = \sigma^2 \cdot iDOP \tag{8}$$

where the iDOP is $\delta HDOP_{max}$.

Next, the data from satellite observation equations are extracted according to the number of faults, the combinations of satellites in which the number of satellites is equal to the number of faults is determined, and the one that results in the fastest decline in accuracy is determined. The most difficult fault/faults to detect are for the fault/faults that exist in that satellite group. BIT can be used with general measurement covariance matrices and is well defined for the assumption of multiple failures. When there is one fault, it can be calculated as:

$$\text{BIT} = \max_i \frac{\|\Delta\hat{x}_{LS}(\mu_v^{(i)})\|^2}{\lambda(\mu_v^{(i)})} = \max_i \frac{\mu_v^{(i)T}N^TN\mu_v^{(i)}}{\mu_v^{(i)}D^TR_v^{-1}D\mu_v^{(i)}} \tag{9}$$

where:

$$N = (A^TR_v^{-1}A)^{-1}A^TR_v^{-1}$$

$$D = I - (A^TR_v^{-1}A)^{-1}A^TR_v^{-1} = I - AN$$

In which $\mu_v^{(i)}$ is the mean of error in the i th observation; it is $\mu_v^{(i)} = [0 \cdots \mu_i \cdots 0]^T$ when only one fault exists. Because different measurement noise leads to different weights, the influence to integrity cannot be simply separated into measurement quality and satellite geometry.

When r failures exist simultaneously, the bias vectors that we attempt to detect will have r nonzero elements. When $r > 1$, the exact direction of the bias vector becomes important. The BIT method uses a worst-case approach again and examines the worst direction of all bias vectors with r nonzero elements.

To find this worst-case direction, we will exploit a standard result from linear algebra. For two positive definite $r \times r$ matrices XTX and YTY, the maximum of the so-called Rayleigh's quotient over all r-vectors equals the largest eigenvalue of (YTY)-1XTX.

$$\text{BIT} = \max_{\mu_v^{(i)}} \frac{\mu_v^{(i)T}N^TN\mu_v^{(i)}}{\mu_v^{(i)}D^TR_v^{-1}D\mu_v^{(i)}} = \max_i \left\{ \lambda_{\max}(Q^{(i)}) \right\} \tag{10}$$

In which:

$$Q^{(i)} = (\tilde{D}^{(i)T}R_v^{-1}\tilde{D}^{(i)})^{-1}\tilde{N}^{(i)T}\tilde{N}^{(i)}$$

where $\tilde{N}^{(i)}$ and $\tilde{D}^{(i)}$ are the sub-matrices that are obtained when we remove all columns that correspond to zero elements in $\mu_v^{(i)}$ from N and D, respectively. $\lambda_{\max}()$ is the maximum eigenvalue of the matrix.

2.4 The Maximum Undetectable Position Bias (MUPB)

The BIT can be interpreted as the slope of the square position error plotted against the non centrality parameter of the distribution of the test statistic; thus, we can calculate the MUPB [3]:

$$\text{MUPB} = \sqrt{\text{BIT} \cdot \lambda_{\min}} \tag{11}$$

where λ_{min} is the detectable minimum noncentrality parameter value with sufficient probability.

Thus, MUPB can be used as integrity metric and can be obtained based on the λ_{min} with the probability, which is sufficient to the integrity of monitoring require different aviation approach phases.

We can see from above definition that the BIT and MUPB consider measurement geometry and measurement quality; thus, they are more suitable for use with multiple constellations and multiple satellites.

3 Data and Analysis Setting

In this paper, we simulated the scenarios with part or all of the satellites of three constellations (GPS, BDS, GLONASS) with 2 or 3 outliers occurring simultaneously; subsequently, we calculated the MUPB and compared it with the Horizontal Alert Level (HAL) of different aviation approach phases and then analyzed the results.

Assuming that the observations are independent,

$$R_v^{-1} = \begin{bmatrix} 1/\sigma_1^2 & 0 & \cdots & 0 \\ 0 & 1/\sigma_2^2 & \cdots & 0 \\ \vdots & \vdots & \ddots & \vdots \\ 0 & 0 & \cdots & 1/\sigma_n^2 \end{bmatrix} \tag{12}$$

In which:

$$\sigma_i^2 = \sigma_{URA,i}^2 + \sigma_{i,user}^2 + \sigma_{i,tropo}^2 \tag{13}$$

where $\sigma_{i,user}$ accounts for multipath and user receiver noise. $\sigma_{URA,i}$ is the standard deviation of the satellite orbit and clock errors. $\sigma_{i,tropo}$ is the standard deviation of the tropospheric delay.

The standard deviation of the receiver noise $\sigma_{i,user}^{GPS}$ is modelled as

$$\sigma_{i,user}^{GPS} = \sqrt{\frac{f_{L1}^4 + f_{L5}^4}{\left(f_{L1}^2 - f_{L5}^2\right)^2}} \sqrt{\sigma_{MP}^2 + \sigma_{Noise}^2} \tag{14}$$

$$\sigma_{MP}(\theta) = 0.13 + 0.53e^{-\frac{\theta}{10}} \tag{15}$$

$$\sigma_{Noise}(\theta) = 0.15 + 0.43e^{-\frac{\theta}{6.9}} \tag{16}$$

where σ_{MP} is the standard deviation of multipath, σ_{Noise} is the standard deviation of user receiver noise, θ is given in degrees and is related to the elevation angle.

The effect of the first order of the ionosphere is eliminated by using the dual-frequency ionosphere-free combination in the integrity calculation. Thus, the remainder term of the ionosphere can be ignored in measurement noise. The calculation of standard deviation of the tropospheric delay $\sigma_{i,\text{tropo}}$ can be simplified to an equation that is only based on the satellite elevation in the user station [6].

$$\sigma_{i,\text{tropo}}(\theta) = 0.12 \frac{1.001}{\sqrt{0.002001 + \left(\sin\frac{\pi\theta}{180}\right)^2}} \tag{17}$$

$\sigma_{\text{URA},i}$ is the standard deviation of the satellite orbit and clock errors, it can be obtained from the User Range Accuracy (URA) in navigation message. In this paper, we make $\text{URA}_{\text{GPS}} = 0.75\text{m}$, $\text{URA}_{\text{BDS}} = 0.75\text{m}$, and $\text{URA}_{\text{GLONASS}} = 1\text{m}$ in the simulations [7].

4 Results

In this simulated segment, RAIM availability analyses were conducted for standalone or integrated systems at 5° intervals of latitude and longitude and an altitude of 100.0 m in the global region ($-89°$ N $-86°$ N, 180° W–180° E) with a 5° masking angle for each 15 min interval in 24 h. Thus, there are 2592 sampling points in the global region, and the BIT and MUPB were calculated in 96 epochs at every sampling point. We calculated the BIT and MUPB of standalone GPS (G) and GLONASS (R), and integrated GPS/BDS (GB), GPS/GLONASS (GR), BDS /GLONASS (BR) and GPS/BDS/ GLONASS (GBR) with 2 outliers. Next, we compared the MUPB with the Horizontal Alert Level (HAL) of different aviation approach phases and analyzed the RAIM availability in those situations. In addition, we performed the same analysis for the case of 3 outliers assumed to occur simultaneously in all the integrated systems.

Table 1. GNSS performance requirements for navigation of the approach phases.

Approach phases	Integrity			Requirement probability	
	Alarm threshold		Alarm time		
	Vertical	Horizontal		Integrity	Continuity
NPA	N/A	1.85–3.7 km	10–15 s	10^{-7}	10^{-4}
LNAV/VNAV	50 m	556 m	10 s	1×10^{-7}–2×10^{-7}/150 s	4×10^{-6}–8×10^{-6}/15 s
LPV-250		40 m			
APV- I	35 m				
APV- II	20 m		6 s		
LPV-200	35 m				
CAT I	10 m				
CAT II/III	<5.3 m	<17 m	<2 s	$<10^{-9}$/150 s	$<4 \times 10^{-6}$/15 s

Table 2 shows that if 2 outliers occur simultaneously, the integrated GPS + GLONASS and GPS + GLONASS + BDS can provide 100% service for the lateral

Table 2. The horizontal availability of systems with 2 outliers (%).

	LNAV	LPV-250~CAT I	CAT II/III
G	99.928	85.150	16.084
R	87.324	7.585	0.001
BR	98.590	73.934	22.121
GB	99.988	98.247	71.836
GR	100.000	99.995	76.507
GBR	100.000	100.000	95.757

Table 3. The horizontal availability of systems with 3 outliers (%).

	LNAV	LPV-250~CAT I	CAT II/III
BR	95.025	48.104	7.563
GB	99.782	90.364	42.370
GR	100.000	99.700	28.969
GBR	100.000	99.963	78.453

navigation of LNAV, and all the RAIM availability of integrated systems are higher than 98%. Only integrated GPS + GLONASS + BDS can 100% meet the lateral navigation requirement of LPV-250 ~ CAT I. The RAIM availability of standalone GLONASS is the minimum (7.585%) in approach phases LPV-250 ~ CAT I. For the requirement of CAT II/III, all the standalone or integrated systems cannot provide good availability in which the best is 95.757%, which is provided by integrated systems of 3 constellations.

Table 3 shows that if 3 outliers occur simultaneously, the combination systems of GPS + GLONASS or GPS + GLONASS + BDS still can satisfy the horizontal integrity requirement of LNAV. In LPV-250 ~ CAT I phases, the horizontal availability of 3 constellations combination is 99.963%, and this value of BDS and GLONASS combination is 48.104%; they are the best and the worst, respectively. In CAT II/III phases, those two values decline to 78.453% and 7.563%, corresponding to the same constellation combinations. From the comparison of Table 2 and Table 3, the horizontal availability of the systems in the scenario with 3 outliers obviously decreased from that in the scenario with 2 outliers.

To analyze the spatial distribution law of the horizontal availability, the median of the MUPB in 96 epochs in a point is selected to represent the MUPB of that point. In the scenario in which 2 outliers occur, Fig. 1 to Fig. 6 show the global spatial distribution of the MUPB median of GNSS systems. Those figures are drawn into equivalent color area according to the MUPB median value.

From Fig. 1 to Fig. 6, when 2 outliers occur simultaneously, the RAIM horizontal availability of standalone GPS or GLONASS is found to be lower, with the interval of

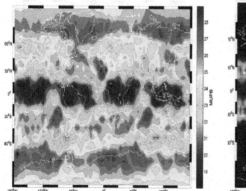

Fig. 1. The spatial distribution of the MUPB median of GPS when 2 outliers occur (the contour interval is 1 m in the interval of 18.014 m–29.858 m).

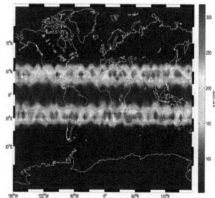

Fig. 2. The spatial distribution of the MUPB median of GLONASS when 2 outliers occur (the contour interval is 10 m in the interval of 52.503 m –327.227 m).

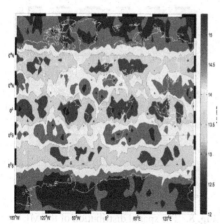

Fig. 3. The spatial distribution of the MUPB median of GPS+ GLONASS when 2 outliers occur (the contour interval is 1 m in the interval of 11.999 m–16.013 m).

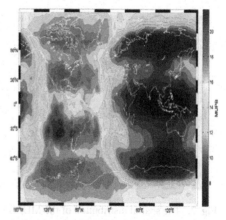

Fig. 4. The spatial distribution of the MUPB median of GPS + BDS when 2 outliers occur (the contour interval is 1 m in the interval of 6.161 m–22.634 m).

MUPB median of those two systems being 18.014–29.857 and 52.503–327.227, respectively. The availability of GLONASS is better at high latitudes and is lower between 30°S–30°N.

The availability of integrated systems has provided great improvement in performance compared to that of standalone systems. With the combination of the systems, the number of satellites increases and its MUPB is significantly reduced. Figure 4 to Fig. 6 show that BDS provides a great contribution to the RAIM integrity in the area (55°S–55°N, 75°E–150°E). The using of BDS can significantly reduce the MUPB values of combined systems in the area.

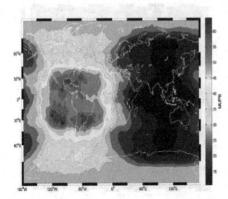

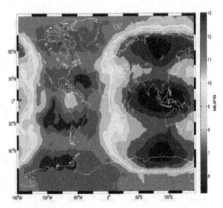

Fig. 5. The spatial distribution of the MUPB median of BDS + GLONASS when 2 outliers occur (the contour interval is 5 m in the interval of 10.724 m–68.789 m).

Fig. 6. The spatial distribution of the MUPB median of GPS + BDS + GLONASS when 2 outliers occur (the contour interval is 1 m in the interval of 5.283 m–13.866 m).

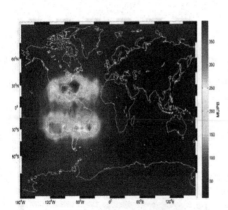

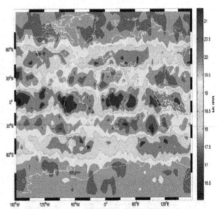

Fig. 7. The spatial distribution of the MUPB median of BDS + GLONASS when 3 outliers occur (the contour interval is 10 m in the interval of 14.086 m–403.300 m).

Fig. 8. The spatial distribution of the MUPB median of GPS + GLONASS when 3 outliers occur (the contour interval is 1 m in the interval of 16.037 m – 22.105 m).

From Fig. 7 to Fig. 10, when 3 outliers occur simultaneously, the MUPB of all the systems are increased relative to that with 2 outliers. The MUPB of combined BDS-GLONASS has the greatest change of the interval of the MUPB median, from 10.724–68.789 to 14.086–403.300; the maximum of the latter is approximately 5.8 times the maximum of former. The increase rate of the MUPB median of other combination systems is approximately 30% to 60%. The spatial distribution of the MUPB median of the systems with 3 outliers is similar to that of the MUPB median of the systems with 2 outliers.

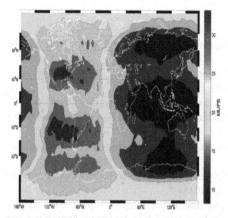

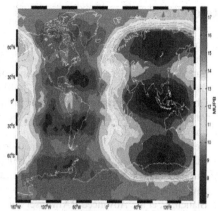

Fig. 9. The spatial distribution of the MUPB median of GPS + BDS when 3 outliers occur (the contour interval is 5 m in the interval of 8.202 m –37.275 m).

Fig. 10. The spatial distribution of the MUPB median of GPS + BDS + GLONASS when 3 outliers occur (the contour interval is 1 m in the interval of 6.572 m–18.512 m).

5 Conclusions

1. The RAIM horizontal availability will be significantly reduced with the increase in the number of outliers.
2. When 2 outliers occur simultaneously, combined GPS + GLONASS + BDS can meet the horizontal integrity requirement of LNAV ~ CAT I globally. Combined GPS + GLONASS can meet the requirement of LNAV globally. The other systems only can provide appropriate service in a partial region. All the combined systems can only meet the requirements of CAT II/III in partial ranges of the globe.
3. When 3 outliers occur simultaneously, the combination systems of GPS +GLONASS or GPS + GLONASS + BDS still can satisfy the horizontal integrity requirement of LNAV. All the combined systems can only meet the requirements of LPV-250 ~ CAT II/III in partial ranges of the globe.

In this paper, we simulated the scenarios of standalone or combined GNSS with 2 or 3 outliers occurring simultaneously and calculated the MUPB in one day in points of the globe. Next, we analyzed the RAIM horizontal availability in scenarios by comparing the MUPB with the HAL of different approach phases. The simulation and analysis of RAIM vertical availability with multiple outliers occurring will be performed in a future study.

Acknowledgments. This work was funded by Guangxi Natural Science Foundation Program (2018GXNSFAA281279).

References

1. Chen, J.P., Xu, Q.F., Liu, X.J.: Analysis of different algorithms for GPS RAIM HPL. J. Surv. Map. Inst. **18**, 1–3 (2001). (in Chinese)
2. Yang, C.S., Xu, X.H., Liu, R.H.: Analysis of algorithm s for G PS RA IM position integrity risk. J. Anhui Univ. (Natl. Sci. Ed). **3**, 76–81 (2011). (in Chinese)
3. Ober, P.B.: New, generally applicable metrics for RAIM/AAIM integrity monitoring. In: 9th International Technical Meeting of the Satellite Division of the Institute of Navigation, pp. 1677–1686 (2010)
4. Hewitson, S., Wang, J.: GNSS receiver autonomous integrity monitoring (RAIM) performance analysis. GPS Solutions, **10**, 155–170 (2006)
5. Xu, J.Y., Yang, Y.X., Li, J.L., et al.: Integrity analysis of COMPASS and other GNSS combined navigation. Sci. China Earth Sci. **56**, 1616–1622 (2013)
6. Blanch, J., et al.: Advanced RAIM user algorithm description: integrity support message processing, fault detection, exclusion, and protection level calculation. In: Proceedings of the 25th International Technical Meeting of The Satellite Division of the Institute of Navigation (ION GNSS 2012), Nashville, TN, pp. 2828–2849 September 2012
7. Blanch, J., Walter, T., Enge, P.: Advanced RAIM system architecture with a long latency integrity support message (2013)

Psychological Semantic Differences of Zhang Juzheng Before and After DuoQing Event

Shuting Li[1,2], Miaorong Fan[1,2], Fugui Xing[1,2], and Tingshao Zhu[1,2(✉)]

[1] Institute of Psychology, Chinese Academy of Sciences, Beijing, China
[2] Department of Psychology, University of Chinese Academy of Sciences, Beijing, China
tszhu@psych.ac.cn

Abstract. After the death of his father, Zhang Juzheng did not follow the rule to mourn his father for 27 months, but continued to work, known as Zhang's DuoQing event. Because the event had a huge influence on the imperial court and Zhang Juzheng himself, the present paper analyzed the differences in written expressions of Zhang Juzheng before and after DuoQing event from a psychological semantic perspective. This paper used a classic Chinese dictionary, CC-LIWC, to analyze Zhang's work (Memorials & letters), and obtained the word frequency of LIWC keywords, then compared their differences. We selected psychological semantically related CC-LIWC keywords for further analysis, including Functional-Personal pronouns, Affective processes, Social process, Cognitive process, Drives process, Personal concerns. The results indicated that only five dimensions of the Third person singular words, the Third person plural words, the Sadness words, the Female words and the Difference words were significantly different among the above keywords. This paper found that Zhang Juzheng was an efficient manager who loved power and work, the development of DuoQing event was beyond Zhang Juzheng's expectation, and the incomprehension and impeachment from his close people made him sad. This paper provides a new method for analyzing one's psychological changes before and after one major event, also provides some reference for modern leaders in terms of psychology and dealing with people.

Keywords: Zhang juzheng's DuoQing event · Psychological semantic analysis · CC-LIWC · Written language

1 Introduction

Zhang Juzheng (1525–1582) was a famous Chinese politician and reformer in the Ming Dynasty, who was appointed as the first grand-secretary of the emperor and the imperial tutor in the first decade of the Wanli reign (1572–1582). During this period, his series of reforms made people live in peace and enjoy their work [1], imperial coffers were well stocked with silver bullion [2], and these remarkable achievements delayed the fall of the Ming Dynasty to some extent. On September 13, 1577 (the 5th year of Wanli), Zhang Juzheng's father died. Following the filial piety of the Ming Dynasty-Dingyou regulation, Zhang Juzheng had to resign to return to his hometown, and mourned his

© Springer Nature Switzerland AG 2021
Q. Zu et al. (Eds.): HCC 2020, LNCS 12634, pp. 161–172, 2021.
https://doi.org/10.1007/978-3-030-70626-5_17

father for 27 months before he could resume his duties to the imperial court. Zhang Juzheng submitted his request to leave three times, but the Wanli emperor refused the application and issued a rescript ordering Zhang Juzheng to stay in office and be loyal to the emperor [1]. The incumbent official whose parents died was deprived of his grief and continued his work, which was called DuoQing QiFu. After Zhang Juzheng accepted the order for DuoQing QiFu from the Wanli Emperor, a series of political storms erupted in the Ming Empire, known historically as Zhang Juzheng's DuoQing event, hereafter referred to as DuoQing event.

Imperial Power Society in Ancient China vigorously promoted the ethics of filial piety, and DuoQing QiFu was usually considered unfilial which was contrary to the ruler's basic policy of ruling the country through a culture of filial piety. The Ming Dynasty was no exception, even the emperors decreed that the standard procedure to mourn the demise of one's parents could not be deprived [2]. "Ming Shi Lu", "History of Ming Dynasty", "Ming Wenhai", "Guoque", "First grand-secretaries of the Cabinet since Jiajing" had recorded and described Zhang Juzheng's DuoQing event, they almost agreed that Zhang Juzheng was authoritarian and arrogant in the event of DuoQing [1–5]. Many modern scholars hold the same view that DuoQing event produced a series of adverse consequences for Zhang Juzheng, such as: moral criticism [6], a decline in personal prestige, posthumous liquidation [7], and the ultimate failure of reform [8]. Because of the significance of DuoQing event to Zhang Juzheng, many researches have been conducted. For instance: Tian Shu analyzed the DuoQing event in the context of the "Da Li Yi" [6]; Han Xiaojie reported the relation between the results of reform and Zhang Juzheng's personality shown in the DuoQing event [9]; Liu Zhiqin argued that Zhang Juzheng's character traits and the reasons for his liquidated after his death through DuoQing event [10]; Tang Shujun et al. discussed the Machiavellian personality of Zhang Juzheng in the case of DuoQing event [11]. However, the researches on Zhang Juzheng's DuoQing event are carried out by qualitative analysis, and quantitative analysis method is rarely used.

Qualitative analysis often extracted a few sentences or paragraphs from Zhang Juzheng's written works to illustrate the author's views or conclusions in the form of examples. Yet the function of written word is far more than that, they are important carrier of emotional information, the emotion words such as *happy*, *sad*, *war*, *reward*, have different ways of mental processing, and they had different psychological semantics [12]. Pennebaker et al. developed a quantitative analysis method, LIWC [13], which was a dictionary based on keyword classification and word frequency statistics. With the dictionary, they explored the expression of written language in various scenes, such as emotion, cognition and language process [13–18]. In one of their studies, they analyzed the contents of emails, the novel "Alice in Wonderland" and Shakespeare's play "King Lear (1606)", etc., then they found the content and manner which authors used in written could reflect personal status, power and could display psychological emotions such as safety and anger [14]. Li Sijia et al. analyzed the categorization and differences in the Big Five personality of the characters in Jin Yong's works [19]; Li Yingyu et al. studied the differences in the gender of translators in written expression [20]; John Sell et al. found that with the passage of time, the authors of Introduction to Psychology have changed in terms of authority and emotion [21]. It is obvious that there are many advantages in analyzing written language by LIWC: 1) it can process a large number of text data, not limited to a few sentences or paragraphs; 2) the entire data analysis process is processed by the program, which is not subject to the researchers' subjective perceptions, and has a high degree of repeatability; 3) the data results are presented in a structured way, and the significance test of differences are provided.

Using Zhang Juzheng's Memorials and letters as written materials, this paper analyzed the psychological semantic differences in Zhang Juzheng's written before and after DuoQing event, especially in the aspects of power achievement motives, emotional processes, personal attention and cognitive processes. Our analysis used the CC-LIWC, a classic Chinese dictionary (Classic Chinese linguistic inquiry and word count) [22, 23], which analyzed the relationship between written language in Ancient Chinese and psychological semantic based on LIWC keywords.

This study found Zhang Juzheng's behavior motivation after DuoQing event from the point of psychological semantic, proposes a new perspective and methodology for studying the personality psychology of the ancients and the psychological impact of major events on the characters.

2   Method

2.1   Data Selection

The DuoQing event was one of the important events of Zhang Juzheng in his career as an official [1, 2], see Table 1. To avoid the influence of the two great events of being the first grand-secretary and dying of illness, we defined the 1st to 4th year of Wanli (1573–1576) as before DuoQing event, and the 6th to 9th year of Wanli (1578–1581) as after DuoQing event.

Table 1. The great events in Zhang Juzheng's political career.

S/N	Year	Great events
1	The 26th year of Jiajing (1547)	Became a metropolitan graduate
2	The 28th year of Jiajing (1549)	Became an official of Han-lin Academy
3	The 33th year of Jiajing (1554)	Asked for back home on the grounds of illness
4	The 38th year of Jiajing (1559)	Returned to the court as an official
5	The 1th year of Longqing (1567)	Became a grand-secretary of Longqing
6	The 6th year of Longqing (1572)	Became the first grand-secretary of Wanli
7	The 5th year of Wanli (1577)	The DuoQing event happened
8	The 10th year of Wanli (1582)	Died of illness

All "Wl" in the following tables in this paper were simplified Chinese Pinyin of Wanli, and the number represented the year.

In this paper, the criteria for selecting Zhang Juzheng's works were as follows:

1) Character works should be formal, easy to understand, and not prone to ambiguity. Zhang Juzheng's collection of "Zhang Taiyue Poetry Collection" was a collection of poetry and prose, it often used allusions and the rhetoric of Fu Bi Xing, which easily interfered with the word segmentation system. Therefore, this paper only collected the Memorials and letters in the "Complete Works of Zhang Wenzhong-gong [24]" for text analysis.
2) The "Complete Works of Zhang Wenzhong-gong" selected in this article was derived from a special edition of the Siku Quanshu-Jibu which was officially compiled by Qing Dynasty, and the original text had high credibility and authority.
3) Through "Ming Shenzong Shilu [1]", "Big Biography of Zhang Juzheng [25]", "A study on the mentality of scholars in the late Ming Dynasty [26]", "Zhang Juzheng and the political situation in the middle and late Ming Dynasty [27]", we confirmed the writing time of Zhang Juzheng's works, and then obtained the complete ancient texts through the "Complete Works of Zhang Wenzhong-gong".

Finally, this paper selected 23 works in 1–4 years of Wanli and 28 works in 6–9 years of Wanli as the data source of ancient Chinese text.

2.2 Data Processing

The tool used in this paper was Classic Chinese Dictionary (hereafter referred to as CC-LIWC), and it was developed by the Chinese Academy of Sciences Institute of Psychology Computational Network Psychology Laboratory based on the SC-LIWC (Simplified Chinese LIWC Dictionary) [28], which a dictionary of classic Chinese word frequency analysis. The CC-LIWC included four parts: the main body of the Python program, the Ancient Classic Chinese Word Dictionary, the LIWC keywords & their frequency, and the corresponding significance difference test.

2.3 Data Analysis

The data analysis and processing flow of this paper was as follows:

1) Ancient Chinese word segmentation. We loaded the CC-LIWC Ancient Chinese Word Segmentation Dictionary, removed the punctuation and spaces in the TXT text, and classified all words by LIWC after obtaining the segmentation results.
2) Statistical indicators calculation.

① Basic statistical indicators: TWC (Total Word Count), LWC (LIWC Word Count), LCR (LIWC Cover Rate, see Eq. 1).

$$LCR = \frac{LWC}{TWC} \tag{1}$$

② Word category and word frequency statistics. Word frequency and proportion of LWC keywords in each category and dimension.

3) Significant difference test, Mann-Whitney U test.

4) In the keyword category of CC-LIWC, we selected psychological semantically related LIWC keywords for analysis, such as Functional-Personal pronouns, Affective processes, Social process, Cognitive process, Drives process, Personal concerns.

3   Results

The basic statistical indicators were shown in Table 2.

Table 2. The basic statistical indicators.

Category name	Wl1–4	Wl6–9
TWC	9015	10516
LWC	6818	7761
LCR	75.6%	73.8%

After excluding the Functional words in Zhang Juzheng's works, there were 43 dimensions keywords. The frequency of *Power* words ($M = 0.131$), *Work* words ($M = 0.088$), *Negative emotion* words ($M = 0.067$), *Space* words ($M = 0.065$) and *Positive emotion* words ($M = 0.063$) were in top five (see Fig. 1). Among the categories of Drives process, Achievement words ($M = 0.043$) ranked 7, Risk words ($M = 0.035$) ranked 10, Affiliation words ($M = 0.0218$) ranked 17, Reward words ($M = 0.022$) ranked18. Social process (Family words, Friends words, Female words, Male words), Personal Concerns (Leisure words, Home words, Religion words, Death words) were all ranked after 26.

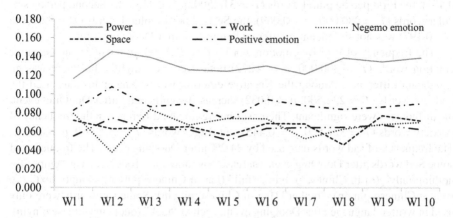

Fig. 1. The top five keywords used by Zhang Juzheng from the 1st to 10th year of Wanli.

In the six categories of keywords we selected, there were significant differences in the five dimensions of keywords before and after DuoQing event, such as Third person singular words, Third person plural words, Sadness words, Female words, Difference words. There were varying increases in Sad words, Female words, and decreased in other dimensional keywords.

Table 3. List of keywords with significant difference.

Category name	Dimension name	M		SD		T-value	p-value
		WI1–4	WI5–9	WI1–4	WI5–9		
Personal pronouns	Third person singular	0.012	0.008	0.008	0.007	215.000	0.015*
Personal pronouns	Third person plural	0.012	0.008	0.008	0.007	233.500	0.033*
Affective processes	Sadness	0.013	0.015	0.008	0.007	232.500	0.032*
Social processes	Female	0.003	0.003	0.007	0.003	242.000	0.038*
Cognitive processes	Difference	0.037	0.025	0.012	0.016	170.000	0.001*

Note: Each number was rounded up to three decimal places, * P < 0.05.

Among the Function-Personal pronouns, the Third person singular words ($T = 215.000$, $p = 0.015$) and the Third person plural words ($T = 233.500$, $p = 0.033$) were significantly different before and after DuoQing event. Compared with the First person singular words, their frequency was higher before DuoQing event and lower after DuoQing event. The frequency of the First person singular words ($T = 327.000$, $p = 0.456$), the First person plural words ($T = 318.500$, $p = 0.315$), the Second person singular words ($T = 280.000$, $p = 0.069$), the Second person plural words ($T = 322.000$, $p = 0.197$) was not significantly different before and after DuoQing event.

The frequency of Positive emotion words ($T = 320.000$, $p = 0.405$) and Negative emotion words ($T = 300.000$, $p = 0.272$) both decreased slightly, but there was no significant difference. Among the Negative emotion words, Zhang Juzheng used the Sadness words ($T = 232.500$, $p = 0.032$) increased significantly after DuoQing event, the differences were significant. There was no significant difference in the frequency of Anxiety words ($T = 325.500$, $p = 0.445$) and Anger words ($T = 275.000$, $p = 0.143$). The frequency of sad words increased by 24.2% after DuoQing event. The frequency of some Sad words after DuoQing event increased by more than 100%, such as "empty/for nothing/only" (tu in Chinese pinyin), "fail" (bai in Chinese pinyin), "empty text" (xu wen in Chinese pinyin), "bother" (fan in Chinese pinyin); Some sad words were only used in written language after DuoQing event, such as "lack" (quefa in Chinese pinyin), "pain" (tong in Chinese pinyin), "accident" (yiwai in Chinese pinyin), "useless" (wuyi in Chinese pinyin).

The Female words ($T = 242.000$, $p = 0.038$) of Social Processes was significantly different before and after DuoQing event, the Family words ($T = 307.000$, $p = 0.315$), the Friends words ($T = 325.000, p = 0.439$), the Male words ($T = 318.000, p = 0.388$) were no significantly different.

After DuoQing event, the Difference words ($T = 170.000, p = 0.001$) in the Cognitive process were significantly reduced, and there were significant differences. There was no significant difference in Insight words ($T = 301.500, p = 0.281$), Causation words ($T = 332.000, p = 0.493$), Discrepancy words ($T = 314.000, p = 0.363$), Tentative words ($T = 325.500, p = 0.445$) and Certainty words ($T = 282.000, p = 0.174$). The frequency of Difference words decreased by 33.5% after DuoQing event. The number of words "but" (er in Chinese pinyin) used decreased from 158 to 90, with a decrease rate of 43.0%.

From the perspective of Dives words, Affiliation words ($T = 284.500, p = 0.186$), Achievement words ($T = 294.500, p = 0.239$), Power words ($T = 300.000, p = 0.272$), Reward words ($T = 295.000, p = 0.242$), Risk words ($T = 303.000, p = 0.290$) were not significant before and after DuoQing event.

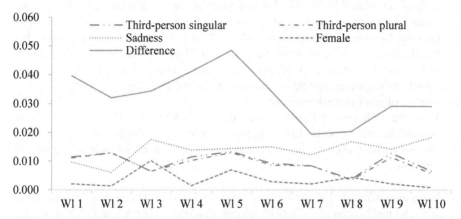

Fig. 2. The frequency of keywords with significant difference from the 1st to 10th year of Wanli.

4 Discussion

By calculating the frequency of CC-LIWC keywords in Zhang Juzheng's works, we quantified the effects of DuoQing event on his written expression. We found that Zhang Juzheng had a strong desire for power, and he was absorbed in his work and seldom entertained himself. He paid less attention to people and events outside his work, such as family, friends, leisure, and death. In the view of obeying the filial piety regulation, Zhang Juzheng was worried that he might lose his position of the first grand-secretary [29], might have no hope of rejoining the cabinet [30]. In his letters to his friends, he admitted that the chosen of DuoQing QiFu was in love with his power and political position, and he didn't dare to retire until the reform was done [24]. Facing doubts and incomprehension after DuoQing event, he constantly expressed heart for constructing his country in letters with friends, such as "reply to General constitution Li Jian'an, reply to River course Lin Yunyuan", etc., and proved to the world that his DuoQing QiFu was beneficial to the country and the people [24, 31].

4.1   Zhang Juzheng was an efficient manager who loves power and work

According to David C. McClelland's social motivation theory, Zhang Juzheng was a man with high power motivation and devoted himself to work [32]. Combined with the ranking of Power words, Work words, Achievement words, Risk words and Affiliation words, we found that Zhang Juzheng was an efficient manager with high power need, moderate achievement need, and low affinity need [33].

1) The act of high power
   After becoming the first grand-secretary, Zhang Juzheng repeatedly compared himself with Yi Yin, said that he was *"words on behalf of the Emperor's words, on behalf of the King's administration* [24]". The emperor of Wanli also said to Zhang that *"I obeyed everything you said"*. So, many scholars believed that *"the punishment of those who opposed DuoQing QiFu was the product of Zhang Juzheng's kidnapping of Emperor Wanli"* [6, 25, 27]. He took pride in the fact that he recommended and quoted the grand-secretaries and hundreds of officials, especially the Cabinet members, the Censorial-supervisory Branch, the Ministries of the Six Boards and other key departments, those were conducive to his holding of power and the development of his work [24]. The above behaviors of Zhang Juzheng occurred both before and after DuoQing event, and there was no significant difference.
2) The act of work and achievement.
   From the 38th year of Jiajing to the 5th year of Wanli (nearly 19 years), Zhang Juzheng had been working in Beijing and never returned to his hometown, Jiangling, Hubei Province, even on the way home to bury his father, he was handling official business.

It was undeniable that Zhang Juzheng's work in his ten-year career as the first grand-secretary, he had accomplished lots of reforms that others dare not and cannot accomplish. See Table 4, his work achievements include but were not limited to: promoted the Kaosei system, popularized the rule of taxation reform, rectified academic administration, destroyed academies, reformed post delivery system, reduced the emperor's vassal [1, 2, 24].

3) The act of low affinity
   People with high affinity cared about the quality of interpersonal relationships and wanted to be liked and accepted by others [33]. Zhang Juzheng who was a person lacked affinity was so devoted to his country that he went out of his way to offend people from the emperor down to the hundreds of officials and students [25–27].

Zhang Juzheng reduced the emperor's expenses and interfered with the emperor's personal life, and gradually accumulated the emperor's aversion to him. He offended incompetent or corrupt officials by Promoting the Kaosei system and popularizing the rule of taxation. He offended royal relatives by cutting off the vassal. He offended those with a vested interest in the post delivery system by reforming the post delivery system. He offended profitmaking officials and people who studied at public expense by closing the Academic Administration in Colleges [25, 34].

Huang Renyu believed that Zhang's control of power relied on his personal political relationship, which was not recognized by the Daming system, and to a certain extent, intensified the party struggle [34]. Zhang Juzheng's hiring strategy to maintain power and lack of affinity were likely to be an important reason for his failure to no one continued his reform policy and his liquidation after his death [7, 35].

Table 4. Zhang Juzheng's behavior before and after DuoQing event.

Category name	Before Duoqing	After Duoqing
Power	Introduced Lv Tiaoyang and Zhang Siwei as grand-secretary into the cabinet	Introduced Ma Ziqiang and Shen Xingshi as grand-secretary into the cabinet
	Held his own ground and put Qi Jiguang and Li Chengliang to guard the frontier	Held his own ground and put Pan Jixun harnessing water conservancy
Work	Worked every day for 19 years	Dealt with official business during the process of returning home to bury his father
Work & Achievement	Promoted the Kaosei system to rectify the administration of officials, started fiscal and tax reform	Popularized the rule of taxation reform by combining all miscellaneous duties into one collected in silver coins instead of real goods
	Rectified the academic administration and study style, reformed the military's government affairs process	Destroyed academies and cut down government students
	Reformed the rule of law and penalties, reformed of post delivery system	Reduced the number of royal relatives, made an exact measurement of the land
	Yelled at the emperor in public. Asked him to reduce his spending	Asked the emperor to recall weaving and reduce the burden on the people. Wrote the edict of guilt on behalf of the emperor without his request
	Disagreed with the Empress Dowager Li's order to build a temple and release prisoners. Punished her father in public	Refused repeatedly to be knighted by the Queen's father

Surprisingly, our data result that Power words ($T = 300.000$, $p = 0.272$) were not significant before and after DuoQing event, did not support the view that Zhang Juzheng's desire for power had expansion after DuoQing event, which was inconsistent with the conclusions of previous research works, [1, 2, 25–27, 35]. Table 4 tells us intuitively that

Zhang Juzheng strictly controlled the personnel power of the cabinet, the Six Boards ministries, Censorial-supervisory and other power centers before and after DuoQing event.

4.2 The Development of DuoQing Event Beyond Zhang's Expectations, and the Incomprehension and Impeachment of His Close People Make Him Sad

According to Pennebaker et al., people over 40 used the First person singular less and paid less attention to themselves; those who used less First person singular and more Third person singular were more authoritative and confident [11]. Based on Zhang's use of First person singular, Third person singular and Third person plural, Zhang Juzheng's attention to himself had not changed significantly before and after DuoQing event, but the moral deficiency of DuoQing event weakened his authority to some extent [6, 8, 25].

The frequency of the Difference words was at the peak on the 5th years of Wanli, which was on the rise from 2nd to 5th years of Wanli and was on a downward trend from 5th to 7th years of Wanli, see Fig. 2. Although the DuoQing event was not in accordance with propriety and etiquette, there were several precedents in Ming dynasty [1, 2]. In many cases, being asked by the emperor for DuoQing QiFu was a symbol of official identity and value [36]. However, Zhang Juzheng's DuoQing event was opposed unprecedented, a disparity that Zhang Juzheng could not understand. In his opinion, these officials did not understand his feeling of forgetting his family to benefit the country, they did not share the responsibility of handling state affairs, and even impeached and slandered himself [24]. Some historians believed that Zhang Juzheng realized that the DuoQing event had gradually evolved into a struggle for power, and his subsequent policies such as the assessment of officials in Beijing and the closure of colleges were related to this [27, 34]. What saddened Zhang Juzheng the most was that these included people close to him, including Minister of Personnel Zhang Han whom he had promoted, his student Wu Zhongxing, his student Zhao Yongxian, his fellow townsman Ai Mu [1, 2].

This paper has the following limitations for the future work. The 51 Memorials and letters we selected in this paper were part of Zhang Juzheng's works. This paper focused on before and after DuoQing event but didn't explain the difference of Zhang Juzheng's affinity before and after his first grand-secretary period, it's necessary to discuss it in the future study.

5 Conclusions

This paper used CC-LIWC to analyze Zhang Juzheng's Memorials and letters, found the differences of written expression from the psychological semantics before and after DuoQing event. The quantitative analysis methods in the paper was more objective than qualitative analysis, its data processing process was automatically completed by the program, and the data results were presented in a structured form.

Zhang Juzheng's motivation of power was always high, but it couldn't be concluded that Zhang Juzheng's power expansion after DuoQing event. The discontent and criticism of many officials over the incident of Zhang Juzheng's DuoQing event was beyond Zhang Juzheng's expectation, and the lack of understanding and impeachment by those close to him made him sad. Zhang Juzheng lacked affinity and offended too many people in the process of handling affairs, the personnel tactics of maintaining his absolute power might be the important reason why his reform policy failed to continue and was liquidated after his death.

This paper provides a new way for analyzing the psychological changes in personalities before and after the major events, cultural, ideological, and psychological changes of the times. It also provides some reference for modern leaders in terms of psychology and dealing with people.

References

1. Lu, M.S.: History of official revision (Ming Dynasty): Ming Emperor Shenzong-Xian emperor's record. 152, 67–82. http://www.wenxue100.com. Accessed 2 Oct 2020
2. Zhang T., et al.: History of official revision (Qing Dynasty). History of Ming Dynasty. The Biography 101-Biography of Zhang Juzheng; the biographic sketches of emperors. http://www.wenxue100.com. Accessed 2 Oct 2020
3. Zongxi, H.: (Ming Dynasty): Ming Wenhai Zhonghua Book Company (1987)
4. Qian, T.: (Ming Dynasty): Guoque Zhonghua Book Company (1958)
5. Shizhen, W.: (Ming Dynasty): First grand-secretaries of the Cabinet since Jiajing Zhongzhou Ancient Books Publishing House (2016)
6. Shu, T.: Zhang Juzheng's DuoQing event and political upheaval from the View of "Da Li Yi". Acad. Res. **000**(003), 109–118 (2017)
7. Wanchun, D.: The contradiction of objective and means on legalization. J. Yangtze Univ. (Soc. Sci.) **28**(05), 36–41 (2005)
8. Zhongtao, F.: On Zhang Juzheng's political activity of Duoqing. J. Yibin Univ. **7**(001), 57–59 (2007)
9. Xiaojie, H.: Politician's personality and reform-on personal factors of Zhang Juzheng's unsuccessful reform. J. Yangtze Univ. (Soc. Sci.) **27**(01), 81–84 (2004)
10. Zhiqin, L., Juzheng's, Z.: Character tragedy. J. Tianjin Normal Univ. (Soc. Sci.), **55**(005), 30–36 (2005)
11. Shujun, T., Yongyu, G.: The analysis of Zhang Juzheng's machiavellism personality. Psychol. Explor. **31**(3), 209–213 (2011)
12. Yang, C., Lin, W.: Processing of emotional information in written language. Chin. Sci. Bull. **063**(002), 148–163 (2018)
13. Pennebaker, J.W., Boyd, R.L., Blackburn, K.: The Development and Psychometric Properties of LIWC2015, TX: University of Texas at Austin. https://doi.org/10.15781/t29g6z, Austin (2015)
14. Pennebaker, J.W.: The Secret Life of Pronouns: What Our Words Say about Us, Translated by Liu Shan. pp. 5–208. Machinery Industry Press, Beijing (2018)
15. Pennebaker, J.W.: Writing about emotional experiences as a therapeutic process. Psychol. Sci. **8**, 162–166 (1997)
16. Pennebaker, J.W., Francis, M.E.: Cognitive, emotional, and language processes in disclosure. Cogn. Emot. **10**, 601–626 (1996)
17. Pennebaker, J.W., King, L.A.: Linguistic styles: language use as an individual difference. J. Pers. Soc. Psychol. **77**, 1296–1312 (1999)
18. Pennebaker, J.W., Mayne, T., Francis, M.E.: Linguistic predictors of adaptive bereavement. J. Pers. Soc. Psychol. **72**, 863–871 (1997)

19. Sijia, Li., Yijun, Y., Shengtao, W., et al.: Writer criticism on Jin Yong: an artificial-intelligence study. Psychol. Tech. Appl. **000**(10), 584–589 (2019)
20. Yingyu, L., Jian, Zhang, Li, Yuan: An empirical study of gender differences in Chinese English translation-quantitative analysis of English translation corpus of Shaanxi literature. Data Culture Educ. **847**(001), 57–58 (2020)
21. Sell, J., Farreras, I.G.: LIWC-ing at a century of introductory college textbooks: have the sentiments changed? Procedia Comput. Sci. **118**(2017), 108–112 (2017)
22. Miaorong, F.: Critical Technology in Ancient Chinese Psychological Semantic Analysis Based on LIWC. Master Dissertation. University of Chinese Academy of Sciences, Beijing (2020)
23. Fugu, X., Tingshao, Z.: Large-scale online corpus based classical chinese integrated dictionary building and word segmentation. J. Chin. Inf. Process. Accept. (2020)
24. Juzheng, Z.: (Ming Dynasty): Complete works of Zhang Wenzhong-gong, pp. 15–458. Commercial Press, Shanghai (1935)
25. Dongrun, Z.: Big Biography of Zhang Juzheng, pp. 82–219. Jilin Publishing Group Co., Ltd, Changchun (2012)
26. Luo, Z.: A study on the mentality of scholars in the late Ming Dynasty, pp. 223–251. Nankai University Press, Tianjin (2006)
27. Qingyuan, W., Juzheng, Z.: and the political situation in the middle and late Ming Dynasty, pp. 467–815. Guangdong Higher Education Press, Guangzhou (1999)
28. Nan, Z., Dongdong, J., Shuotian, B., et al.: Evaluating the validity of simplified chinese version of LIWC in detecting psychological expressions in short texts on social network services. PLoS ONE **11**(6), 1–15 (2016)
29. Xie, X.: (Qing Dynasty): Ming Tong Jian, vol. 66, pp. 516. Yuelu Publishing House, Changsha (1990)
30. Yinglin, F.: (Late Ming and early Qing Dynasty): Mingshu Biography of Zhang Ju-zheng, vol. 150. http://www.guoxuedashi.com/guji/zx_6665528ocwg/. Accessed 2 Oct 2020
31. Xiong, S.: Discussion on Zhang Jiangling with Friends, pp. 182–184. Shanghai Bookstore Publishing House, Shanghai (2007)
32. Hou, Y.: Foundations of Social Psychology. 3rd edn., pp. 215–216. Peking University Press, Beijing (2013)
33. McClelland, D.C., Boyatzis, R.E.: The leadership motive pattern and long-term success in management. J. Appl. Psychol. **67**(6), 737–743 (1982)
34. Fuli, M., Ruide, C.: The Cambridge history of China (The Ming Dynasty, 1368–1644), Translated by Zhang Shusheng, Huang Mo, Yang PINQUAN, Siwei, Zhang Yan, etc., vol. 1, pp. 570–571. China Social Sciences Press, Beijing (1992)
35. Houan, G., Shu, T.: Anatomy of Zhang Juzheng's power. Gansu Soc. Sci. **000**(02), 67–72 (1989)
36. Kesheng, Z.: On the mourning system of civil officials in the ming dynasty. J. Southwest China Normal Univ. (Hum. Soc. Sci. Ed.) **32**(5), 48–52 (2006)

Digital Rights Management Platform Based on Blockchain Technology

Wenan Tan[1,2(✉)], Xiao Zhang[1], and Xiaojuan Cai[1]

[1] College of Computer Science and Technology, Nanjing University of Aeronautics and Astronautics, Nanjing 211106, Jiangsu, China
wtan@foxmail.com
[2] School of Computer and Information Engineering, Shanghai Polytechnic University, Shanghai 201209, China

Abstract. Traditional copyright protection technologies are unable to adapt to the current digital era of the Internet, and some problems such as difficult in digital works confirmation copyright, complex copyright transactions, copyright protection and infringement monitoring. Although Digital Rights Management (DRM) can provide certain security for digital copyrights, it still cannot effectively solve these problems. The decentralization, non-tampering, traceability and other characteristics of the blockchain are of great help to address the above problems. This paper proposes a scheme combining blockchain technology with digital rights management, and builds a digital rights management platform based on blockchain. In this paper, we propose platform system architecture based on blockchain technology by using the open source Hyperledger Fabric framework as the underlying blockchain development platform and combining smart contracts, IPFS systems, timestamps and other technologies, and have implemented the functions of the four core modules of the platform, which brings new breakthroughs for digital copyright protection technology.

Keywords: Blockchain · Smart contracts · IPFS · Copyright transaction

1 Introduction

In the current digital age, various digital works have appeared and published on the Internet. The process of confirmation copyright of traditional digital works is too complicated and cumbersome. When the copyrights of digital works are confirmed, the infringers have already processed the content of the digital works, which makes it impossible to verify the infringement and damages the interests of copyright owners. Therefore, it is easy to infringe copyright, but difficult to confirmation copyright, copyright protection, monitor infringements and protect privacy.

The traditional digital rights management platform is a centralized system, and almost all data information is managed by the administrator of the management center, because the administrator has absolute control and lack of power constraints, it is prone to hacker attacks, information leakage, or data tampering [1]. The introduction of

© Springer Nature Switzerland AG 2021
Q. Zu et al. (Eds.): HCC 2020, LNCS 12634, pp. 173–183, 2021.
https://doi.org/10.1007/978-3-030-70626-5_18

blockchain technology and the use of its "decentralized" subversive design concept and its data tamper resistance, traceability and other advantages can perfectly make up for the shortcomings of traditional digital rights management technology.

This paper designs a digital rights management platform based on blockchain that stores the copyright information and transaction information of digital works, which can not only prevent information data from being tampered with, but also can be traced at any time. The contributions of this paper are: 1) Using smart contracts to replace manual inquiry and review of information can shorten the confirmation copyright and copyright transactions cycle and prevent these information from being tampered with. 2) Combining blockchain technology with digital rights management technology can bring new breakthroughs to digital copyright protection technology.

2 Related Work

2.1 Overview of Blockchain

Blockchain is a chained data structure that combines data blocks in a chronological order by connecting them sequentially, and uses cryptography to ensure that the data can-not be tampered with or forged [2]. Generally, a mature blockchain should have the following characteristics: decentralization, traceability, transparency and credibility, among which decentralization is the core advantage of blockchain technology. (1) Decentralization: The blockchain network is a distributed system composed of many nodes, and each node has a high degree of autonomy. The use of P2P network ena-bles each node to have the same network rights, eliminating centralized service nodes and avoiding data information leakage and tampering. (2) Traceability: Blockchain stores data in chain, and uses timestamps, consensus mechanisms and other technical means to realize the functions of non-tampering and traceability of data. For example, after the transaction information is transmitted to the blockchain network through the P2P network, the consensus algorithm is used to reach consensus on each node. If the consensus can be reached, the smart contract information is stored on each node and stamped with timestamp, which realizes the full traceability of data information and solves the problem of copyright forensics to a certain extent. (3) Transparency and credibility: In a decentralized system, each user obtains complete information, thereby realizing information transparency. The consensus algorithm ensures the consistency among all nodes when the transaction is finally confirmed, so the information of the entire system is credible.

2.2 Smart Contract

The concept of smart contract was first proposed by Nick Szabo [3] in 1994, and it was not actually implemented until the block 2.0 stage. Blockchain provides a credible execution environment for smart contracts, which expand applications for blockchain. The typical representative of Blockchain 2.0 is Ethereum. The biggest advantage of Ethereum is the smart contract. Once it is set, as long as the execution conditions are met, it will be automatically executed without the participation of a third party. This is also an important reason why the blockchain is called "decentralized". It allows us to perform traceable, irreversible and secure transactions and fully automated processes without a third party.

2.3 IPFS

IPFS (InterPlanetary File System) [4] is a point-to-point distributed file system. Its aim is to replace HTTP used in the past 20 years. IPFS replaces HTTP-based domain name addressing with content-based addresses, it will generate a unique hash identifier for each file content, and users can find the corresponding file by hash value. When the same hash value file appears, the system will automatically delete it which saves storage space. The essence of the blockchain is a distributed ledger, and one of its bottlenecks is insufficient file storage. The storage method of IPFS is fragmented distributed storage, which cannot be attacked by hackers and files are not easy to be lost. From a certain perspective, IPFS solves many problems caused by Internet centralization, such as high cost, low efficiency, information redundancy and security.

2.4 Hyperledger Fabric

Hyperledger Fabric is a platform that provides distributed ledger solutions. The goal is to implement a common underlying framework of Permissioned Chain, in order to apply to different occasions, a modular architecture is used to provide switchable and extensible components, including consensus algorithms, encryption security, smart contracts, monitoring services and other services [5]. This paper also uses the Fabric framework as the underlying blockchain development platform. It provides SDK and GRPC connection API interfaces for applications, so that the same communication interface as the internal model is used, and fewer ports are opened to avoid some malicious attacks. Member services provide registration and login, identity management and transaction audit functions to limit access rights to ensure the security of system login. In the Fabric system, the smart contract is the chaincode [6]. The chaincode service provides a smart contract execution engine, which provides a deployment environment for Fabric's contract code programs.

2.5 Related Technology Analysis

Blockchain has the characteristics of decentralization, tamper resistance, traceability, transparency and credibility, which can solve the problem of multi-party trust. It has been widely used in supply chain finance [7], logistics chain [8], Internet of Things [9], public welfare [3] and other fields, which has greatly reduced collaboration costs and improved collaboration efficiency within and between industries.

Combining blockchain technology with digital rights management technology will start a new chapter for digital works, which is a new starting point for it. Meng et al. [10] proposed a scheme of copyright management based on digital watermarking and blockchain, but when digital works are infringed, the watermark data cannot be used as effective evidence for infringement. Xu et al. [11] proposed a digital rights management scheme for network media based on blockchain. Although this scheme can use these functions of the blockchain to realize effective copyright management, transaction management, and user behavior management network media, it ignores storing all huge network digital media works on the blockchain, which will lead to the decline of the blockchain's ability to process data. Ding et al. [12] designed a digital copyright protection and transaction system based on blockchain technology, using smart contracts and other technologies to provide registration, tracking, authentication, query, and transaction services for digital content copyright.

These researches provide some enlightenment for blockchain technology in digital rights management. Namely, this paper proposes a scheme based on the block-chain digital rights management platform. Combining smart contract, IPFS, Fabric and other technologies, it can effectively solve the problems of copyright confirma-tion, infringement monitoring, copyright protection and evidence collection.

3   Analysis and Design on Digital Rights Management

The introduction of blockchain technology into the digital rights management plat-form realizes online infringement monitoring of digital works and safeguards the interests of copyright owners. The main functions of the platform are user registra-tion, user login, copyright application, transaction application, transaction inquiry, etc. By storing copyright-related information and copyright transaction information to each node in the Fabric blockchain and stamping it with timestamps, the information can be monitored in real time to avoid problems such as difficulty in infringement monitoring. Combining the IPFS system with the blockchain can make up for the bottleneck of the insufficient storage memory of the blockchain and prevent the copyright information from being tampered with or lost.

3.1   Business Analysis

All users register and log in first on the platform. Note: The user needs to perform real-name authentication before confirmation copyright and transaction application. After the administrator passes the verification, he can perform other operations. Figure 1 shows the process of confirmation copyright of digital works in a digital rights management platform based on blockchain technology.

The creator first uploads the digital work to the copyright management platform to duplicate checking, if it passes the test, the digital work is stored in the IPFS system [13]. IPFS is responsible for storing encrypted digital works and returning the hash value of the work as an index for querying digital works. If not, the work will be re-turned directly. Then the creator temporarily stores the personal information, creation time and other information in the smart contract for copyright application through the platform [14]. After being approved by a third-party regulatory agency, the smart contract automatically packages the copyright-related information on the blockchain, stores it in the blockchain and stamps with a timestamp [15], which is convenient for later copyright protection.

The creator can also transfer or authorize the work whose copyrights have been confirmed, and can fill in transaction information for different demands, and then the information will be reviewed by a third party after the contract is constructed. If approved, the transaction information will be stored in the smart contract of the transaction application and transmitted to each node of the blockchain through the P2P network. Otherwise, you need to fill out the transaction information again. Figure 2 shows the process of copyright

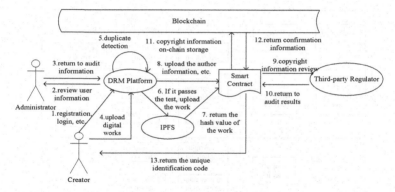

Fig. 1. The process of confirmation copyright of digital works

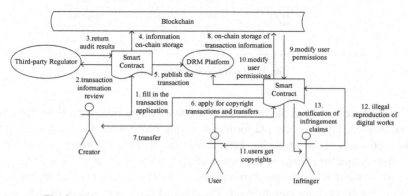

Fig. 2. The process of copyright transaction and copyright protection

transaction and copyright protection in the digital rights management platform based on blockchain.

When users want to use a certain work, they can choose different types of transactions according to their own needs, and then perform related transaction operations through smart contracts. Once the transaction is completed, the transaction information is automatically packaged and stored on the blockchain. Then it will be broadcast to other nodes to update the status data of the blockchain and modify the corresponding permissions of the user, and finally the user will obtain the copyright of the work. If users illegally reprint or use digital works, each node of blockchain will fail to verify which indicates that infringement has occurred [16]. The smart contract directly sends the infringement compensation notice to the infringing user.

3.2 System Architecture

The digital rights management platform based on blockchain designed in this paper adopts Hyperledger Fabric framework as the underlying blockchain development platform. As shown in Fig. 3, it is mainly composed of three parts, which can be divided into the blockchain bottom platform, business layer and application layer.

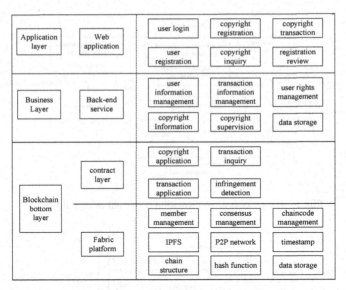

Application layer	Web application	user login	copyright registration	copyright transaction
		user registration	copyright inquiry	registration review
Business Layer	Back-end service	user information management	transaction information management	user rights management
		copyright Information	copyright supervision	data storage
Blockchain bottom layer	contract layer	copyright application	transaction inquiry	
		transaction application	infringement detection	
		member management	consensus management	chaincode management
	Fabric platform	IPFS	P2P network	timestamp
		chain structure	hash function	data storage

Fig. 3. System architecture design

The blockchain bottom platform is mainly divided into two parts: Fabric platform and the contract layer. Fabric platform is not only responsible for the processing of the basic business of the whole blockchain, but also provides technical support for the whole system platform (such as chaincode service, IPFS, consensus service, etc.). The realization of the functions of the web application and business layers also depend on the support of the blockchain bottom layer. The contract layer is to realize the function of system through the chaincode. Chaincode is responsible for providing data on-chain and query interface, which is mainly used in copyright and transaction application, copyright detection and other aspects [17].

The business layer is a bridge between Web applications and Blockchain bottom layer, and mainly provides functions such as user information management, copyright transaction management, and other functions. As a back-end service, the business layer is responsible for providing Restful interfaces to Web applications and interacting with them through API interfaces provided by Fabric, which also provides effective evidence for the copyright protection of digital works.

The application layer uses html +css +JavaScript as the front-end framework to provide creators, users, and administrators with a good web interface. Its main functions include copyright registration, copyright inquiry, copyright transaction, etc. The Web service module can interact with the data through the RESTful API provided by the business layer to realize the management of related information. When a user submits a demand application, not only the application layer and the contract layer interface to interact with the blockchain platform, but also need to use the GRPC interface to query or invoke the chaincode, realizing the functions of confirmation copyright, transaction copyright, copyright protection of digital works.

4 System Function Design

In this digital rights management platform, user information management module, digital work confirmation copyright module, digital copyright transaction module, and digital work copyright protection module are the four major modules of the platform.

4.1 User Information Management Module

New users can directly choose to register on the web interface and perform real-name authentication. After the administrator approves the user information, the information is stored in the database. Every time a user logs into an account, the system will automatically detect whether the user's account login information is consistent with the information stored in the database at the time of registration. If they are consistent, the login is successful; otherwise, the login fails.

4.2 Digital Work Confirmation Copyright Module

The creator uploads the digital work to the copyright management platform for duplicating checking, and only if the duplicate checking rate is lower than that specified by the platform, can it be certified as an original work. If it passes the test, the digital work is stored in the IPFS in the Fabric blockchain through the API interface, and then the hash value of the work is returned. If not, the work will be returned directly. The copyright information (Table 1) is temporarily stored in the chaincode.

Table 1. Copyright information sheet

Field name	Data type	Name
Author	String	Creator name
Copy type	String	Copyright type
Creative_time	String	Creation time
Submit_time	String	Upload time
Data_hash	String	Work hash value
Title	String	Title

After the approval of the third-party regulatory agency, the copyright-related information will be automatically packaged and stored in the Fabric blockchain. The hash value of the digital work is returned to the creator as the unique confirmation copyright code through the chaincode, which is the only proof of the original copyright protection.

Storing Copyright Information, a contract algorithm that stores copyright Information into the Fabric blockchain, is described below.

Algorithm 1 Storing Copyright Information Algorithm

Input: Author, copyright type, creation time, work hash value, title of digital work, upload time and other copyright information.

Output: Return the confirmation copyright code to the creator.

Step:

1. Obtain copyright information such as author, copyright type, creator time, work hash value, etc. and review it by a third-party regulatory agency.

2. Monitor the approval event.

3. If it approved, the chaincode will automatically package the copyright information on the chain and store it on the Fabric blockchain.

4. The unique confirmation copyright code of the work is sent by the chaincode to the creator.

5. If not, the work will be returned.

4.3 Digital Copyright Transaction Module

Creators can transfer or authorized transactions for works with confirmation copy-right, and fill in transaction information (including transaction type, price, etc.) ac-cording to different demands of users (Table 2). After being approved by a third par-ty, it can be transmitted to fabric blockchain through P2P network.

Table 2. Copyright transaction information sheet.

Field name	Data type	Name
Tx_type	String	Transaction type
Agreement_content	String	Agreement content
Contract_address	String	Contract address
Author_account	String	Creator account
Payment_money	Int	Payment amount

When users use the content of a digital work, the blockchain will automatically trigger the relevant code of the chaincode. The nodes on the Fabric network can trace back to the blockchain according to the user's address. If the records in the chain meet the requirements for obtaining the copyright of the work, it will be open to the user for use. If not, you can choose to purchase the copyright of the work.

The user first checks the information of the corresponding chaincode according to his own requests, and then transfers the amount specified in the chaincode to the creator's account and performs identity verification. When the transaction is completed, the blockchain system will automatically trigger the chaincode to package the transaction information and send it to each node in the Fabric network to update the blockchain data status of the whole network and modify the permissions of relevant users, and finally the users will obtain the rights and interests of the work. Copyright Transaction Algorithm for users to conduct copyright transactions is described below.

Algorithm 2 Copyright Transaction Algorithm

Input: Copyright ID, transaction type, creator account, payment amount.
Output: The user gets the copyright of the work.
Step:
1. If the user wants to use the work, the system will first judge whether the user has purchased the copyright of the work.
2. If not purchased, the relevant code of the chaincode is automatically triggered, and the user can view the corresponding chaincode according to his own needs and obtain the content of the agreement (creator account, payment amount, etc.).
3. The user transfers the specified amount to the creator's account.
4. After the transaction succeeds, the chaincode will automatically package and store the transaction records to each node on the Fabric block chain and modify the user permissions.
5. Finally, the user obtains the copyright of the work.
6. If purchased, the user can directly use the work.

4.4 Digital Works Copyright Protection Module

When the user illegally reprints or uses the content of a work, each node in the Fabric network will fail to verify. From the two characteristics of the collision resistance and one-way irreversible of the hash function, it can be seen that the digital work and the hash value of the work should have a one-to-one correspondence. If the same content with different hash values occur when verifying block nodes in the Fabric network, it means that infringement has occurred. Since the blockchain records all copyright usage, so creators can look at the copyright record and status to find evidence of infringement by backtracking the block hash value on the blockchain. Finally, the chaincode directly sends the infringement compensation notice to the infringing user.

5   Comparison with Traditional Platforms

Compared with the traditional digital rights management platforms, this platform has the following advantages:

1) Easily confirmation copyright. By introducing blockchain technology, the content and creator information of digital works can be packaged and stored in the block

through smart contract and combined with timestamps, so that the copyright can be confirmed when entering the chain. In the operation process, unlike the cumbersome copyright registration process of traditional platforms, this platform shortens the cycle of confirmation copyright and ensures that every piece of copyright information will not be tampered with and can be traced back at any time which can avoid problems such as difficulty in proof on traditional platforms.

2) The transaction process is simple and safe. In the traditional platform, users need to submit the transaction application first, and then can perform operations such as transfer transactions after approval. The whole transaction process is too cumbersome. In this paper, information query and audit, transfer transaction and other work are all automatically completed by smart contract. For example, if the transfer address or amount is inconsistent during the transfer process, the smart contract will automatically terminate the transaction. The whole transaction process do not need the intervention of the third party to avoid information leakage or data tampering.

3) Relieve the pressure of insufficient memory in blockchain. In this paper, IPFS is responsible for storing encrypted digital works and returning the hash value of the work as an index for querying digital works, which can prevent the slow response of the whole system due to the rapidly growth of storage occupancy of the blockchain.

6   Conclusion and Outlook

This paper applies the blockchain to the digital rights management and combines it with the Fabric system, IPFS, smart contract and other technologies. It not only addresses the problems of traditional copyright management, such as difficult to confirmation copyright, copyright transaction and protection, but also alleviates the problem of insufficient memory in the blockchain, and improves system performance.

Although the platform established in this paper can solve some important prob-lems of copyright protection of digital works, further work can be carried out in the following aspects: (1) Improve privacy protection. Copyright and transaction information are broadcast to each node on the blockchain, and the information is still visible to each node, which can further strengthen privacy protection. (2) Improve system scalability. This paper is divided into two levels of review by administrators and third-party regulatory agencies. It is possible to consider how to completely replace their responsibilities with smart contracts to automatically carry out digital copyright confirmation and transactions. (3) Improve system functions. The design of this platform is at an early stage, and the realization of functions can be continuously updated and improved in conjunction with new requirements in operation and maintenance.

References

1. Thomas, T., Emmanuel, S., Subramanyam, A.V., Kankanhalli, M.S.: Joint watermarking scheme for multiparty multilevel DRM architecture. IEEE Trans. Inf. Forensics Secur. **4**(4), 758–767 (2009)
2. Chen, W.W., Cao, L., Shao, C.H.: Blockchain based efficient anonymous authentication scheme for VANETs. J. Comput. Appl. **40**(10), 2992–2999 https://kns.cnki.net/kcms/detail/51.1307

3. Wang, P., et al.: Smart contract-based negotiation for adaptive QoS-aware service composition. IEEE Trans. Parallel Distr. Syst. **30**(6), 1403–1420 (2019)
4. Chen, Y., Li, H., Li, K., Zhang, J.: An improved P2P file system scheme based on IPFS and Blockchain. In: 2017 IEEE International Conference on Big Data (Big Data), Boston, MA, pp. 2652–2657 (2017)
5. Mazumdar, S., Ruj, S.: Design of Anonymous Endorsement System in Hyperledger Fabric. IEEE Transactions on Emerging Topics in Computing (2019)
6. Yamashita, K., Nomura, Y., Zhou, E., Pi, B., Jun, S.: Potential risks of hyperledger fabric smart contracts. In: IEEE International Workshop on Blockchain Oriented Software Engineering (IWBOSE), Hangzhou, China, pp. 1–10 (2019)
7. Du, M., Chen, Q., Xiao, J., Yang, H., Ma, X.: Supply chain finance innovation using Blockchain. IEEE Trans. Eng. Manage. **67**(4), 1045–1058 (2020)
8. Perboli, G., Musso, S., Rosano, M.: Blockchain in logistics and supply chain: A lean approach for designing real-world use cases. IEEE Access, **6,** 62018–62028 (2018)
9. Yao, H., Mai, T., Wang, J., Ji, Z., Jiang, C., Qian, Y.: Resource trading in blockchain-based industrial internet of things. IEEE Trans. Ind. Inform. **15**(6), 3602–3609 (2019)
10. Meng, Z., Morizumi, T., Miyata, S., Kinoshita, H.: Design scheme of copyright management system based on digital watermarking and blockchain. In: IEEE 42nd Annual Computer Software and Applications Conference, pp. 359–364 (2018)
11. Xu, R., Zhang, L., Zhao, H., Peng, Y.: Design of network media's digital rights management scheme based on blockchain technology. In: IEEE 13th International Symposium on Autonomous Decentralized System., pp. 128–133 (2017)
12. Ding, Y., Yang, L., Shi, W., Duan, X.: The digital copyright management system based on Blockchain. In: 2019 IEEE 2nd International Conference on Computer and Communication Engineering Technology, pp. 63–68 (2019)
13. Poudel, K., Aryal, A.B., Pokhrel, A., Upadhyaya, P.: Photograph ownership and authorization using blockchain. In: 2019 Artificial Intelligence for Transforming Business and Society, pp. 1–5 (2019)
14. Tan, W.A., Wang, H.: Scheme and platform of trusted fund-raising and donation based on smart contract. J. Comput. Appl. **40**(05), 1483–1487 (2019)
15. Liang, W., Zhang, D., Lei, X., Tang, M., Li, K., Zomaya, A.: Circuit copyright blockchain: Blockchain-based homomorphic encryption for IP circuit protection. IEEE Transactions on Emerging Topics in Computing (2020)
16. Liu, Z.Y., Liu, X.Z.: Application of blockchain in the field of digital rights. Cyberspace Secur. **10**(12), 36–45 (2019)
17. Han, P., Sui, A., Jiang, T., Gu, C.: Copyright Certificate Storage and Trading System Based on Blockchain. In IEEE International Conference on Advances in Electrical Engineering and Computer Applications (AEECA), Dalian, China, pp. 611–615 (2020)

A Deep Hybrid Neural Network Forecasting for Multivariate Non-stationary Time Series

Xiaojian Yang and Xiyu Liu[✉]

Business School of Shandong Normal University, Jinan, China
xyliu@sdnu.edu.cn

Abstract. In the field of financial time series prediction, multivariate time series is increasingly considered as the input of the prediction model and non-stationary time series have always been the most common data sets. However, the processing efficiency is low but the cost is high whenthe traditional methods is used for modeling non-stationary time series. This paper is aimed at providing the methodological guidance for building low-cost models for modeling multivariate non-stationary time series. By building a univariate CNN, a multivariate CNN, a non-pooling CNN (NPCNN), a CNN-LSTM and a NPCNN-LSTM, we conducted a series of comparative experiments. We found that multivariate non-stationary time series is not complex enough, the pooling operation will lose the useful information and the LSTM layer can weaken this negative effect. Meanwhile, convolutional layers and LSTM layers can improve the prediction accuracy. Adding the LSTM to prediction models can make models have better performance in short-term prediction.

Keywords: CNN · LSTM · Multivariate non-stationary time series · Pooling layers

1 Introduction

In the stage of rapid development of the world economy, financial time series forecasting is a problem that has been widely discussed [1]. However financial time series always present non-stationary. How to find ways to build a low-cost model able to directly predict the future trends of financial time series will be the research hotspot of academia and financial circles [2].

The convolutional neural network (CNN) is a hierarchical network, which can be said to be a generalization of the traditional neural network. In the field of financial time series forecasting, the CNN is attracting more and more attention, because of it's robustness and adaptivity in solving complex problems in the real world [3, 4].

In the field of time series forecasting, many models have been proposed [5]. But most models are oriented towards modeling univariate time series. When faced with data like financial time series, it's not enough to only considering the historical values of the predictive variable. Moreover, when processing non-stationary time series, traditional

© Springer Nature Switzerland AG 2021
Q. Zu et al. (Eds.): HCC 2020, LNCS 12634, pp. 184–190, 2021.
https://doi.org/10.1007/978-3-030-70626-5_19

time series models usually require preprocessing of data. Extensive manual operations lead to the higher operation cost of the model.

In this way, the main distribution of this paper is: first, in the field of time series, for non-stationary time series forecasting, the CNN model is introduced, and it's hoped that the prediction accuracy can be improved by taking multivariate time series data as the input. Second, we report the effects of the pooling layer in multivariate time series processing. Finally, we present a hybrid neural network. This approach offers a competitive model for fitting multivariate non-stationary time series.

2   State of the Art

The earliest theoretical method for time series forecasting is the mathematical modeling method based on statistic [6, 7]. With the development of the theory of Machine learning, many related methods have been applied in the time series analysis [8]. Grudnitski et al. applied the artificial neural network to stock market prediction and found it was suitable for processing complex, noisy data in financial market [9].

LeCun et al. published a paper in the 1990s, and established the modern structure of CNN. They designed a multilayer artificial neural network called Lenet-5 that could categorize handwritten Numbers [10]. Cao et al. apply CNN in the financial field to analysis financial time series, and studies have proved that the better prediction can be obtained from the CNN [11].

Moreover, aiming at the existing problems of traditional RNN, Hochreiter put forward Long Short-Term Memory (LSTM) model [12]. Visin Francesco et al. used 4-directional RNN (LSTM unit) to replace CNN for image processing.

CNN and LSTM both have good performance in multivariate data processing and broad application in financial time series forecasting. But CNN is usually used to process univariate time series. In this paper, the influences of the convolution and the pooling layer were discussed in multivariate time series analysis, and LSTM's ability to learn multivariate data features is applied to multivariate time series forecasting.

3   Time Series

3.1   Financial Time Series

Financial time series do not have stable statistical characteristics and shows highly non-linear characteristics [13]. Periodic wave motion usually exists in financial time series. Meanwhile, generally there is an upward trend in the financial time series. So financial time series usually presents the non-stationary characteristics. Although financial time series presents very complex characteristics, it is not irregular [14].

3.2   Multivariate Time Series

Depending on the number of observation variables in the system, financial time series can be divided into univariate time series and multivariate time series. For a multivariate time series $X = \{D_1^m, D_2^m, ..., D_n^m\}$, $D_i^m = \{d_{i1}, d_{i2}, ..., d_{im}\}$, $i \leq n$, d_{ij} represents the attribute value of the jth dimension variable at time i, the length is n, and there are m dimensions.

4 Methodology

4.1 The Convolutional Layer and the Pooling Layer for Modeling the Non-stationary Multivariate Time Series

First, we built a CNN financial time series forecasting model. The model structure designed in this paper is based on Lenet-5. Because Lenet-5 determines the most basic architecture of CNN: convolutional layers, pooling layers and full connection layers, it is one of the most representative CNN models.

We use the Lenet-5 network structure directly, because identifying the best structure of the CNN is not our goal in this section and Lenet-5 has the ability to process multi-dimensional data. The model structure we designed consists of 2 convolutional layers, 2 pooling layers, 2 full connection layers and a linear output layer. Then, we just create a big enough size of the lag variables so that they represent the full pattern [5]. So, the model input data structure is 100 * 4. After each convolutional layer is a pooling layer of 2 * 2 structure with maximum pooling operation, the amplitude of downsampling size is 2. The size of the convolution kernel is uniformly set as 3 * 3. In the first convolutional layer, there are 16 convolution kernels. The number of convolution kernels in each subsequent layer is twice that of the previous layer. With a batch_size of 256, the network parameters were updated every 256 samples, and the number of training (epoch) was 500. We end up with a linear output from the output layer of the model. For convenience, we refer multivariate CNN as CNN in the rest of the paper. In addition, we built a univariate CNN model by changing the number of neurons in the input layer to 1.

Then, on this basis we built a NPCNN. The pooling layers is removed from the CNN. Because of this, the fully connected part will have a different number of input neurons. And the other parameters are the same as those of the CNN.

4.2 CNN + LSTM Model for Modeling the Non-stationary Multivariate Time Series

Based on the above discussion, a hybrid neural network prediction model based on LSTM method was proposed: CNN - LSTM model. The model consists of two parts. The first part is the CNN method used to extract features from the original input vector, the second part is the LSTM model used to process corresponding attributes. The hidden layer representation of the last layer of the CNN will be input into the LSTM model as a summary and representation of the financial market environment.

The parameters of the first part are the same as parameters in the CNN above. In the build process of the LSTM neural network, there are many parameters that need to be set. Among them, there is no rule of thumb for hiding levels and time steps. In this paper, after hyparameter optimization, the number of hidden layers is set to 4 and the time step is set to 100.

Then, similarly, we removed the pooling layers from the CNN - LSTM. Except that the number of input neurons in the fully connected part has changed, the other parameters are the same as those of the CNN - LSTM.

5 Experimental Results and Analysis

5.1 Experimental Data and Experimental Environment

The experimental data used to trained and tested models in this paper are from the Dow Jones Industrial Average (DJIA). 4,520 samples of daily frequency data of the DJIA from January 3, 2000 to December 29, 2017 are selected. The Daily trading data for the DJIA includes the daily Open, High, Low and Close. So the trend of time series is influenced by many inducing factors, this four variables are considered in this article, and we set the closing price as the predicted variable.

To avoid the influence of dimension, the original time series data was linearly transformed before feature extraction and mapped to the interval [0,1], and we divided the data in two set: training set (80% of the data point), and test set (20% of the data point). The normalization formula is

$$x_{norm} = \frac{x - x_{min}}{x_{max} - x_{min}} \tag{1}$$

In this paper, the tensorFlow open source platform is used as the deep learning platform, and Python 3.6 is used to write the experimental program. Meanwhile, some third-party libraries are used, such as Talib to calculate technical indicators and Keras to build the network structure. The experimental operating system is Windows 8.

5.2 Experimental Result

In order to determine if the prediction accuracy can be improved by taking multivariate time series data as the input, we made the following comparative experiment. Figure 1 shows the predicted results of the multivariate CNN model constructed in this paper. Figure 2 and 3 show the predicted results of univariate CNN model and the BP neural network model, considering only the historical closing price as input data. The CNN model in Fig. 2 just changes the number of input neurons in the input layer, compared with the multivariate CNN model constructed in this paper. And the results of this experience are shown in Table 1.

As can be seen from the table and figures above, considering multivariate time series as the input can improve the precision of forecasting. When modeling the non-stationary time series, the multivariate CNN model can get better prediction performance than univariate CNN models and the traditional neural network model.

Then, we compared the prediction performance of CNN and NPCNN and the only difference between this two models is whether the pooling layer is included. Figure 4 show the prediction results of the NPCNN models and Table 1 shows the result of performance indicators. NPCNN has better performance. In other words, when modeling the multivariate non-stationary time series, the pooling layers will lose the too much useful information.

Similarly, Fig. 5 and 6 show the prediction results of CNN - LSTM and NPCNN - LSTM. Table 1 summarizes the performance of this two models. Compared with CNN and NPCNN, the hybrid models both are batter methods for modeling multivariate non-stationary time series. And we found it was difficult to say which one would have been better. They were very equal, and the NPCNN - LSTM model exhibited slightly advantages over the CNN - LSTM. CNN - LSTM is also good choice with only slightly worse MSE and RMSE values. So, although multivariate time-series data are more complex than unitivariate data, it's not enough. The pooling layer still may lose partial useful information. But LSTM neural network has stronger ability to process features than full connection layers. This ability may weaken the negative effect of the pooling layer.

Fig. 1. The prediction result of the multivariate CNN

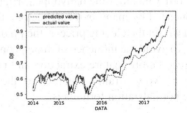

Fig. 2. The prediction result of the univariate CNN

Fig. 3. The prediction result of the BP neural network

Fig. 4. The prediction result of the multivariate NPCNN

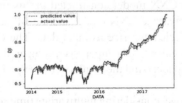

Fig. 5. The prediction result of the CNN + LSTM

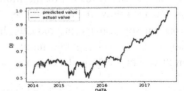

Fig. 6. The prediction result of the NPCNN + LSTM

As shown in the Fig. 7, CNN and NPCNN have did poorly in the short-term prediction. And the hybrid models have a far better effectiveness overall. We think LSTM is helpful to short-term prediction of multivariate non-stationary time series.

Table 1. The prediction performance of models.

models	MSE
BP neural networks	0.0519
Univariate CNN	0.0012
Multivariate CNN	5.7928e−04
NPCNN	3.4639e−04
CNN-LSTM	1.9924e−04
NPCNN-LSTM	1.3463e−04

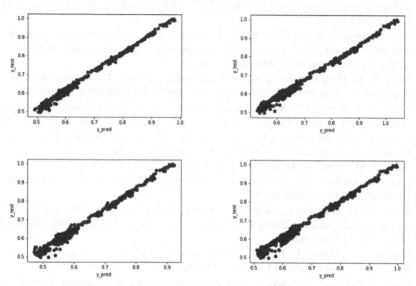

Fig. 7. The closing prices predicted by the CNN, NPCNN, CNN - LSTM, NPCNN - LSTM compared to the actual closing price of DJI

6 Conclusion

We expect to build models that can directly process multivariate non-stationary time series and get some useful conclusions that are helpful to the future research.

The experimental results indicate that CNN can directly process multivariate non-stationary time series without any complicated data preprocessing, and can get better prediction performance than univariate models. And the pooling operation still have negative effects in modeling multivariate non-stationary time series. So, multivariate time series is not complex enough to require the use of pooling layers in the prediction models. Moreover, we found that the introduction of LSTM can enhance the performance of predicted model. Meanwhile, it is worth mentioning that the negative effect of pooling layers is weaker than we think for CNN - LSTM. LSTM can weaken this effect. So, for modeling multivariate non-stationary time series, the introduction of convolution layers and LSTM should be encouraged, but pooling layers shouldn't be used.

We provided a theoretical and experimental basis for building predicted models for time-series data, but did not explore the optimal parameters of the model. On the other hand, on other data sets, the validity of the conclusions remain to be tested.

References

1. Sezer, O.B., Gudelek, M.U., Ozbayoglu, A.M.: Financial time series forecasting with deep learning: a systematic literature review: 2005–2019. Appl. Soft Comput. **90**, 106181 (2020). https://doi.org/10.1016/j.asoc.2020.106181
2. Yan, W.: Toward automatic time-series forecasting using neural networks. IEEE Transactions on Neural Networks and Learning Systems **23**(7), 1028–1039 (2012). https://doi.org/10.1109/TNNLS.2012.2198074
3. Zhao, B., Huanzhang, L., Chen, S., Liu, J., Dongya, W.: Convolutional neural networks for time series classification. Journal of Systems Engineering and Electronics **28**(1), 162–169 (2017). https://doi.org/10.21629/JSEE.2017.01.18
4. Zhou, D.-X.: Deep distributed convolutional neural networks: Universality. Analysis and Applications **16**(06), 895–919 (2018). https://doi.org/10.1142/S0219530518500124
5. Liu, S., Ji, H., Wang, M.C.: Nonpooling convolutional neural network forecasting for seasonal time series with trends. IEEE Trans. Neural Networks Learn. Syst. **31**(8), 2879–2888 (2020)
6. De Gooijer, J.G., Hyndman, R.J.: 25 years of time series forecasting. Int. J. Forecast. **22**(3), 443–473 (2006). https://doi.org/10.1016/j.ijforecast.2006.01.001
7. Tseng, F.-M., Tzeng, G.-H., Hsiao-Cheng, Y., Yuan, B.J.C.: Fuzzy ARIMA model for forecasting the foreign exchange market. Fuzzy Sets Syst. **118**(1), 9–19 (2001). https://doi.org/10.1016/S0165-0114(98)00286-3
8. Längkvist, M., Karlsson, L., Loutfi, A.: A review of unsupervised feature learning and deep learning for time-series modeling. Pattern Recogn. Lett. **42**, 11–24 (2014). https://doi.org/10.1016/j.patrec.2014.01.008
9. Grudnitski, G., Osburn, L.: Forecasting S&P and gold futures prices: an application of neural networks. J. Futures Market **13**(6), 631–643 (2010)
10. Lecun, Y., Bottou, L.: Gradient-based learning applied to document recognition. Proc. IEEE **86**(11), 2278–2324 (1998)
11. Cao, J., Wang, J.: Stock price forecasting model based on modified convolution neural network and financial time series analysis. Int. J. Commun. Syst. **32**(12), e3987 (2019). https://doi.org/10.1002/dac.3987
12. Hochreiter, S., Schmidhuber, J.: Long short-term memory. Neural Comput. **9**(8), 1735–1780 (1997). https://doi.org/10.1162/neco.1997.9.8.1735
13. Brock, W.A.: Nonlinearity and complex dynamics in economics and finance. Econ. Dev. Cult. Change **53**(3), 760–762 (1988)
14. de Ricardo, A., Araújo, N.N., Oliveira, A.L.I., de Silvio, R., Meira, L.: A deep increasing–decreasing-linear neural network for financial time series prediction. Neurocomputing **347**, 59–81 (2019). https://doi.org/10.1016/j.neucom.2019.03.017s

Risk Assessment of Flood Disasters in Hechi City Based on GIS

Min Yin[✉], Chengcheng Wei, Juanli Jing, and Xiaoqian Huang

College of Geamatics and Geoinformation,
Guilin University of Technology, Guilin 541004, China
416721782@qq.com

Abstract. Flood disasters are one of the meteorological disasters that occur frequently and cause heavy losses. Hechi City is located in a low-latitude zone with a typical subtropical monsoon climate. Water resources are unevenly distributed in time and space, and floods often occur. The flood risk assessment of Hechi City has strong practical significance. This paper collects three phases of remote sensing image data and many years of meteorological and attribute data, builds a flood disaster risk assessment model based on the flood formation mechanism, uses Analytic Hierarchy Process (AHP) to determine the weight of each index factor, and calculates each evaluation index data with GIS technology Analyze, through the GIS spatial analysis function, superimpose the analysis results of flood hazard, disaster-generating environment sensitivity, and disaster-bearing body vulnerability in Hechi City to obtain the comprehensive risk distribution of flood disaster in Hechi City, showing Hechi City The flood disaster risk of the northeast region and the southwest region is gradually reduced to the central region. It is necessary to strengthen the actual disaster prevention and mitigation work in high-risk areas in the northeast region.

Keywords: Risk assessment · Flood disaster · GIS spatial analysis

1 Introduction

Flood disaster is one of the most frequent natural disasters in China, which can cause the largest direct economic loss among all the natural disasters in the world. In recent years, the frequent flood disasters in Guangxi have a great impact on the social and economic production. Hechi is one of the high incidence areas of flood disaster in Guangxi. The flood season from April to September every year is the frequent flood disaster period in Hechi. The floods caused by the precipitation in the flood season during this period had a great impact on the socio-economic development of Hechi. The combination of remote sensing image and geographic information system technology can effectively evaluate and predict flood disaster, and provide effective technical support for flood disaster.

In recent years, many scholars at home and abroad have used remote sensing images and Geographic Information System (GIS) techniques to analyze flood-related factors such as precipitation, land form, soil properties, meteorological conditions and river

© Springer Nature Switzerland AG 2021
Q. Zu et al. (Eds.): HCC 2020, LNCS 12634, pp. 191–201, 2021.
https://doi.org/10.1007/978-3-030-70626-5_20

drainage capacity. The flood hazard is evaluated by statistical method and GIS spatial analysis [1–3]. Among the methods of Flood Disaster Risk Assessment, there are principal component analysis method, hydraulic numerical simulation method, analytic hierarchy process method and so on. These methods can be used to build assessment models, generate flood risk assessment maps, or simulate flood processes to display flood risk grades [4–9].

The study of Flood Disaster Risk Assessment by different methods is mostly aimed at the flood disaster risk assessment of a specific area at a certain time, few of the long-term series, multi-period of the comprehensive index and dynamic simulation of risk indicators comparative analysis. Based on the data from 2011 to 2019 in Hechi, this paper uses the analytic hierarchy process (AHP) to determine the assessment factors and their weights, and constructs a flood risk assessment model. Then, the risk of flood disaster in Hechi is evaluated scientifically.

2 Study Area Description

Hechi belongs to the Guangxi Zhuang Autonomous Region. It is located in the northwestern border of Guangxi Zhuang Autonomous Region and the southern foot of the Yunnan-Guizhou Plateau. It is an important transportation channel from the southwestern region to coastal ports. There are 11 counties in Hechi. The geographical location is

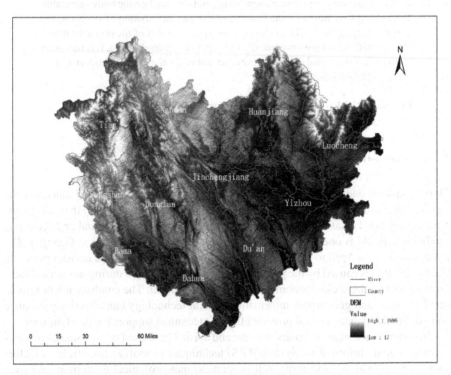

Fig. 1. The summary of Hechi City

between 106°34′~109°09′ east longitude and 23°41′~25°37′ north latitude. 228 km east-west length, 260 km north-south width, the city's land area is 33,500 square km, mainly in mountainous areas, diverse terrain, large karst area, widely distributed, it covers 66% of the area and is a typical karst. Hechi is a typical sub-tropical monsoon climate at low latitudes. Plenty of sunshine and rainfall all the year round. There are more than 630 rivers, of which more than 50 major rivers, rivers a total length of 5,130 km. The two major river basins for Hongshui River and Longjiang River. And Hongshui River is the main Trunk River of XiJiang River. The summary of the study area is shown in Fig. 1.

3 Data and Approach

3.1 Data Source and Preprocessing

The remote sensing image data of this research area are derived from LANDSAT7 data of 2013, 2015 and 2018 from the geospatial data cloud (https://www.gscloud.cn/). Using mosaics, clipping to obtain the remote sensing images of Hechi. Using infrared and near-infrared band to calculate the NDVI index of the study area. The situation of land use types in 2011–2013, 2014–2016 and 2017–2019 were obtained by monitoring classification.

The DEM data selects the GDEMDEM 30m resolution digital elevation data. On the basis of DEM data, terrain slope, elevation standard deviation and river network density are derived as index factors in the evaluation model.

The vector data of administrative division boundaries and water network basins comes from the BIGEMAP map downloader. The data format is shape format, and the default coordinates are CGS_WGS1984.

The social and economic data mainly involve the administrative land area, total population, GDP and farmland area of 11 counties in the study area, all of which are from Guangxi Statistics Bureau and some statistical yearbooks.

Data of annual precipitation, monthly precipitation and maximum daily precipitation of 11 stations in the study area from 2011 to 2019 were collected from China Meteorological Data Bureau (https://data.cma.cn/). According to the time characteristics of flood disaster, the annual average data from May to August in Hechi is counted as the index factor in the evaluation model.

3.2 Research Methods

Index Selection. The topography, climate and hydrology of Hechi have an important impact on the risk assessment of flood disaster. Therefore, the risk assessment model of flood disaster is constructed by 9 indexes from three aspects: Hazard factors, environmental sensitivity of disaster-inducing environment and vulnerability of disaster-bearing body, The specific index system of risk assessment model is shown in Table 1.

Table 1. Flood disaster risk assessment index system

Target level	Standard level	Program level
Flood disaster risk assessment model	Hazard factor	Multi-year average monthly precipitation in flood season
		Annual average precipitation
		Multi-year average maximum daily precipitation
	Disaster-pregnant environment	Elevation standard deviation
		River network density
		Normalized vegetation index
	Disaster-bearing body	Population density
		gross product
		Farmland area

Weight Determination. The Analytic Hierarchy Process (AHP) was proposed by Professor T.L. Saaty of the United States in the 1970s. This method is based on the correlation of various factors to carry out hierarchical division and finally construct a judgment matrix to determine the weight of each evaluation index factor. In this paper, AHP is used to determine the weight of each index factor of flood disaster in Hechi City, such as the risk of disaster causing factors, the sensitivity of disaster pregnant environment and the vulnerability of disaster bearing body. And they are all passed the consistency test. The weight of each indicator is shown in Table 2.

Table 2. Index factor weight value

Criterion layer	Weight	Index layer	Weight	Combination weight
Hazard risk	0.58	Monthly precipitation in flood season	0.32	0.1856
		Annual precipitation	0.12	0.0696
		Maximum daily precipitation	0.56	0.3248
Disaster-pregnant environment risk	0.31	Elevation standard deviation	0.60	0.1860
		River network density	0.28	0.0868
		Normalized vegetation index	0.12	0.0372
Vulnerability of disaster bearing body	0.11	Population density	0.54	0.0594
		gross product	0.16	0.0176
		Farmland area	0.30	0.0330

Risk Assessment Model. Construct a flood disaster risk assessment model according to the index weight:

$$\text{Comprehensive risk} = F\,(\text{risk, sensitivity, vulnerability}) \qquad (1)$$

According to the results of the model, the study area is divided into 5 levels according to the natural discontinuity point classification method: low risk, low risk, and medium risk, Higher risk, high risk.

4 Results and Analysis

4.1 Risk Analysis of Hazards

For flood disaster, the annual average precipitation, annual average precipitation and annual average maximum daily precipitation are selected as index factors. There are several possible causes of flood disaster: The impact of rainstorm characteristics, rainfall distribution.

Impact of Rainstorm Characteristics. As shown in Fig. 2, the annual precipitation in Hechi is basically above 1100 mm, and the annual precipitation from 2011 to 2019 shows an upward trend with a large increase, the change rate of precipitation was 53.666 MM/A, which reached two peaks in 2015 and 2017, the values were 1851.760 mm and 1792.965 mm respectively.

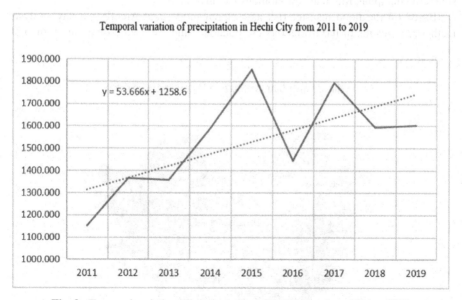

Fig. 2. Temporal variation of precipatation in HeChi City from 2011 to 2019

Rainfall Distribution. Based on the precipitation data from 2011 to 2019, this paper uses ArcGIS interpolation method to interpolate the annual average precipitation, monthly average precipitation and maximum daily precipitation in flood season, the results are divided into 5 grades: little, less, medium, more, and many according to the natural break point method as shown in Fig. 3.

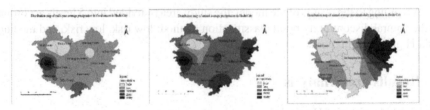

Fig. 3. Distribution map of precipitation in Hechi

From the spatial distribution point of view, the precipitation distribution of the stations in the study area shows the following trend: The average monthly precipitation in flood season in Hechi is much more than that in other time in a years. The precipitation in western Donglan County, northern Bama County and southern Dahua Yao Autonomous County and southern Duan County is more widely distributed, while the precipitation in Jinchengjiang District, Yizhou district, Luocheng Mulao Autonomous County, northern Duan County, northern Dahua Yao Autonomous County and northern Huanjiang Maonan Autonomous County is medium. Rainfall in Nandan County, Tian'e County and southern Huanjiang Maonan Autonomous County is low.

Based on the flood disaster risk assessment model and ArcGIS overlay analysis method, the results of hazard factors are obtained, and the hazard distribution map of

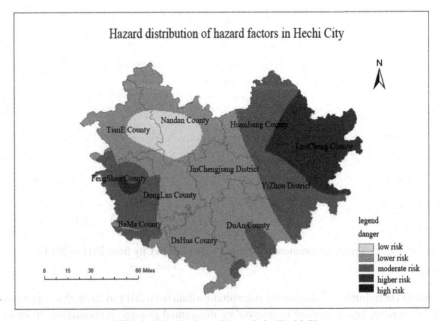

Fig. 4. Distribution of hazards in Hechi City

hazard factors is obtained by natural break point classification method, as shown in Fig. 4. Driven by disaster-causing index factors, the risk of flood disaster is higher in the east and southwest, and the risk of flood disaster is higher in Luocheng Mulao Autonomous County, Huanjiang Maonan Autonomous County, Yizhou, Fengshan County and Bama counties. The risk in Jinchengjiang District, Dahua Yao Autonomous County and Du'an counties is moderate, while the risk in Tian'e County and Nandan County counties is low.

4.2   Environmental Sensitivity Analysis

The topography, river network density and vegetation coverage of the study area are the main factors affecting the sensitivity of the disaster-pregnant environment. Therefore, the altitude standard deviation, river network density and the normalized vegetation index (NDVI) are selected as the disaster-pregnant environment sensitivity for the index factor.

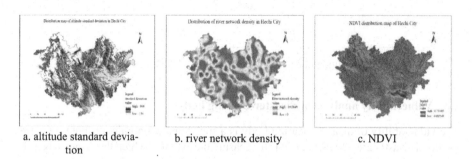

a. altitude standard devia-   b. river network density   c. NDVI
tion

Fig. 5. Environmental sensitivity analysis

Based on the analysis of the results of Hechi Elevation Standard Deviation (Fig. 5(a)) and NDVI (Fig. 5(C)), it can be seen that Hechi is distributed from east to west in a flat plain to a mountainous area, the terrain is high in the west and low in the east, and the west is inclined from the outside to the river valley. Like the analysis of river network density (Fig. 5(b)) shows that the distribution of river networks in the north-east and north-west is relatively dense, but the topography in the north-west is low in the middle and high all around. When the water level rises, it will not spread out rapidly, whereas the topography in the north-east is flat, as the water level rises, the flood waters spread outwards.

When analyzing the sensitivity of the disaster-pregnant environment in Hechi, the three index factors of altitude standard deviation, river network density and normalized vegetation index (NDVI) were weighted and superimposed. The weight values were based on Table 2, and the analysis results are shown in Fig. 6.

According to the distribution map of environmental sensitivity to disasters, the environmental sensitivity of disasters in Hechi is decreasing from the southeast to the North-Western Territory. The risk of floods in the southeast is higher, and more precautions should be taken in normal times.

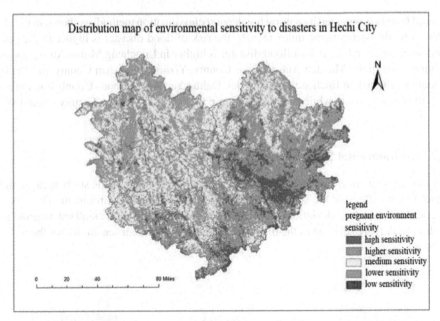

Fig. 6. Distribution of environmental sensitivity to disasters in Hechi City

4.3 Vulnerability analysis of disaster-bearing body

The vulnerability of disaster-bearing body is a social and economic attribute reflecting the flood disaster area. In essence, it is to measure the extent to which the disaster-bearing body is affected by the disaster-causing factors and the disaster-bearing environment. The analysis is based on 2011–2019 Hechi Statistical Yearbook data from both social and economic attributes, respectively, the population density, farmland area and gross domestic product (GDP) were selected as the evaluation index factors to assess the vulnerability, as shown in Fig. 7.

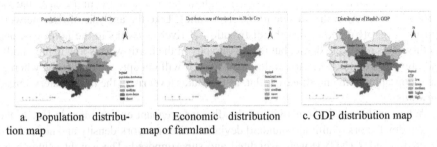

a. Population distribu-   b. Economic distribution   c. GDP distribution map
tion map   map of farmland

Fig. 7. Vulnerability analysis of disaster-bearing body

According to the Flood Risk Assessment Model, the vulnerability distribution map of the disaster-bearing bodies in Hechi was obtained, as shown in Fig. 8. The vulnerability of the disaster-bearing bodies in Hechi is increasing from the north-western territory to the southeast. The areas with high vulnerability and high vulnerability include Dahua Yao Autonomous County and Yizhou districts. The moderately vulnerable areas include Bama, Duane, Jinchengjiang District and Luocheng Mulao Autonomous County; the Donglan County are less vulnerable; and the low vulnerable areas include Fengshan County, Tian'e County, Nandan County and Huanjiang County.

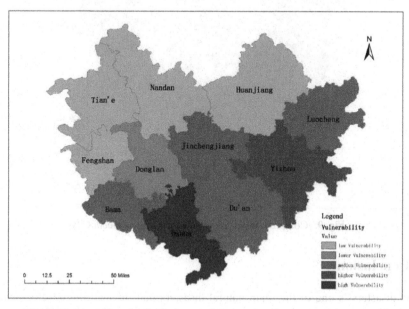

Fig. 8. Vulnerability distribution map of disaster-bearing bodies in Hechi City

4.4 Risk Assessment of Flood Disaster

According to the Comprehensive Risk Assessment model of flood disaster, the risk result of the disaster-causing factors, the result of the environmental sensitivity of the pregnant disaster and the result of the vulnerability of the disaster-bearing body of the flood disaster in Hechi were calculated by using the GIS spatial overlay technology, according to the natural break point classification method, the comprehensive risk grade distribution map of flood disaster in Hechi was obtained, as shown in Fig. 9.

It can be seen from Fig. 9 that Luocheng Mulao Autonomous County is a high-risk area, with Dahua Yao Autonomous County, Bama and Fengshan County being the higher-risk area, mainly because they are located along rivers and the main rivers of Luocheng Mulao Autonomous County are Longjiang Tributaries. The main tributaries of the higher-risk areas are the Hongshui River river and the Panyang River, where the river network is dense, flat and sensitive, the population distribution is concentrated and the social wealth is relatively concentrated, and the vulnerability is high, there's going to be a huge loss.

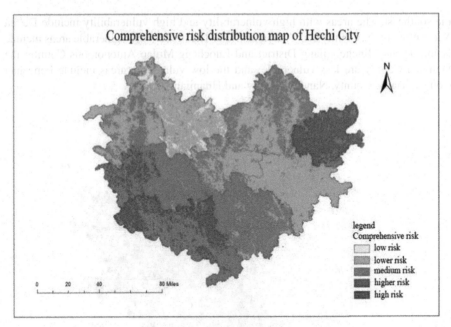

Fig. 9. Comprehensive risk distribution map of Hechi City

5 Conclusion

Flood disasters is one of the main natural disasters that threaten the development of human society and economy. Hechi is located in the Northwest Frontier of Guangxi and is constantly affected by flood disasters. An effective assessment of flood risk in Hechi can provide a good technical support for flood prevention and mitigation in Hechi. In this paper, the risk assessment model of flood disaster is established by using analytic hierarchy process (AHP). Remote sensing images and thematic data from Hechi were collected, and the study area was evaluated using remote sensing and GIS spatial analysis techniques. The results show that Luocheng County of Hechi City is a high-risk area, located along the river, with flat terrain and concentrated population, which has high risk, and needs to strengthen the response measures. For the prediction of flood disaster, this paper has not yet involved, and will carry out systematic research in the next step.

Acknowledgments. This work was funded by Guangxi Natural Science Foundation Program (2018GXNSFAA281279).

References

1. Haruyama, S., Ohokura, H., Simking, T., et al.: Geomorphological zoning for flood inundation using satellite data. GeoJournal **38**(3), 273–278 (1996)
2. Kim, Y.O., Seo, S.B., Jang, O.J.: Flood risk assessment using regional regression analysis. Nat. Hazard **63**(2), 1203-1217 (2012)

3. Anding, P., Huiping, L., Bishan, C., et al.: Preliminary study on the risk assessment of flood disaster in Guangzhou. J. Nat. Disasters (4), 23–28 (2010)
4. Shukun, L., et al.: Numerical simulation of flood evolution in Xiaoqinghe flood diversion area. Adv. Water Sci. (3), 188–193 (1991)
5. Elkhrachy, I.: Flash Flood Hazard Mapping Using Satellite Images and GIS Tools: a case study of Najran City, Kingdom of Saudi Arabia(KSA). Egyptian J. Remote Sens. Space Sci. 18(2), 261–278 (2015)
6. Yipeng, W., Jiesong, L., Liyuan, L.: Application of analytic hierarchy process in the comprehensive evaluation of agricultural meteorological disasters in Fujian province. J. First Mil. Med. Univ. 25(4) (2009)
7. Xing, Z., Hui, C., Juxin, W.: Application of analytic hierarchy process in determining weight coefficients of evaluation indexes. J. Trop. Meteorol. 25(4) (2009)
8. Yiming, W., Liangju, J., Chenghu, Z., Qing, W., Jiren, L.: Research on flood disaster assessment system. Catastrophe Sci. 12(3), 1–5 (1997)
9. Xingjun, L., Zhaopei, Z., Yuhui, Z., et al.: Water resources security assessment in Shanxi Province based on analytic hierarchy process. J. Ludong Univ. (Nat. Sci. Ed.) 30(2), 162–166 (2014)

Cultural Symbol Recognition Algorithm Based on CTPN + CRNN

Yanru Wang[✉]

School of Economics, Jilin University, ChangChun 130012, China
wangyanru@jlu.edu.cn

Abstract. This paper proposes a cultural symbol recognition algorithm based on CTPN + CRNN. The algorithm uses the improved VGG16 + BLSTM network to extract the depth features and sequence features of the text image, and uses the Anchor to locate the text position. Finally, the task of cultural symbol recognition is carried out through the CNN + BLSTM + CTC deep network, The algorithm is an ideal cultural symbol recognition scheme in terms of the recognition efficiency and accuracy.

Keywords: Machine learning · Deep neural network · Cultural symbol recognition

1 Introduction

In recent years, text symbols symbolizing national culture can be seen everywhere. With the rapid development of artificial intelligence, people have higher and higher requirements for the convenience and intelligence of life and work. The research of intelligent cultural symbol recognition represented by Chinese and English characters is of great significance for people to exchange and spread information more quickly. This paper proposes an improved CTPN + CRNN recognition algorithm by comparing the traditional text symbol detection and recognition algorithms. The algorithm uses the improved VGG16 + BLSTM network to extract the depth features and sequence features of the text image, and uses the Anchor to locate the text position, Finally, the task of cultural symbol recognition is carried out through the CNN + BLSTM + CTC deep network, The algorithm is in the recognition efficiency and In terms of accuracy, they are all ideal cultural symbol recognition schemes.

2 CTPN + CRNN Recognition Algorithm Principle

2.1 CTPN Text Detection Principle

The CTPN model combines the characteristics of RNN and CNN to improve the accuracy and reliability of detection. Among them, CNN is used to extract the depth features of pixels, and RNN is used for feature recognition of sequences [1]. The whole process is mainly divided into six steps:

© Springer Nature Switzerland AG 2021
Q. Zu et al. (Eds.): HCC 2020, LNCS 12634, pp. 202–208, 2021.
https://doi.org/10.1007/978-3-030-70626-5_21

1) Input $3 \times 600(h) \times 900(w)$ image, use VGG16 for feature extraction, get the feature of conv5_3 (the third convolutional layer of the fifth block of VGG) as the feature map, The size is $512 \times 38 \times 57$;

2) Make a sliding window on this feature map, the window size is 3×3, $512 \times 38 \times 57$ becomes $4608 \times 38 \times 57$ (512 is expanded by 3×3 convolution);

3) Input the features corresponding to all windows in each row into the two-way recurrent network, each memory layer has 128 hidden layers, $57 \times 38 \times 4608$ becomes $57 \times 38 \times 128$, Reverse-LSTM also gets $57 \times 38 \times 128$, and the final result after merging is $256 \times 38 \times 57$;

4) Input the output of LSTM to the fully connected layer (FC layer), which is a 256×512 matrix parameter, and get the result of $512 \times 38 \times 57$;

5) The features of the fully connected layer are input to the three softmax layers. The first 2k vertical coordinate and the third k side-refinement are used to regress the position parameters of k anchors, and the second 2k scores represents the category information of k anchors (whether it is a character);

6) Using the algorithm of text construction, the obtained small character boxes are merged into text sequence boxes. The main idea of the text construction algorithm is: every two adjacent candidate regions form a pair, and merge different pairs until they are all merged.

The CTPN network mainly uses the deep convolution network of VGG16 for convolution operation, so we can only select the top five convolutional layers of VGG16, and the convolutional layers finally output a series of feature map after the five-layer convolution operation.

LSTM is the most common RNN network in practical application, but in the CTPN framework, we select BLSTM network to replace LSTM [2]. BLSTM, it is bidirectional recurrent neural network. Actually, it is composed of two LSTM networks with opposite directions. The purpose of such improvement is actually to enable the model to better understand the characteristics of sequences in the light of the text context.

Because of the CTPN is aimed at lateral text detection. Naturally, the set of anchor are same in width, and placed in a vertical direction. Anchor width and heights: width = [16], heights = [11, 16, 23, 33, 48, 68, 97, 139, 198, 283].

The way to locate the text position is to access the RPN network after the convolutional layer of "FC". The RPN in here is similar to the Faster R-CNN mentioned in the introduction, and it is divided into two branches:

(1) The left branch is used for bounding box regression. Since fc feature map is equipped with 10 Anchor for per point, at the same time, only return regression center y coordinate and 2 values of height, so the rpn_bboxp_red has 20 channels.

(2) The right branch is used for softmax to classify Anchor.

When an Anchor is obtained, CTPN will do the following, similar to Faster R-CNN:

(1) Softmax judges the Anchor contains text whether or not, that is to say, it will select the main Anchor with the largest score of Softmax

(2) The Bounding box regression modifies the center y coordinates and height of the Anchor containing the text.

Notice: unlike Faster R-CNN, Bounding box regression in here doesn't correct the Coordinates of the center and width of Anchor. The specific regression method is shown in Formula (1):

$$v_c = \left(c_y - c_y^a\right)/h^a, \quad v_h = \log(h/h^a)$$
$$v_c^* = \left(c_y^* - c_y^a\right)/h^a, \quad v_h^* = \log(h^*/h^a) \tag{1}$$

Where $v = (v_v, v_h)$ is the coordinate of the regression prediction, $v = (v_{v*}v_{h*})$ is Ground Truth, c_y^a and h^a are the center y coordinate and height of Anchor. After the Anchor is processed by the above-mentioned softmax and the directional bounding box regression, a set of vertical bars next to a dense text proposal (a small block of text that is part of a line of text) will be obtained, with the same width but different heights, depending on which height of the Anchor the text at that point is adapted to. Subsequently, the text positions can be obtained by connecting the text proposal together with a text line construction algorithm.

In the previous step, the text pre-selection frames have been obtained, and then the large and small pre-selection frames are combined into a large text detection frame by a text line construction algorithm.

The core of CTPN network training is to determine the loss function, and the expression of the CTPN loss function is shown as a formula [3] (2):

$$Loss(s_i, v_j, o_k) = \frac{1}{N_s} \sum_i L_s^{cls}(s_i, s_i^*) + \frac{\lambda_1}{N_v} \sum_j L_v^{reg}(v_j, v_j^*) + \frac{\lambda_2}{N_o} \sum_k L_o^{reg}(o_k, o_k^*) \tag{2}$$

As can be seen, the Loss is divided into 3 parts:

(1) Anchor softmax loss: the loss is used to supervise and learn whether each Anchor contains text, indicating whether it is Groud truth;
(2) Anchor y coordinate regression loss: this loss is used to supervise the offset of the Boundary box regression y direction of each Anchor containing text;
(3) Anchor x coordinate regression loss: this loss is used to supervise the offset of the Boundary box regression x direction of each Anchor containing text;

The training process is that the predicted value is obtained by forward propagation, and the error is calculated and propagated back layer by layer. The network is trained in the direction with the minimum gradient until the parameters can make the network model converge.

2.2 CRNN Text Recognition Principle

The CRNN can feed input images of different sizes and produce predictions of different lengths. It runs directly on coarse-grained tags (such as words), and you don't need to specifically mark each individual element (such as a character) during the training phase. In addition, we have improved the structure of the CRNN network by removing the full connection layer at the end of the previous infrastructure. This has the advantage of simplifying the network model and greatly reducing the computation scale. In addition, since the core operation layer of the convolution and memory layer has not changed, it has little impact on the feature extraction ability of the network.

The CRNN text recognition architecture includes three parts [5]:

(1) Convolution layer, which uses CNN to learn pixel features of a text image;
(2) Recurrent layer, using BLSTM to predict the label distribution of the feature sequence obtained from the previous layer;
(3) Transcription layer, CTC is used to remove the repeated labels and blank labels in the features obtained from the upper layer and integrate them into the final result.

Combined with the knowledge of our software architecture, after inputting text images, the CRNN framework is actually a pipeline processing or stream processing process for text images. First, pixel feature learning is conducted through CNN, followed by sequence feature learning through BLSTM, and finally, CTC is integrated and transcribed into a recognition result, such a processing flow looks very simple and clear on the macro level. In addition, CNN and RNN can be combined to train the loss function, because although the two networks have different structures, they have formed a close relationship between output and input in the framework, so it is feasible to directly train the CRNN as a whole.

In the CRNN model, Feature extraction is realized by improved CNN network. The improved CNN network removes the full connection layer of traditional CNN network, It is mainly divided into convolution and pooling.

The mathematical definition of convolution is divided into continuous definition and discrete definition [4]:

Definition of continuous type:

$$(f * g)(n) = \int_{-\infty}^{+\infty} f(x)g(n-x)dx \tag{3}$$

Discrete definition:

$$(f * g)(n) = \sum_{x=-\infty}^{\infty} f(x) \cdot g(n-x) \tag{4}$$

In this paper, we use the convolution operator similar to the above discrete type for CNN convolution operation. We set the convolution kernel as 3×3 and the step parameter as 2 to carry out the convolution process.

The maximum pooling operation of 2×2 is used here, That is, every four pixels select a maximum pixel value as their feature, which can greatly simplify the operation scale of the back network and avoid over fitting phenomenon. After convolution pooling, the pixel matrix can be transmitted to blstm as a pixel feature matrix for sequence annotation.

The deep bidirectional recurrent neural network is built on the convolution layer, as the circulation layer. The cyclic layer predicts the feature sequence $x = x_1, x_2, x_3, \cdots x_T$, and the label distribution y_t of each frame x_t in.

Transcription is the process of converting each output of RNN into the final text sequence. Mathematically, transcription is based on prediction per frame to obtain the tag sequence with the highest probability. In practice, I used the conditional probability defined in the CTC layer. According to each frame, to predict, $y = y_1, y, y_3, \cdots y_T$, Probability of the tag sequence L, ignoring the position of each tag in L. Therefore, when the negative log likelihood probability is used as the objective function of the training network, the complexity is reduced.

Crnn network training, $\aleph = \{I_i, l_i\}_i$ is the training set, I_i is the training image, l_i is the real tag sequence. The goal is to minimize the loss function:

$$O = -\sum_{I_i, l_i \in X} \log p(l_i | y_i),\tag{5}$$

y_i is the sequence generated from l_i by the cyclic layer and convolution layer. The network layer and circulation layer are trained at the same time, and the forward calculation is carried out layer by layer. The final results are compared with the real tag sequence, and the parameters are updated in the negative direction of the loss function parameter gradient, Each gradient descent use random samples, that is, random gradient descent (SGD). Then the back propagation completes the iterative process until it converges.

2.3   Process Flow of Text Recognition Based on CTPN and CRNN

The overall framework of text detection and recognition is shown in the figure, After inputting the text image, the image is preprocessed by grayscale binarization and denoising, and then the recognized text is output through the above-mentioned text detection framework and text recognition framework [6] (Fig. 1).

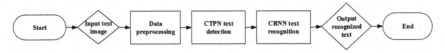

Fig. 1. Flow chart of text detection and recognition framework.

In the text detection stage, input the pre-processed text image data, use the first five convolutional layer blocks of VGG16 for depth feature extraction, and then send it into the two-way LSTM for text sequence feature recognition. After learning the pixel characteristics and sequence characteristics of the text [7], Each point is detected and located by a group of equal width anchors to obtain text small pieces of text proposals. After that, text lines construction algorithm is used to connect text proposals in the same area into text lines. Finally, the text lines determine the location of each text area to determine the text detection box (Figs. 2 and 3).

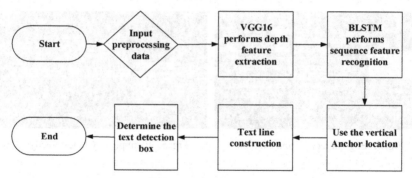

Fig. 2. Flow chart of CTPN text detection framework

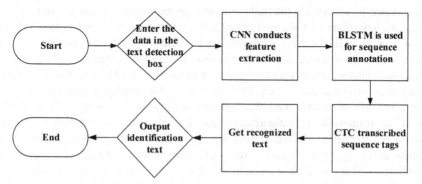

Fig. 3. Flow chart of CRNN text recognition framework

In the process of text recognition, only the text image data in the previous stage of text detection box is input, and the text feature is extracted through standard CNN network (convolution layer+maximum pooling layer) [8,9], and then it is sent to bidirectional LSTM network for sequence tagging. After that, sequence tags are transcribed through CTC layer, mainly to remove blank tags and illegal duplicate tags, and finally the recognized text sequence is obtained, and output it.

3 Algorithm Experiment

Based on the recognition algorithm proposed in this paper, we develop an intelligent cultural symbol recognition experiment app, the APP is also used for actual text content detection and recognition. The text detected successfully will generate a preview diagram and mark each word with a blue text box. After the successful recognition, the text content will be printed on the right side of the screen. The APP has completed the detection and recognition of English text well, and the recognition of punctuation marks and Numbers is also effective. As shown in Figs. 5 and 6, after obtaining the target text image using the experimental app, the English and Chinese are recognized. The recognition results are more accurate (Fig. 4).

Fig. 4. English recognition example   **Fig. 5.** Example of Chinese recognition

4 Conclusion

In this paper, by comparing the traditional text symbol detection and recognition algorithms, we have done a lot of research on the basic theory of machine learning and deep neural network in recent years, especially for the machine learning methods in the field of text detection and text recognition, an improved recognition algorithm of ctpn + crnn is proposed. This algorithm uses the improved VGG16 + BLSTM network to extract the depth feature and sequence feature of text image, and uses Anchor to locate text position, Finally, the task of cultural symbol recognition is carried out through CNN + BLSTM + CTC deep network, This algorithm is an ideal text recognition scheme at present in terms of recognition efficiency and accuracy. In the following work, we will further improve the recognition rate and accuracy of the algorithm, and gradually improve the algorithm based app to make it a practical cultural symbol recognition software.

References

1. Ma, J., Shao, W., Ye, H., et al.: Arbitrary-Oriented Scene Text Detection via Rotation Proposals (2017)
2. Liao, M., Shi, B., Bai, X., et al.: TextBoxes: A Fast Text Detector with a Single Deep Neural Network. The Association for the Advancement of Artificial Intelligence, San Francisco, California, USA, pp. 4161–4167 (2017)
3. He, W., Zhang, X.Y., Yin, F., et al.: Deep Direct Regression for Multi-oriented Scene Text Detection. In: IEEE International Conference on Computer Vision, Venice, Italy, pp. 745–753 (2017)
4. He, T., Huang, W., Qiao, Y., et al.: Text-attentional convolutional neural network for scene text detection. IEEE Trans. Image Process. 25(6), 2529–3254 (2016)
5. Wojna, Z., Gorban, A., Lee, D.S., et al.: Attention-based extraction of structured information from street view imagery. In: International Conference on Document Analysis and Recognition, Kyoto, Japan, pp. 844–850 (2017)
6. Zheng, L., Yang, Y., Tian, Q.: SIFT meets CNN: a decade survey of instance retrieval. IEEE Trans. Pattern Anal. Mach. Intell. 40(5), 1224–1244 (2018)
7. Uijlings, J., van de Sande, K., Gevers, T., et al.: Selective search for object recognition. Int. J. Comput. Vis. 104(2), 154–171 (2013)

Spatial-Temporal Variation and Future Changing Trend of NDVI in the Pearl River Basin from 1982 to 2015

Caixia He[1], Bingxin Ma[1], Juanli Jing[1,2(✉)], Yong Xu[1], and Shiqing Dou[1]

[1] College of Geomatics and Geoinformation, Guilin University of Technology, Guilin 541006, Guangxi, China
2003080@glut.edu.cn
[2] Guangxi Key Laboratory of Spatial Information and Geomatics, Guilin 541006, Guangxi, China

Abstract. Based on the GIMMS NDVI3g data from 1982 to 2015, this paper uses trend analysis, Mann-Kendall test and R/S analysis methods to explore the spatial- temporal variation and future trends of NDVI in the Pearl River Basin, in order to provide basis for the improvement of ecological problems in the area. The results show that: (1) In terms of time, the NDVI of the Pearl River Basin showed a fluctuating upward trend as a whole from 1982 to 2015, and it was extremely significant ($P < 0.01$); (2) In terms of space, the vegetation coverage was generally better in 34 years; the low and medium vegetation coverage areas (NDVI < 0.6) only accounted for 2.78% of the study area, which were concentrated in the Pearl River Delta region in the lower reaches of the Pearl River Basin; (3) In the trend analysis, NDVI showed an increasing trend in 34 years, and it accounted for 81.44% of the study area, mainly distributed in the upper reaches of the Yunnan-Guizhou Plateau and the middle reaches of the area; (4) In the future trend, the area with H > 0.75 accounts for 51.71%, indicating that the vegetation coverage will continue to increase in the future.

Keywords: NDVI · Trend analysis · Mann-Kendall test · Hurst index

1 The Introduction

Vegetation is a natural link connecting soil, atmosphere and water. It plays an important role in the energy exchange, biogeochemical cycle and hydrological cycle of the land surface, and is an important part of the ecosystem [1, 2]. In recent decades, under the combined influence of changes in natural conditions and human activities, global warming has had an unprecedented impact on the earth's environment and has become one of the major problems facing contemporary humans [3]. As an important part of the ecosystem, vegetation is very sensitive to climate change [4]. Understanding the dynamics of vegetation and its response to climate change is very important for revealing the behavioral mechanism of terrestrial ecosystems, predicting future vegetation growth and thus guiding environmental management [5], As the most commonly used indicator to characterize vegetation status, the Normalized Vegetation Index (NDVI) is closely related to vegetation coverage, growth status, biomass and photosynthesis intensity, and is an effective indicator for monitoring changes in regional vegetation and ecosystems [6].

The Pearl River Basin is located in the subtropical humid area of southern China, and its upstream is at the junction of the Yunnan-Guizhou Plateau and Guangxi. It is a concentrated distribution area of karst landforms in my country. The large population and acute human-land conflicts have led to vegetation destruction and soil erosion, resulting in a very serious phenomenon of rocky desertification. It is of great significance to analyze the spatiotemporal changes of vegetation in the Pearl River Basin. The current research period is relatively short. In view of this, this paper analyzes the characteristics of the spatial-temporal variation and future changing trend of NDVI in the Pearl River Basin based on the long-term series of GIMMS NDVI data sets from 1982 to 2015, supplemented by trend analysis, Mann-Kendall test, and R/S analysis to provide a scientific basis for the changes in the ecosystem of the Pearl River Basin in China.

2 Overview of the Research Area

The Pearl River basin is located between $102°14'–115°53'$ E and $21°31'–26°49'$ N, with a length of about 2320 km and a drainage area of about 446,800 km^2. The basin spans Yunnan, Guizhou, Guangxi, Guangdong, Hunan, and Jiangxi provinces. The terrain is high in the northwest and low in the southeast. The topography is complex, it is connected to the Nanling Mountains to the north, the South China Sea to the south, the Yungui Plateau to the west, and the delta impact plain to the southeast. Hills and basins alternate.

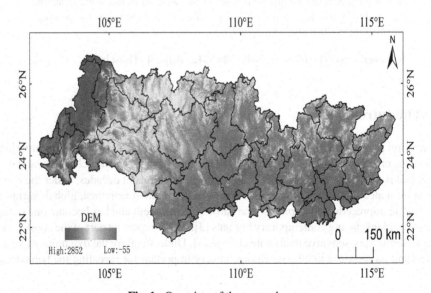

Fig. 1. Overview of the research area.

The study area is located in tropical and subtropical monsoon climate regions, with obvious monsoon climate. The average temperature is 14–20 °C and the average rainfall is 1200–2200 mm in the Pearl River Basin. The rainfall distributes uneven throughout the year and decreases from east to west (Fig. 1).

3 Data and Methods

3.1 Data Source and Preprocessing

The GIMMS NDVI3g data set used in the research comes from the Goddard Space Flight Center (GSFC) (https://ecocast.arc.nasa.gov/data/pub/gimms/3g.v1/). The dataset spans from July 1981 to December 2015, with a spatial resolution of 8 km and a temporal resolution of 15 days. The downloaded file is in nc4 format. Each nc4 file contains half a year of NDVI data, and the monthly NDVI data is divided into the first half of the month and the second half. This study uses Maximum Value Composites (MVC) to obtain monthly NDVI data, seasonal NDVI data and annual NDVI data.

The SRTM (Shuttle Radar Topography Mission) DEM data used in the study comes from the geospatial data cloud (https://www.gscloud.cn/search) with a spatial resolution of 90 m.

3.2 Research Methods

Trend Analysis Method. Linear regression analysis can simulate the changing trend of each pixel in the raster, and then reflect the spatiotemporal pattern of NDVI in the study area [7], the calculation process is as follows:

$$S = \frac{n \sum_{i=1}^{n} i \times NDVI_i - \sum_{i=1}^{n} i \sum_{i=1}^{n} NDVI_i}{n \sum_{i=1}^{n} i^2 - \left(\sum_{i=1}^{n} i\right)^2} \tag{1}$$

In the formula, S is the slope of the regression trend from 1982 to 2015, n is the total number of years in the study period, i is the year, and $NDVI_i$ is the maximum value of NDVI in the i-th year. If S > 0, it means that the NDVI value changes with time during the study period, and the increase trend is more obvious; S < 0, it means that the NDVI value changes with time and shows a downward trend.

Mann-Kendall (MK) Test. This article uses the Mann-Kendall (MK) test. MK is also called the non-distribution test. It is a non-parametric rank test method. Its advantage is that it does not require a specific distribution test on the data and is not affected by a few extreme values [8]. MK can be used to test the changing trend of long-term series in various regions, and quantitatively reflect the significance of the changing trend. The calculation formula is as follows:

$$\text{Suppose } \{NDVI_i\}, \quad i = 1982, \ 1983, \ \ldots, \ 2015,$$

The following formula is the calculation formula to define the test statistic Z:

$$Z = \begin{cases} \frac{S-1}{\sqrt{V(S)}}, & S > 0 \\ 0, & S = 0 \\ \frac{S+1}{\sqrt{V(S)}}, & S < 0 \end{cases} \tag{2}$$

The above formula means:

$$S = \sum_{i=1}^{n-1} \sum_{j=i+1}^{n} f(NDVI_j - NDVI_i) \tag{3}$$

$$f(NDVI_j - NDVI_i) = \begin{cases} 1, & NDVI_j - NDVI_i > 0 \\ 0, & NDVI_j - NDVI_i = 0 \\ -1, & NDVI_j - NDVI_i < 0 \end{cases} \tag{4}$$

$$\text{Variance}: V_{(S)} = n(n-1)(2n+5)/18 \tag{5}$$

In the formula, $NDVI_i$ and $NDVI_j$ are the annual NDVI values of the pixel in year i and j in the study area, respectively. f is a sign function, and n is the length of time to study vegetation changes. Z adopts a two-sided test with a value range of $(-\infty, +\infty)$. By looking up the normal distribution table, when $|Z| > 1.96$ is selected in this study, it passes the 95% significance test.

R/S Analysis. The R/S analysis method is also called rescaled range analysis [9], which was proposed by the British hydrologist Hurst when studying the hydrological time series data of the Nile [10], and is widely used in time series data. In the study, the calculation process is as follows:

Given a time series $\{NDVI_{(t)}\}$, $t = 1, 2,..., n$. For any positive integer p, define the mean sequence:

$$NDVI_{(p)} = \frac{1}{p} \sum_{t=1}^{p} NDVI_{(t)} \tag{6}$$

Define the cumulative difference sequence $NDVI_{(t,p)}$:

$$NDVI_{(t,p)} = \sum_{\substack{t=1 \\ 1 \le t \le p}}^{p} (NDVI_{(t)} - NDVI_{(p)}) \tag{7}$$

Define the range $R_{(p)}$:

$$R_{(p)} = \frac{Max}{1 \le t \le P} NDVI_{(t,p)} - \frac{Min}{1 \le t \le P} NDVI_{(t,p)} \tag{8}$$

Define the standard deviation sequence $S_{(p)}$:

$$S_{(p)} = \sqrt{\frac{1}{p} \sum_{t=1}^{p} (NDVI_t - NDVI_p)^2} \tag{9}$$

Hurst has established the following relationship through long-term practice summary:

$$\frac{R_p}{S_p} = (\alpha p)^H \tag{10}$$

In the formula, H is the Hurst exponent, which is obtained by fitting in the double logarithmic coordinate system $\{\ln(p),\ \ln(R_{(p)}/S_{(p)})\}$ using the least square method.

The H value is between 0 and 1. When $0 < H < 0.5$, the time series has anti-persistence character, indicating that the future vegetation changing trend is opposite to the changing trend in the past 34 years. The closer the H value is to 0, the stronger the anti-persistence of the time series. When $H = 0.5$, the time series is a random series, and the vegetation changing trend cannot be judged. When $0.5 < H < 1$, the time series has persistence character, indicating that the future vegetation changing trend is consistent with the changing trend in the past 34 years.

4 Result Analysis

4.1 Spatial and Temporal Characteristics of NDVI

Temporal Characteristics of NDVI. It could be seen from Fig. 2 that the NDVI fluctuated between 0.74 and 0.8 during 34 years, and the overall trend was upward, with a growth slope of 0.01/10a and R^2 is 0.3903. The changing trend of NDVI was significant ($P < 0.01$), indicating that the policy of returning farmland to forest and grassland had achieved results in the study area. The large fluctuations period of NDVI were 1989–1995 and 1998–2002, and the fluctuation range reached 0.04, which had a greater impact on the growing trend of NDVI from 1982 to 2015. After 2002, the volatility decreased and showed a steady growth trend.

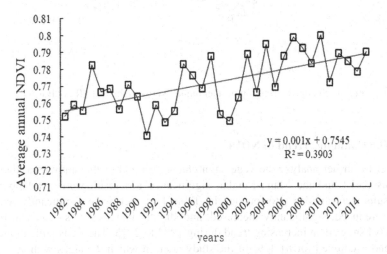

Fig. 2. Annual average NDVI from 1982 to 2015 in the Pearl River Basin.

Spatial Distribution of NDVI. The average value of NDVI was calculated from 1982 to 2015 in the study area and the spatial distribution of annual NDVI was obtained. As shown in Fig. 3, NDVI was divided into 4 grades, which were low vegetation cover areas

(NDVI < 0.4), medium vegetation cover areas (0.4 < NDVI < 0.6), high vegetation cover area (0.6 < NDVI < 0.8) and very high vegetation cover (NDVI > 0.8). The overall vegetation cover was relatively good during the 34a in the study area. The medium-low vegetation cover area only accounted for 2.78% of the study area. It was concentrated in the Pearl River Delta region and the vegetation types were mainly crops, where the terrain was low and the water resources was abundant. The high vegetation area accounted for 64.14%, which was distributed in various regions of the Pearl River Basin. The very high vegetation cover area (NDVI > 0.8) accounted for 33.07%, mainly distributed in the middle and upper reaches of Guangxi Hechi and Baise. The main landforms were mainly pure and relatively pure carbonate rocks (limestone), the altitude was between 0 and 800 m, and the vegetation was mainly open shrubs and mixed with evergreen coniferous forests and most grasslands.

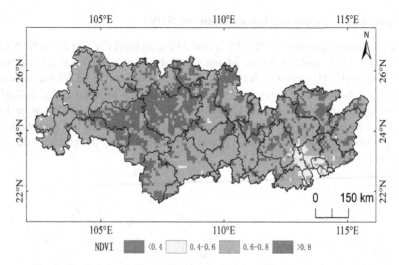

Fig. 3. Spatial distribution of interannual NDVI in the Pearl River Basin.

4.2   The Changing Trend of NDVI

In order to further analyze the vegetation changing trend of the study area, the trend analysis result and the Mann-Kendall significance test result were overlapped and the results were divided into four levels (Table 1), which were significantly reduced, slightly reduced, significantly increased, and slightly increased. As shown in Fig. 4, the NDVI showed an increasing trend during 1982 to 2015. The areas with an increasing trend accounted for 81.44% of the study area, in which the areas with significant increase accounted for 49.50%, mainly distributed from the middle to upper reaches of the Yunnan-Guizhou Plateau. The regional topography exhibited significant variation, mostly dominated by karst landforms with obvious differences in vegetation types. The areas with slight increase accounted for 31.94%, scattered throughout the basin. The areas with a decreasing trend accounted for 18.56%, among them, the areas with significant decrease accounted for 4.44% and slight decrease accounted for 14.12%, mainly distributed in the Pearl River Delta and its surrounding areas. The economy has been highly developed, which led to urban expansion.

Table 1. NDVI change trend table in the study area

	Significantly reduced S < 0 IZI >1.96	Slightly reduced S < 0 IZI <1.96	Significantly increased S > 0 IZI >1.96	Slightly increased S < 0 IZI <1.96
Year	4.44%	14.12%	49.50%	31.94%

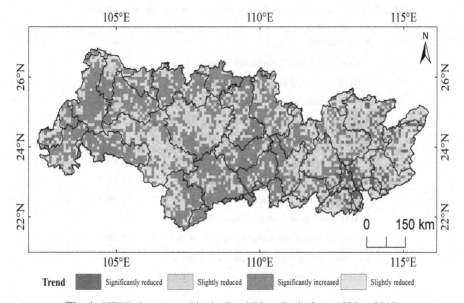

Fig. 4. NDVI change trend in the Pearl River Basin from 1982 to 2015.

4.3   The Future Trend of NDVI

The above analysis mainly reflects the characteristics of the spatial and temporal changes of vegetation of the study area in the past 34 years, and its future changing trend is not yet clear, for this reason, the results of NDVI changes from 1982 to 2015 are superimposed with the results of R/S analysis to reveal the future changing trend of the study area. The results are showed in Fig. 5. In this study, when $H < 0.5$, indicating an anti-persistence character; when H is larger than 0.5 but smaller than 0.75, indicating a weak persistence; when $H > 0.75$, indicating a strong persistence. The average H of the study area was 0.749. Of which the areas with H value smaller than 0.5 accounted for 2.65%, the areas with H value between 0.5 and 0.75 accounted for 45.64%, and the areas with H value greater than 0.75 accounted for 51.71%, it shows that the vegetation cover change in the future is more sustainable. As shown in Table 2 and Fig. 6, the vegetation cover will show an upward trend in the future. The areas with increasing trend in the future accounted for 80.91% of the study area, of which the areas with strong continuous and significant increase (34.11%) were mainly distributed in the middle upstream area. Followed by areas with weak persistent and insignificant increase (19.91%), scattered in various provinces, the areas with decreasing trend in the future accounted for 19.09% of the study area, of which the areas with weak persistent and insignificant decreases (9.04%) occupied the largest proportion. The areas with a decreasing trend in the future were mainly concentrated in the downstream of the Pearl River Delta. Above regions are characterized as densely population and intensively human activities.

Table 2. Future changing trend of vegetation cover in the Pearl River Basin

Variation type	Year
1 Strong sustained significant reduction	2.98%
2 Weak sustained significant reduction	1.22%
3 Anti-continuous and significant increase	0.11%
4 Strong persistent without significant reduction	4.05%
5 Weak persistent without significant reduction	9.04%
6 Anti-sustaining does not increase significantly	1.71%
7 Strong sustained significant increase	34.11%
8 Weak continued to increase significantly	15.48%
9 Anti-continuous and significant reduction	0
10 Strong sustained without significant increase	10.58%
11 Weak lasting without significant increase	19.91%
12 Anti-sustainability is not significantly reduced	0.83%

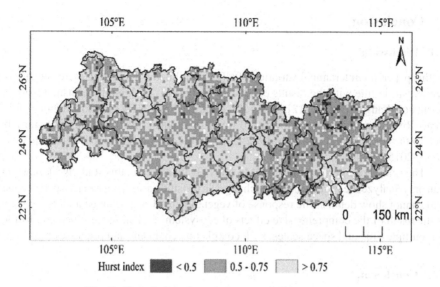

Fig. 5. Hurst index of vegetation cover of the study area.

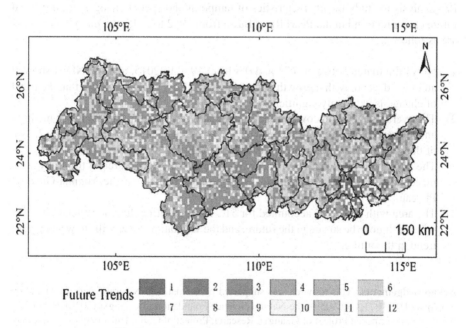

Fig. 6. Future changing trend of vegetation cover.

5 Conclusion

5.1 Discussion

In this paper, the interannual variation of NDVI exhibited a fluctuating increasing trend, which was in line with the results conducted by Jing J L [11], but inconsistent with the results of Wang Z L et al. [12]. Wang Z L addressed that NDVI in most regions of the Pearl River Basin showed a downward trend. The reason for the differences may be due to the study period was only 1982–2003, while the study period of this paper was 1982–2015.

The spatial resolution of the GIMMS NDVI3g data used in this study is 8 km, which can only analyze the vegetation cover changes in the study area from a macro perspective, and cannot show details. The response of vegetation to climate change and its feedback is the result of the comprehensive effects of ecosystems and climate influencing factors on multiple time and space scales, with complex mechanisms and processes [13].

5.2 Conclusion

Based on GIMMS NDVI3g data, this paper uses trend analysis, Mann-Kendall test and R/S analysis to study the characteristics of temporal and spatial changes of NDVI and future changing trend in the Pearl River Basin from 1982 to 2015, the main conclusions are as follows:

1) NDVI fluctuated between 0.74 and 0.8 from 1982 to 2015, and the NDVI showed an upward trend, with a growth rate of 0.01/10a. This situation shows that the trend of change is extremely significant ($P < 0.01$).
2) The overall vegetation cover in the study area was relatively good, and the medium and low vegetation coverage areas with NDVI less than 0.6 accounted for only 2.78% of the study area, which were concentrated in the Pearl River Delta.
3) The areas with increasing trend accounted for 81.44% of the area of the study area, mainly distributed from the middle to the upper reaches of the Yunnan-Guizhou Plateau.
4) The area with $H > 0.75$ accounted for 51.71%, indicating that the vegetation cover will continue to be strong in the future, and the vegetation cover will show an upward trend in the future.

Acknowledgements. This research was supported by the national Natural Science Foundation of China (41801071;42061059); the NaturalsScience Foundation of Guangxi (2020JJB150025); Bagui Scholars Special Project of Guangxi; Research Project of Guilin University of Technology (GUIQDJJ2019046).

References

1. Li, Z., Yan, F.L., Fan, X.T.: NDVI variation in Northwest China and its relationship with temperature and precipitation . J. Remote Sens. **03**, 308–313 (2005)

2. Piao, S., Fang, J., He, J.: Variations in vegetation net primary production in the Qinghai-Xizang Plateau, China, from 1982 to 1999. Climatic Change **74**(1–3), 253–267 (2006)
3. Landman, W.: Climate change 2007: the physical science basis **92**(1), 86–87 (2010)
4. Gottfried, M., Pauli, H., Futschik, A., et al.: Continent-wide response of mountain vegetation to climate change **2**(2), 111–115 (2012)
5. Zhao, J., Huang, S., Huang, Q., et al.: Time-lagged response of vegetation dynamics to climatic and teleconnection factors, 189 (2020)
6. Wang, X.L., Hou, X.Y.: Change of normalized vegetation index (NDVI) and its response to extreme climate in coastal areas of China from 1982 to 2014. Geogr. Res. **38**(04), 807–821 (2019)
7. Dai, Z.J., Zhao, X., Li, G.W., et al.: Spatial-temporal variations in NDVI in vegetation growing season in Qinghai based on GIMMS NDVI3g.v1 in past 34 years. Pratacultural Sci. **35**(4), 713–725 (2018)
8. Vicente-Serrano, S.M., Beguería, S., López-Moreno, J.I.: A multi scalar drought index sensitive to global warming: the standardized precipitation evapotranspiration index. J. Clim. **23**, 1696–1718 (2010)
9. Rao, A.R., Bhattacharya, D.: Hypothesis testing for long-term memory in hydrologic series **216**(3), 183–196 (1999)
10. Hurst, H.: Long-term storage capacity of reservoirs. Trans. Am. Soc. Civil Eng. **116**, 770–779 (1951)
11. Jing, J.L., Wang, Y.F., Yin, M.: Spatiotemporal variation of vegetation cover in Pearl river basin based on SPOT4-VGT. J. Guilin Univ. Technol. **34**(04), 711–716 (2014)
12. Wang, Z.L., Chen, X.H., Zhang, L., et al.: Temporal and spatial evolution characteristics of precipitation in Pearl river basin in recent 40 years. Hydrology **06**, 71–75 (2006)
13. Bachelet, D., Neilson, R.P., Lenihan, J.M., et al.: Climate change effects on vegetation distribution and carbon budget in the United States. Ecosystems **4**, 164–185 (2001)

An Improved Collaborative Filtering Algorithm Based on Filling Missing Data

Xin Zhou[1] and Wenan Tan[1,2(✉)]

[1] College of Computer Science and Technology, Nanjing University
of Aeronautics and Astronautics, Nanjing 211106, Jiangsu, China
wtan@foxmail.com
[2] School of Computer and Information Engineering,
Shanghai Polytechnic University, Shanghai 201209, China

Abstract. At present, most of the literature uses error metrics to evaluate the performance of recommendation algorithms. However, especially in Top-N recommendation tasks, the accuracy metrics can better show the pros and cons of the recommendation algorithm. Researching traditional recommendation algorithms have a big problem is that the data of recommendation system is always sparse. In order to solve the problem and improve the accuracy metrics, we add a filling matrix when predicting the rating. And then, we considering that different users have different scoring preferences, the user and item biases are added to the loss function. Finally, we use the improved alternating least square method as the optimization method to update the filling matrix in the iterative process. The experiment compared four recommendation algorithms and the results show that, in terms of accuracy metrics, the accuracy of the improved algorithm has been improved many times. In addition, since the filling of the rating matrix needs to consider both the item and the user, we compared the algorithm that considers both and only considers one. The improved algorithm that considered both has an improvement of about 1%.

Keyword: Recommendation algorithm · Alternate least square method · Accuracy metrics · Matrix factorization

1 Introduction

From the emergence of Arpanet in the United States in 1969, the Internet began to enter civilian life and the Internet became universal in the 21st century. The huge number of users generates a large amount of data every day. The ocean of data is rich in information and value. While satisfying the needs of users, the huge amount of data imposes a great burden on user to acquisition and screening. The problem of information overload has also arisen: the audience has "bottleneck" and "obstacles" in the process of transforming a large amount of information, unable to internalize it into the knowledge they need and absorb it effectively, resulting in "overload" [1].

The emergence and development of collaborative filtering recommendation algorithms have achieved good results in some social fields. Through the user's hobbies

Q. Zu et al. (Eds.): HCC 2020, LNCS 12634, pp. 220–226, 2021.
https://doi.org/10.1007/978-3-030-70626-5_23

and historical behaviors, products or information that may be of interest to the user can be extracted from the massive amount of information and recommended to the user. Collaborative filtering recommendation algorithms are mainly divided into item-based collaborative filtering [2], user-based collaborative filtering [3], and model-based collaborative filtering algorithms [4, 5]. This paper aims to improve and innovate collaborative filtering algorithms based on matrix factorization.

The rest of this paper is organized into following sections. Section 2 reviews some related work. Section 3 shows the details of our proposed loss function for filling missing values and rating prediction approach, and the optimization process of parameters. Section 4 presents the experimental results. Finally, Sect. 5 concludes this paper and points out the future work.

2 Related Work

At present, the performance of the recommended system is often judged by the error metrics (RMSE and MAE) [6–9]. MAE and RMSE calculate the average error between the predicted rating and the true value to illustrate the accuracy of the algorithm's prediction. However, in the opposite direction When a user recommends Top-N recommendations for items that may interest him, the error of the prediction rating does not play a big role, and the accuracy metrics (Precision and Recall) can better represent the performance of Top-N recommendations. Cremonesi [10] and others first proved that there is no necessary connection between error metrics and accuracy metrics, which laid the foundation for subsequent research. Hu Y [11]and others first applied the alternating least square method to implicit feedback, and added an attention mechanism to the loss function, and made significant progress in implicit recommendation. Zhang S [12] proposed to divide user trust relationships into strong and weak trust relationships on the basis of social relationships, and obtain weighted social relationships through the SVD algorithm. Finally, it was verified on the experimental data set and the results showed that there are higher than other methods. Accuracy. Wen J [13] used dance pictures as the background and combined social behavior and depth matrix decomposition to predict the dancer's interest in candidate images. The experimental results show that when the data set is very sparse, the proposed model has More significant effect.

3 Recommendation Algorithm

3.1 Loss Function

Square Error Loss. The loss function is used as a learning criterion in the experimental algorithm, showing the performance of the predicted and the true. And the performance of the algorithm is evaluated by minimizing the loss function.

The loss calculation method in the experimental algorithm uses square error loss (L2 loss):

$$L = \min \sum_{u \in U, i \in V} \left(r_{ui} - \hat{r}_{ui} \right)^2 \tag{1}$$

Where r is the true value and $\hat{r}$ is the predicted rating. The prediction rating is calculated as follows:

$$\hat{r}_{ui} = u_u v_i^T + b_u + b_i \tag{2}$$

Among them $\hat{r}_{ui}$ is the predicted rating of user u for item i, p_u is the vector of the u-th row of the user feature matrix; q_i is the vector of the i-th row of the item feature matrix; b_u is the paranoia of user u for rating prediction; b_i is the paranoia of item i for rating prediction.

Filling Missing Data. In order to reduce the sparseness of the matrix, we use the method of filling the sparse matrix. Since the rating is related to the item and the user, we add two rough filling matrices about the user and the item. The final loss function is as follows:

$$L = \min \frac{1}{2} \sum_{u \in P, i \in Q} \left(r_{ui} - u_u v_i^T + bu_{ui} + bi_{ui} \right)^2 + \lambda \left(\|u_u\|^2 + \|v_i\|^2 + \|bu\|^2 + \|bi\|^2 \right) \tag{3}$$

Where bu_{ui}, bi_{ui} is the added filling matrices, λ is the regularization coefficient and $\|u_u\|^2 + \|v_i\|^2 + \|bu\|^2 + \|bi\|^2$ is the regularization term.

3.2 Improved Alternating Least Squares Method

After the loss function calculates the error between the predicted result that the model output and the true value, the parameters in the model are optimized by the method of minimizing the loss function. In this paper, we use the improved alternating least squares method to optimize the parameters.

To update u_u, v_i, bu_{ui}, bi_{ui}:

$$\begin{cases} u_u = (V^T V + \lambda E)^{-1} (V^T R_i - V^T Bu_i - V^T Bi_i) \\ v_i = (U^T U + \lambda E)^{-1} (U^T R_u - U^T Bu_u - U^T Bi_u) \\ bu_u = \frac{1}{\lambda+1} (R_i - Bi_i - U^T v_i) \\ bi_i = \frac{1}{\lambda+1} (R_u - Bu_u - u_u^T V) \end{cases} \tag{4}$$

Alternating least squares method is a special case of least squares method. When the least squares method needs to deal with multiple parameter variables, take two parameter variables as an example. First fix the first parameter and solve the other; then fix the other parameter and solve the first parameter is to achieve convergence of results through alternate iterations. Relative to the gradient descent method, the alternate least squares method often requires fewer iterations to achieve convergence, and the results are more accurate.

The comparison found that in the commonly used gradient descent method, the optimization of parameters is only for the scored data. Although the missing values are filled in the data preprocessing stage, the missing values are not optimized in the optimization process, which leads to poor performance in the subsequent calculation of accuracy and regression rate. Therefore, the improved alternating least squares method is used as the optimization method of the parameters in the experimental algorithm.

3.3 Algorithm Description

The steps of the improved recommendation algorithm considering project users are as follows:

1. Convert the data set into a user-item matrix, initialize the regular term sparseness to 0.1, the maximum number of iterations is 50, and the potential dimension is 10;
2. Randomly generate user feature matrix and item feature matrix U, V, and user preference matrix and item preference matrix Bu and Bi;
3. Calculate the error by formula (3);
4. Update the parameters by formula (4) to determine whether the maximum number of iterations is reached, if the maximum number of iterations is not reached, return to step 3; if the maximum number of iterations is reached, output U, V;
5. Calculate the prediction rating by formula (2), and generate the Top-N recommendation sequence based on the prediction rating.

4 Experimental Design and Result Analysis

4.1 The Experimental Data Set

The data set used in the experiment is the rating provided by MovieLens about the user's movie. The data set includes 943 users' records of 100,000 ratings of 1,683 movies. In the rating record, each user's rating data for a movie includes each user's id (userId), each movie's id (movieId), user rating (rating) and timestamp (timestamp), where the user's rating is 1 to 5 [10]. It is also worth mentioning that the MovieLens dataset is a fairly sparse scoring matrix, with 93.7% of the data not being scored.

4.2 Evaluation Metrics

In this paper, we generate a recommendation list containing N recommended items for each user. In order to evaluate the accuracy of the generated recommendation list, we introduce the accuracy evaluation parameter for the recommendation list.

Precision and regression rate Recall are defined as followed:

$$\text{Pr} ecision@N = \frac{|C_{N,\text{rec}} \cap C_{\text{adopted}}|}{N} \tag{5}$$

$$\text{Recall}@N = \frac{|C_{N,\text{rec}} \cap C_{\text{adopted}}|}{|C_{\text{adopted}}|} \tag{6}$$

Among them, $C_{N,\text{rec}}$ is the recommendation given by us, and C_{adopted} is the recommendation item of the user in the test set. In the actual calculation, the accuracy and regression rate are obtained by calculating the average value of all users in the recommendation system.

Average precision (AP)

$$AP@N = \frac{\sum_{k=1}^{N} \text{Pr} ecision@k \times rel(k)}{N} \tag{7}$$

The mean average precision (MAP) is the mean of the average precision of all users.

4.3 Experimental Analysis

The experiment compares the traditional recommendation algorithm based on matrix factorization (MF), the matrix factorization recommendation algorithm that integrates the implicit behavior of users and items (SVD++), the improved recommendation algorithm that only considers users (SingleBiasSVD), and the improved recommendation algorithm that considers both users and items (BiasSVD).

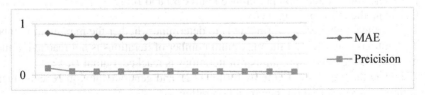

Fig. 1. MAE and Precision in SVD + +

As shown in Fig. 1, the precision convergence curve of SVD++ is obvious. Since the optimization function only optimizes scoring data, although MAE can achieve good results, the accuracy is surprisingly low when recommending items of interest to users, so the improvement on the basis of improving the optimization algorithm, considering the scoring habits of different users and the quality of different items, the user bias and item bias are added to the loss function, which makes the convergence speed and accuracy great promote.

As shown in Fig. 2, biasSVD is inferior to the other three algorithms in terms of convergence speed because of the optimization of two more parameters, but in terms of final accuracy, it is better than the classic SVD++ and MF algorithms. The experiment shows that the accuracy has a great improvement that reaches more than 31%. Compared with the SingleBiasSVD algorithm, the accuracy is also 1% improvement.

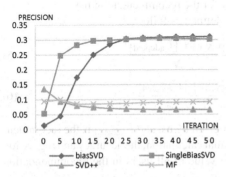 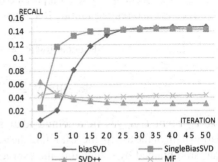

Fig. 2. Comparison of the precision of the four algorithms

Fig. 3. Comparison of the recall of the four algorithms

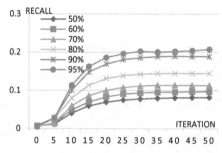

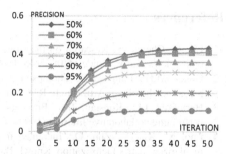

Fig. 4. The relationship between set size and recall

Fig. 5. The relationship between training set size and precision

Figure 3 shows the improvement of the recall of the four algorithms as the number of optimization iterations increases. Like the accuracy curve, biasSVD and SingleBiasSVD have a very significant improvement in accuracy compared with SVD++ and MF. The highest value of biasSVD is 14.6%. Around the same time, due to the optimization of two more parameters, the convergence speed of biasSVD has been reduced, which is about 0.5% improvement compared with SingleBiasSVD. Figure 4 shows the changes in the Recall values obtained by different recommendation algorithms when the training set is 50%, 60%, 70%, 80%, 90% and 95%; Fig. 5 shows the changes in the Recall value When the training set is 50%, 60%, 70%, 80%, 90%, and 95%. Figure 4 shows that as the number of test sets increases, the regression rate has a steady increase, which means with the increase of test machines, the generated recommendation sequence performs well on the test set; but comparing with Fig. 5, it is found that the increase of the training set reduces the accuracy rate. It is speculated that the increase of the training set introduces wrong samples that are not of interest to users. Resulting in errors when extracting feature information in matrix factorization training.

5 Summary

This paper establishes a collaborative recommendation algorithm based on filling missing value. The algorithm solves the problem of sparse scoring matrix by adding user-item matrices. In order to optimize the filling matrix, we propose an improved alternating least squares method. Having observed that different users and items have different preferences, we added bias when predict the rating. Experiments have proved that the proposed algorithm has a great improvement in accuracy. In the addition of filling matrix, BiasSVD is better.

Acknowledgments. The paper is supported in part by the National Natural Science Foundation of China under Grant No. 61672022, No. 61272036 and No. U1904186, Key Disciplines of Computer Science and Technology of Shanghai Polytechnic University under Grant No. XXKZD1604.

References

1. Liu, S., Cheng, Q.: Research on the strategy of personalized recommendation system for online news. Softw. Guide **16**(08), 146–148 (2017)
2. Goker, M.H., Langley, P., Thompson, C.A.: A personalized system for conversational recommendations. J. Artif. Intell. Res. **21** (2004)
3. Herlocker, J.L., Konstan, J.A., Borchers, A., Riedl, J.: An algorithmic framework for performing collaborative filtering (1999)
4. Luo, X., Xia, Y., Zhu, Q.: Incremental collaborative filtering recommender based on regularized matrix factorization. Knowl.-Based Syst. **27**, 271–280 (2012)
5. Liu, D.: A new item recommend algorithm of sparse data set based on user behavior analyzing. In: 12th International Conference on Signal Processing (ICSP), pp. 1377–1380 (2014)
6. Tan, W., Qin, X., Wang, Q.: A hybrid collaborative filtering recommendation algorithm using double neighbor selection. In: Tang, Y., Zu, Q., Rodríguez García, J.G. (eds.) HCC 2018. LNCS, vol. 11354, pp. 416–427. Springer, Cham (2019). https://doi.org/10.1007/978-3-030-15127-0_42
7. Wu, Y., Dubois, C., Zheng, A.X., Ester, M.: collaborative denoising auto-encoders for Top-N recommender systems. In: ACM International Conference on Web Search & Data Mining (2016)
8. Ocepek, U., Rugelj, J., Bosnić, Z.: Improving matrix factorization recommendations for examples in cold start. Expert Syst. Appl. **42**(19), 6784–6794 (2015)
9. He, X., Liao, L., Zhang, H., Nie, L., Hu, X., Chua, T.S.: Neural Collaborative Filtering, pp. 173–182 (2017)
10. Cremonesi, P., Koren, Y., Turrin, R.: Performance of recommender algorithms on top-n recommendation tasks. In: Proceedings of the fourth ACM conference on Recommender systems, Barcelona, Spain, pp. 39–46. Association for Computing Machinery (2010)
11. Hu, Y., Koren, Y., Volinsky, C.: Collaborative filtering for implicit feedback datasets. In: The 2008 Eighth IEEE International Conference on Data Mining (2008)
12. Zhang, S., Liu, X., Jiang, Y., Zhang, M.: A novel fine-grained user trust relation prediction for improving recommendation accuracy. In: 2016 International Conference on Advanced Cloud and Big Data (2016)
13. Wen, J., She, J., Li, X., Mao, H.: Visual background recommendation for dance performances using deep matrix factorization. ACM Trans. Multimed. Comput. **14**(1), 1–19 (2018)

Diagnosis Method of Alzheimer's Disease in PET Image Based on CNN Multi-mode Network

Shupei Wu[1,2] and He Huang[1(✉)]

[1] Intelligent Agriculture Engineering Laboratory of Anhui Province,
Institute of Intelligent Machines, Chinese Academy of Sciences, Hefei 230031, China
hhuang@iim.ac.cn
[2] University of Science and Technology of China, Hefei 230026, China

Abstract. Developing a correct diagnosis of Alzheimer's disease (AD) is a challenging task. Positron emission tomography (PET) is a good method to help doctors assist in the diagnosis of AD. In recent years, artificial intelligence methods such as machine learning have been widely used in image analysis and judgment and medical auxiliary diagnosis. The current methods are mainly to manually extract image features from medical images and then train classifiers to judge AD, or use deep learning, neural networks for end-to-end AD classification, most methods only use a single-mode method, and the classification effect is limited. This paper proposes a multi-mode network structure based on CNN to classify and diagnose AD. The network is mainly divided into three parts: CNN-based multi-scale deep-level feature extraction module, image texture feature extraction module, and SVM-based feature integration classification module. The network fully combines the advantages of the two modes of manual feature extraction and neural network. Compared with single mode feature extraction, this method has higher accuracy and has a good performance on the classification and diagnosis of AD.

Keywords: Alzheimer's disease diagnosis · Multi-mode · PET image · Multi-scale CNN

1 Introduction

Alzheimer's disease (AD) is a major neurodegenerative encephalopathy and the main cause of Alzheimer's disease worldwide [1]. It is estimated that the global prevalence of dementia is as high as 24 million, and as the world's population ages, this number will increase, reaching 66 million by 2030 and 115 million by 2050. The main increase will occur in low- income and middle-income countries, where more than 70% of people with dementia will live by 2050 [2, 3]. For AD, there is currently no effective treatment method, it is very meaningful to develop a treatment method to delay the disease under the condition of early diagnosis.

The computer-aided diagnosis method for Alzheimer's disease is based on image processing technology. With the development of computer diagnostic technology, such

© Springer Nature Switzerland AG 2021
Q. Zu et al. (Eds.): HCC 2020, LNCS 12634, pp. 227–237, 2021.
https://doi.org/10.1007/978-3-030-70626-5_24

as positron emission tomography (PET), a functional medical imaging method, it can help doctors diagnose neurodegenerative diseases at an early stage. A positron-emitting radionuclide (tracer) with biologically active molecules, such as an analog of glucose (18) F-fluorodeoxyglucose ((18) FDG), is introduced into the body. Use related imaging equipment to image the concentration of the tracer, and different tracer concentrations represent different glucose uptake activities to indicate the metabolic activity of local tissues [4, 5]. However, before the diagnosis is confirmed, results from tests and various sources must be collected, and this process can be complicated and time-consuming. Also, there is a certain degree of uncertainty in the results of the diagnosis, such as the commonly used diagnosis category ("possible" or "probable"). At the same time, detecting low-metabolism areas in PET is usually performed by medical doctors with expertise in Neuro assessment. It is necessary to understand the typical changes in PET brain images and the distribution of brain areas (patterns) that indicate specific neurological diseases. This "manual" interpretation is still subjective and cannot be replicated, nor can it be done quantitatively [6]. In the current artificial intelligence era, machine learning and deep learning have been applied in many fields, especially in the medical and health fields. Although the method of integrating them into the clinical process has not been developed and verified, great efforts have been made. Deep learning applies to many diseases and imaging types, such as breast cancer detection by mammography, lung nodule detection by CT, and hip osteoarthritis classification by radiography [7–10]. The application of machine learning technology in complex discovery patterns, such as those found in functional PET imaging of the brain, has just begun to be explored.

In recent years, for the related analysis of PET brain images, various pattern recognition methods have been studied, to summarize and analyze the pattern characteristics of Alzheimer's disease for computer-aided diagnosis (CAD), and help doctors treat patient's diagnosis. A method of manually extracting features from related regions to classify PET images into AD is proposed [11]. In this method, a total of 116 feature extraction regions of the image are divided, and features such as histogram entropy in these regions are calculated as the features of the region before the features are sent to the support vector machine (SVM) and random forest classifier, use the characteristic curve to sort the distinguishability of relevant regions of interest and take the top 21 regions. In the literature [12], a semi-supervised method is proposed, which extracts 286 features from the cortical grouping map of Automatic Anatomical Labeling (AAL), and uses random manifold learning and affinity regularization to process labeled and unlabeled data. A certain integration is used to classify AD. The literature [13, 14] proposes a boosting classification method that mixes classifiers and performs feature selection and classification at the same time. At the same time, it proposes the optimal combination of classifiers, and each basic classifier uses a different feature subset. In [1], three levels of features including statistical features, connectivity features, and graph-based features are extracted, and a similarity-driven ranking method is proposed to decompose connectivity features into three different feature sets. This reduces the dimensionality of features and increases the diversity of features.

Deep learning has been widely used in the field of medical imaging. Some articles help AD classification by extracting potential features. Cheng et al. [4] constructed a cascaded convolutional neural network to learn multi-layer image features hierarchically, and these features were fused to classify AD. Vu et al. [15] combined sparse auto-encoder (SAE) and convolutional neural network (CNN) for multi-modal learning.

Most of the above methods extract features from images or only use neural networks for end-to-end training. The single training mode has certain shortcomings, and further research is needed to improve the ability to classify AD.

This paper proposes a CNN-based multi-modal classification method to learn the shallow texture features of the image and the deep features of the neural network and merge these features into the SVM classifier to perform AD classification on PET brain images. First, extract texture features such as LBP. Second, build CNN networks of different image sizes and cascade them to adapt to different resolution images and integrate the features learned from multiple CNNs. Third, combine texture features And CNN extracts the features for fusion and sends them to the SVM classifier for image classification. The method proposed in this paper combines the wide applicability of traditional texture features, the convenience of the CNN network structure, and the advantages of learning hidden features. The rest of this article is organized as follows. The second part introduces the classification method in detail. The third part reports and analyzes the experimental results. Finally, the summary of this work is shown in the fourth part.

2   Design of Model Architecture

In this part, we will introduce in detail the classification method of Alzheimer's disease (AD) based on brain PET images in this work. PET image is a functional imaging method that is often used to assist doctors in diagnosing AD and other diseases. To improve the classification effect of images, we proposed a method of CNN-based multi-mode fusion to classify brain PET images for Alzheimer's disease.

Our method innovatively combines the advantages of deep learning and traditional feature extraction. As shown in Fig. 1, the network structure mainly includes three parts: CNN-based multi-scale deep-level feature extraction module, image texture feature extraction module, and SVM-based feature integration classification module. First of all, the multi-scale neural network based on CNN can adapt to the inconsistency of different image resolutions, and extract the deep semantic information of training images from it, and based on image texture features, a feature with good generalization, it can well reflect the characteristics of an image category, and effectively prevent the classifier from over-learning the image features of the training set, so that the classifier maintains a good generalization performance. Finally, the features of these two patterns are fused and sent to SVM to complete the classification and diagnosis of Alzheimer's disease (AD). We then introduce each module in detail.

2.1   Image Preprocessing

Since the resolution and size of the collected image data are not the same, the direct use of these data will bring obstacles to subsequent training. Therefore, we resample and scale the original image data to keep the size of the image data consistent. At the same time, considering that our training sample data is small and image classification requires

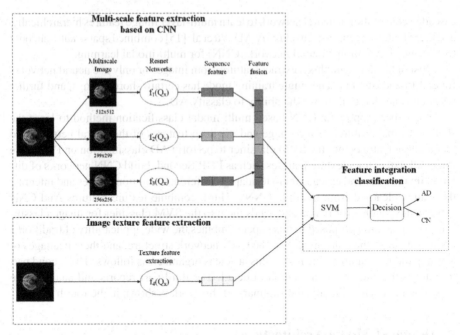

Fig. 1. Network structure diagram of the proposed classification method

high accuracy, we need to reduce the content loss after image scaling. After comparing the nearest neighbor interpolation algorithm, the bilinear interpolation algorithm, and the bicubic interpolation algorithm, we found that the bicubic interpolation algorithm is the best in our scheme, so we use the bicubic interpolation algorithm to preprocess the image.

2.2 Texture Feature Extraction

As a significant visual feature, texture features not only do not depend on color or brightness, but also include the arrangement and organization of the surface structure of things, showing the connection of contextual content, and reflecting the visual features of the repeated occurrence of homogeneity in the image. In this article, we mainly use the histogram statistical method, Gray Level Co-occurrence Matrix (GLCM), and Local Binary Pattern (LBP) to extract texture features of the image.

Statistical Characteristics of the Histogram. The brain PET images of patients with Alzheimer's disease and healthy people are sampled, and their grayscale histograms are drawn. As shown in Fig. 2, we can observe that there are certain differences in the grayscale histograms of the brain PET images of the AD patients. Therefore, we extract the image grayscale first-order statistical features based on the image grayscale histogram, mainly including regions pixel value average, peak value, and pixel energy value.

Based on GLCM Texture Features. Gray-level co-occurrence matrix (GLCM) extracts the spatial correlation of image texture gray levels, establishes a gray-level co-occurrence

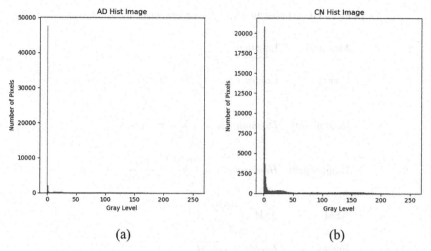

Fig. 2. Alzheimer's patient (AD) and normal gray histogram statistics

matrix based on the distance and direction between pixels, and then calculates the co-occurrence matrix to obtain some eigenvalues of the matrix to represent them respectively some texture characteristics of the image. The gray-level co-occurrence matrix can reflect the comprehensive information of the image gray-level in the direction, adjacent interval, and change range. The GLCM calculation formula is as follows:

$$P(i,j|d,\theta) = \frac{P(i,j|d,\theta)}{\sum_i \sum_j P(i,j|d,\theta)} \tag{1}$$

Where $P(i,j|d,\theta)$ expresses the probability of the couple pixels at θ direction and d interval. When θ and d are determined, is showed by Pi, j. Distinctly GLCM is a symmetric matrix. Its level is determined by the image gray-level [17–19]. Haralick [19, 20] defined 14 statistical features from the grey-level co-occurrence matrix for texture classification.

Here we use the six statistical features, and the specific calculation methods are shown in Table 1. In the equation, level means No. of distinct gray levels in the image, and Pi, j is the probability of each pixel value.

Based on the LBP Texture Feature. In this article, we use the texture feature extraction method that combines LBP and GLCM, and use the rotation-invariant LBP operator to calculate the texture image. After obtaining the LBP image, we extract the texture information feature of the LBP image.

3 Multi-scale Feature Extraction Based on CNN

In the previous preprocessing step, we resample the image and scale it to a uniform size for training, but it will cause the image to lose a certain amount of information during the scaling process. Based on this, we designed a multi-scale feature extraction network

Table 1. GLCM Measures with their Mathematical formulation

Measures	Equation
Contrast	$Contrast = \sum_{i,j=0}^{levels-1} P_{i,j}(i-j)^2$
Dissimilarity	$Dissimilarity = \sum_{i,j=0}^{levels-1} P_{i,j}|i-j|$
Homogeneity	$Homogeneity = \sum_{i,j=0}^{levels-1} \frac{P_{i,j}}{1+(i-j)^2}$
ASM	$ASM = \sum_{i,j=0}^{levels-1} P_{i,j}^2$
Energy	$Energy = \sqrt{ASM}$
Correlation	$Correlation = \sum_{i,j=0}^{levels-1} P_{i,j}\left[\frac{(i-u_i)(j-u_j)}{\sqrt{(\sigma_i^2)(\sigma_j^2)}}\right]$

based on CNN. By extracting image features at multiple scales, the semantic information of the image is increased and the accuracy of image classification is improved.

First, we scale the image to different sizes, and perform certain image enhancement and standardization. Each image size is fed into the convolutional neural network. In this work, we use a basic structure like Resnet and make certain changes to it to adapt to our classification task. For each scaled image, we train it separately. At the same time, to speed up the training of the network, we use network migration to migrate the originally trained Resnet-network to our model and fine-tune it. Considering that the shallow network layer of the neural network learns the similarity of the image content on different data sets, while the deep network layer has a large difference, so the network parameters of the first few layers of our neural network are frozen, and only the parameters of the later fully connected layer is updated. After all the networks are trained, we send the images to the network, extract the deep semantic information of the images, and get the final sequence features.

According to the universal approximation theorem, a single-layer feedforward network is sufficient to represent any function as long as sufficient capacity is given. However, this layer may be very large, and the network and data are prone to overfitting. However, the deepening of the network cannot be achieved through simple stacking of layers. Due to the notorious vanishing gradient problem, deep networks are difficult to train. Because the gradient is propagated back to the previous layer, repeated multiplication may make the gradient infinitely small. As a result, as the number of layers of the network becomes deeper, its performance tends to saturate and even begin to decline rapidly. Before ResNet appeared, there were several ways to deal with the vanishing gradient problem, but none of them really solved this problem.

In our proposed approach, we use resnet34 as our underlying deep network.

The network consists mainly of four layers of convolution modules, each using a 3 × 3 convolution kernel. As shown in Fig. 3, is a diagram of our network structure.

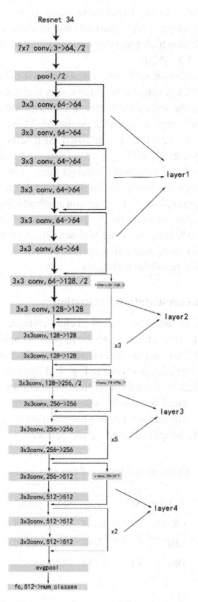

Fig. 3. Resnet34 network structure diagram

4 Experiments and Result

4.1 Experimental Setup and Data Set

The brain PET images used in the experiment came from normal data from elderly volunteers, including Alzheimer's (AD) patients and healthy people. The data set has a total of 2,000 pictures, all of which have the same length and width but are not uniform in size. The size range is 220–520.

To evaluate the classification performance, we used 4 different indicators, namely classification accuracy (ACC), sensitivity (SEN), specificity (SPE), and area under the curve (AUC). The higher the value, the better the corresponding method. Specifically, ACC is the proportion of samples correctly predicted. SEN represents the proportion of AD or MCI samples that are correctly classified. SPE refers to the proportion of NC samples that are correctly classified [1].

To better verify our method, we mainly conduct three experiments, namely: the evaluation of texture feature representation, the evaluation of the deep learning network, and the final feature fusion evaluation. All evaluation methods, except for special instructions, adopt the SVM method to evaluate the performance. We use 1600 images as the training set and another 400 images as the test set. Due to the limited number of data samples, we use a 10-fold cross-validation technique to evaluate the performance and repeat it 10 times to reduce possible deviations.

4.2 Texture Features Representation Evaluation

For the three extracted texture features: statistical characteristics of the histogram, GLCM texture, and LBP texture features, to evaluate the impact of each feature on classification, we respectively used SVM to evaluate the classification effect of a single texture feature. The results are shown in Table 2. As shown in Fig. 4. It can be seen that the performance of texture features on indicators such as classification accuracy is not very good. Among them, the histogram statistical feature is better than the other texture features in all aspects. In this result, we will add weight to histogram statistical features in the feature fusion later to increase the weight to improve the performance of the overall classifier.

Table 2. Experimental results of different texture features

Feature	ACC	SEN	SPE	AUC
GlCM	75.33	80.34	70.25	83.54
LBP	69.17	67.67	70.53	73.26
Hist	84	83.33	84.67	86.32

4.3 Deep Learning Features Representation Evaluation

In Sect. 2, we proposed a CNN-based multi-scale feature extraction method. To verify our proposed method, we respectively compare the classification performance of images at different scales and compare them with our method.

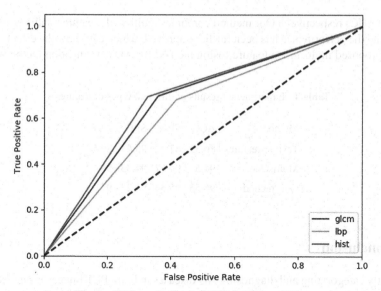

Fig. 4. AUC comparison of GlCM, LBP, and Hist

During the experiment, to speed up the training, we initialized the network parameters and assigned the trained Resnet-network parameters to our experimental network. At the same time, considering the small amount of data in our training data set, overfitting is prone to occur when using CNN, so our network uses dropout, L1 and L2 regularization methods [21]. The experimental results are shown in Table 3. It shows that, compared to the single-scale CNN classification network, the method we propose has a good performance on each evaluation index.

Table 3. Experimental results of different image scales

Scale size	ACC	SEN	SPE	AUC
256	93.5	93.7	92.3	94.5
299	92.34	91.67	94	94.23
512	89	88	90	91.24
Multiscale-fusion	94.5	94.67	94.33	96

4.4   Feature Fusion Evaluation

In this part, we compare different types of features and merge features of the same category. The results are shown in Table 4. Overall, the classification effect of deep learning is better than texture features. The fusion of different texture features can improve the effect relative to a single feature by 2.33% (ACC), 3.67% (SEN), 1.0% (SPE), and

3.24% (AUC) respectively. Our method combines multi-scale features with texture features, and the performance has been further improved, which can show the effectiveness of our proposed multi-mode feature fusion method for AD classification methods.

Table 4. Experimental results of different types of features

Method	ACC	SEN	SPE	AUC
Texture features	86.33	87	85.67	89.56
Multiscale	94.5	94.67	94.33	96
Our method	96.53	95.34	96.23	97

5 Conclusion

Correctly categorizing and diagnosing AD diseases in brain PET images is challenging. It has fewer data samples. Simply using traditional image processing techniques or deep learning methods for diagnosis has certain defects. Based on this, this article is an innovative AD classification method based on multi-class feature representation is proposed. First, extract the image-related texture features step by step. Then, based on a similar Resnet-network, learn features from different scales of the image to generate deep-level features of the image. Finally, comprehensively consider the above two categories of features and send them to SVM for processing Image classification.

This research proposes an innovative classification method, which combines the advantages of traditional image processing and deep learning methods. To meet the requirement of high accuracy in real applications, we need further work to improve our method of diagnosing Alzheimer's disease.

References

1. Pan, X., Adel, M., Fossati, C., Gaidon, T., Guedj, E.: Multilevel Feature Representation of FDG-PET Brain Images for Diagnosing Alzheimer's Disease. IEEE J. Biomed. Health Inf. **23**, 1499–1506 (2019)
2. Reitz, C., Brayne, C., Mayeux, R.: Epidemiology of alzheimer disease. Nat. Rev. Neurol. **7**, 137–152 (2011)
3. Wortmann, M.: Dementia: a global health priority - highlights from an ADI and world health organization report. Alzheimer's Res. Ther. **4**, 40 (2012)
4. Cheng, D., Liu, M.: Classification of Alzheimer's disease by cascaded convolutional neural networks using PET Images. In: Wang, Q., Shi, Y., Suk, H.-I., Suzuki, K. (eds.) MLMI 2017. LNCS, vol. 10541, pp. 106–113. Springer, Cham (2017). https://doi.org/10.1007/978-3-319-67389-9_13
5. Garali, I., Adel, M., Bourennane, S., Guedj, E.: Region-based brain selection and classification on pet images for Alzheimer's disease computer aided diagnosis. In: 2015 IEEE International Conference on Image Processing (ICIP), pp. 1473–1477. Quebec City (2015)

6. Serag, A., Wenzel, F., Thiele, F., Buchert, R., Young, S.: Optimal feature selection for auto-mated classification of FDG-PET in patients with sus-pected dementia. In: Medical Imaging 2009, Florida, United States (2009)
7. Dhungel, N., Carneiro, G., Bradley, A.: A deep learning approach for the analysis of masses in mammograms with minimal user intervention. Med. Image Anal. **37**, 114–128 (2017)
8. Setio, A., et al.: Pulmonary Nodule detection in CT images: false positive reduction using multi-view convolutional networks. IEEE Trans. Medi. Imaging **35**, 1160–1169 (2016)
9. Xu, L., Wu, X., Chen, K., Yao, L.: Multi-modality sparse representation-based classification for Alzheimer's disease and mild cognitive impairment. Comput. Methods Programs Biomed. **122**, 182–190 (2015)
10. Xue, Y., Zhang, R., Deng, Y., Chen, K., Jiang, T.: A preliminary examination of the diagnostic value of deep learning in hip osteoarthritis. PLoS ONE **12**, e0178992 (2017)
11. Garali, I., Adel, M., Bourennane, S., Guedj, E.: Region-based brain selection and classification on pet images for Alzheimer's disease computer aided diagnosis. In: IEEE International Conference on Image Processing, pp. 1473–1477 (2015)
12. Shen, L., Xia, Y., Cai, T.W., Feng, D.D.: Semi-supervised manifold learning with affinity regularization for Alzheimer's disease identification. In: International Conference of the IEEE EMBS, p. 2251 (2015)
13. Silveira, M., Marques, J.: Boosting Alzheimer disease diagnosis using PET images. In: International Conference on Pattern Recognition, pp. 2556–2559 (2010)
14. Cabral, C., Silveira, M.: Classification of Alzheimer's disease from FDG-PET images using favourite class ensembles. In: Engineering in Medicine and Biology Society, pp. 2477–2480. IEEE (2013)
15. Vu, T., Yang, H., Nguyen, V., Oh, A., Kim, M.: Multimodal learning using convolution neural network and sparse autoencoder. In: IEEE International Conference on Big Data and Smart Computing, pp. 13–16. Jeju, South Korea (2017)
16. Keys, R.: Cubic convolution interpolation for digital image processing. IEEE Trans. Acoustics Speech Signal Process. **29**, 1153–1160 (1981)
17. Kong, F.: Image retrieval using both color and texture features. In: 2009 International Conference on Machine Learning and Cybernetics, pp. 2228–2232. Hebei, China (2009)
18. Haralick, R., Shanmugam, K., Dinstein, I.: Textural features for image classification. IEEE Trans. Syst. Man Cyber. **SMC-3**, 610–621 (1973)
19. Haralick, R.: Statistical and structural approaches to texture. Proc. IEEE **67**, 786–804 (1979)
20. Nikoo, H., Talebi, H., Mirzaei, A.: A supervised method for determining displacement of gray level co-occurrence matrix. In: 7th Iranian Conference on Machine Vision and Image Processing, pp. 1–5, 16–17. (2011)
21. Srivastava, N., Hinton, G., Krizhevsky, A., Sutskever, I., Salakhutdinov, R.: Dropout: a simple way to prevent neural networks from overfitting. J. Mach. Learn. Res. **15**(1), 1929–1958 (2014)

Research on Multi-floor Path Planning Based on Recommendation Factors

Yanru Wang[1](✉) and Haoyang Yu[2]

[1] School of Economics, Jilin University, Changchun 130012, China
wangyanru@jlu.edu.cn
[2] The 54th Research Institute of CETC, Shijiazhuang 050081, China

Abstract. This paper proposes a path planning method for indoor multi-floor. By combining the Dijkstra algorithm, using intelligent recommendation predicted value as the weight of multi-floor path node, we give a multi-floor path plan based on the recommended factor. Compared with the traditional indoor path planning method, the method not only solves the problem of multi-floor planning in the interior, but also has the characteristics of personalized recommendation.

Keywords: Recommendation algorithm · Indoor path planning · Indoor navigation

1 Introduction

At present, in terms of outdoor navigation, both at home and abroad have been quite mature, which can provide accurate positioning for people to travel outside, and provide a series of travel plans, such as AutoNavi map and Baidu map in China, which can provide users with specific plans to reach their destinations. However, once entering the multi-floor interior, such as museums, shopping malls and other indoor public places, outdoor navigation based on GPS satellite positioning becomes inadequate. Meanwhile, compared with outdoor navigation with accurate positioning, multi-floor indoor navigation is developing very slowly, so it is an important direction of navigation service development at present. Compared with two-dimensional outdoor paths, multi-floor paths have three-dimensional characteristics, and path planning is the core content of multi-floor indoor navigation. The application scenario of multi-floor path planning studied in this paper is indoor public places such as museums and shopping malls. Therefore, users' personal preferences should be taken into account when solving the problem of multi-floor path planning. At present, the research on indoor path planning mainly considers the influence of multi-floor problem and positioning problem on path planning to provides the users with the shortest path to the destination. However, there are few studies on how to plan according to users' interests and characteristics. In this paper, the predicted value of intelligent recommendation is taken as the recommendation factor and converted it into the weight of multi-floor path nodes, to realize the goal that planning the routes with individual characteristics for users.

© Springer Nature Switzerland AG 2021
Q. Zu et al. (Eds.): HCC 2020, LNCS 12634, pp. 238–245, 2021.
https://doi.org/10.1007/978-3-030-70626-5_25

2 Multi-floor Path Planning Method Based on Recommendation Factors

2.1 Recommendation Algorithm

In this paper, the recommendation of cultural relics in museums is taken as the research object, and the data set adopted in the algorithm is the collected cultural relics label data set and user behavior data set. According to the data set of user behavior and cultural relic label data set, the tagging algorithm is improved. The input of the algorithm is the cultural relic label data set and the user behavior data set, which can be converted into the cultural relic label matrix and user-cultural relic matrix for calculation after processing.

Step 1: Select the specified target user u_r, obtain the historical data of the user u_r, and extract the user-cultural relic behavior feature vector.

Step 2: According to the user-cultural relic behavior feature vector and culturalrelic-label matrix, the user's selection probability value vector for the label set is calculated through formula (1), which is used as the user's label selection feature [1].

$$P(u_r, L_p) = \frac{\sum_{I_i \in I(u_r)} \sum_{l_x \in L(I_i)} \delta(l_x, L_p)}{\sum_{I_i \in I(u_r)} S(I_i)} \tag{1}$$

Step 3: According to the culturalrelic-label matrix and the tag selection probability value vector calculated in Step 2, calculate the user u_r's culturalrelic-labelselection probability value matrix for all cultural relics IT through formula (2). This matrix is the mapping of the user's tag selection characteristics on the cultural relic label set, which can only explain the user's behavioral characteristics.

$$P(u_r, L_p) = \begin{cases} \frac{\sum_{I_i \in I(u_r)} \sum_{l_x \in L(I_i)} \delta(l_x, L_p)}{\sum_{I_i \in I(u_r)} S(I_i)} L_p \in L(I_i) \\ 0 \quad L_p \notin L(I_i) \end{cases} \tag{2}$$

Step 4: Select the probability value matrix according to the cultural relic-label calculated in Step 3, and calculate the user u_r's selection score value [2] for all cultural relics through formula (3).

$$R(u_r, I_i) = \sum_{L_p \in LB} \sum_{l_x \in L(I_i)} \delta(l_x, L_p) P(u_r, L_p) \tag{3}$$

Step 5: Sort the vector of the selected score value in Step 4, and select the first n cultural relics as the collaborative filtering sample cultural relics, and also as part of the recommendation result at the end.

Step 6: The sample cultural relics obtained in Step 5 are taken as the target cultural relics of collaborative filtering. Calculate the similarity of each sample cultural relic to all cultural relics through similarity formula (4), and select the nearest cultural relics of each sample m. Finally, there are altogether n · m neighboring cultural relics.

$$sim(I_i, I_j) = \frac{\sum_{l_p \in LB} (P_{I_i l_p} \cdot P_{I_j l_p})}{\sqrt{\sum_{l_p \in LB} P_{I_i l_p}^2} \cdot \sqrt{\sum_{l_p \in LB} P_{I_j l_p}^2}} \tag{4}$$

Step 7: We need to make a choice of the nearest neighbor cultural relics calculated in step 6. According to formula (5), the predicted value [3] of the user u_r's selection of the nearest $n \cdot m$ neighbor cultural relics can be calculated, then exclude duplicate neighbors and the first k cultural relics can be selected as the predicted recommended cultural relics by sorting out the nearest neighbors.

$$H(r, S_{ij}) = \bar{r} + \frac{\sum_{j=1}^{n} (r_i - \bar{r}) \cdot S_{ij}}{\sum_{j=1}^{n} S_{ij}} \tag{5}$$

Step 8: Take the n sample cultural relics obtained in step 5 and the k predicted recommended cultural relics obtained in step 7 as the final recommendation results, and finally provide the user with () recommendation results and the recommended labeling reasons [4, 5].

2.2 The Recommendation to Multiple Floor Path Nodes

This paper uses museum indoor path planning as Application scenario, bases on the algorithm of Sect. 2.1 to calculate the recommendation predicted value of the cultural relics, and through the selection of Top-N to get the recommendation list [6]. Then, determine the recommendation predicted value of cultural relics as the weight of indoor path nodes, and took the exhibition hall with cultural relics as the node of the path, the indoor corridor and the hall without cultural relics as the path.

For the sake of concisely discussion, this paper makes summation calculation on the score value of the recommended cultural relics according to the exhibition room where the cultural relics are located. If the node set of indoor path is P, $N(P) = n$, then the recommended prediction set of cultural relics in the node can be expressed as $H(p_i) = \{h_j | h_j \in H, j = 1, 2, 3 \ldots N(p_i)\}$, where, $i \in (0, n), p_i$ represents the specified path node [7] in the node set, $H(p_i)$ represents the prediction score set of the node p_i, H is the complete set of the prediction score set, $N(p_i)$ is the number of prediction scores in the node p_i, and the predicted cumulative value of the node p_i can be expressed as:

$$C_{p_i} = \sum_{j=1}^{N(p_i)} h_j \tag{6}$$

Formula (6) is the indoor path node of accumulative value calculation method, it is worth noting that the complete set H of predicted rating define it as a selection list of Top-N after recommended prediction, rather than the set of all cultural relics predicted rating.

According to formula (6), the predicted accumulation value of nodes with recommended cultural relics can be calculated. In order to facilitate data observation and standardize data processing, the Min-Max method carries out linear transformation on the original data [8], and the method is as follows:

$$ND = \frac{OD - Min}{Max - Min} \tag{7}$$

Where, ND represents the data value after standardization, OD represents the original data value, Min represents the minimum value in the original data, and Max represents the maximum value in the original data. When the original data is the minimum value, the normalized value becomes 0. When the original data is the maximum value, the normalized data becomes 1. It can be seen that the Min-Max method reduces the original data linearly to the interval of [0, 1].

In this paper, the shortest path is calculated through the weight of path nodes and the distance between nodes. The greater the recommendation degree of path nodes, the smaller the weight of nodes should be. In addition, for non-recommended path nodes, it should be unified into a fixed value, and the value should be greater than that of all recommended nodes [9]. Formula (7) cannot meet the above requirements, so it needs to be improved as follows:

$$ND = \frac{Max - OD}{Max} + 1 \tag{8}$$

Formula (8) adjusts the Min-Max method, when $OD = Min$, ND is a decimal in the interval of [1, 2]; when $OD = Max$, $ND = 1$. At this point, the weight of nodes that are not recommended can be set to 2. In this way, for all nodes in the path, the greater the weight, the smaller the recommendation degree will be. If the weight is 2, it will be a normal node that is not recommended.

2.3 Multi-floor Path Planning Model Construction

Dijkstra algorithm is good at calculating the shortest path from the single source point to the remaining points. It calculates the shortest path according to the increasing path length, and the final result is the shortest distance from the starting point to the remaining points. In 2.1, the recommended nodes of indoor path can be obtained through the calculation of the recommended cultural relics. In order to combine the recommended nodes with the path planning, some changes need to be made to Dijkstra algorithm to meet the requirements of path planning in this chapter.

Based on Dijkstra algorithm, two node sets are introduced, one is the complete set U of nodes of the current floor, and the other is the recommended node set R. The recommended nodes, which are the core of the recommended path, should to be applied to all of the planned path as intermediate nodes, and in order to realize the floor of the connection, the stairs node should be as the layer of the end of the path planning and the starting point of the next layer of planning path, so the current floor nodes includes not only the exhibition hall, corridor and other path nodes in the floor, but also the stairs, elevators and other nodes that can connect the upper and lower floors, then separately record the nodes with the function of connecting the upper and lower floors, such as stairs and elevators, as set S_t. According to the above principles and combined with Dijkstra algorithm, the process of multi-floor path planning algorithm proposed in this paper with recommendation factor is described as follows:

(1) The adjacency matrix arcs represents a directed graph of a floor and arcs[i][j] is the weight of the edge from the node v_i to v_j. If there is no direct adjacency between two nodes, the value is set as ∞. A set E represents a set where the end point of the shortest path has been found from the starting point v. The vector D[i] represents the shortest distance from the starting point v to the corresponding node v_i.

(2) Select the node v_j to make

$$D[j] = Min\{D[i]|v_i \in V - E\} \tag{9}$$

Where, the node v_j represents the end point of one of the currently known shortest paths starting from the starting point v of the path. Let:

$$E = E \cup \{j\} \tag{10}$$

(3) $D[k]$ is the length of the shortest path from the starting point to any node v_k in the sets $V - E$, if:

$$D[j] + arcs[j][k] < D[k] \tag{11}$$

Modify the value of

$$D[k] = D[j] + arcs[j][k] \tag{12}$$

(4) Assuming that the number of nodes in the graph structure is n, the shortest distance from the starting point to the remaining points on the graph can be obtained by following the steps (2) and (3) $n - 1$ times.

(5) Steps (2), (3) and (4) are the main process of Dijkstra algorithm search and calculation. After calculation, the shortest path from a single node to other points can be obtained. All nodes in set R are calculated by Dijkstra algorithm, that is, the shortest paths of all recommended nodes to other nodes can be obtained, then mark the number of nodes in set R is m.

(6) Plan the path according to the shortest path set of the recommended nodes obtained in step (5). First find the node r_1 in the set R that is closest to the starting point v, mark the nodes in collection R as used nodes, then find the nearest node r_2 to the node r_1 according to the shortest path set of r_1, marked r_2 as a used node, then find the next nearest recommendation node according to the shortest path set. In this way, the sequence of nodes in set R is the recommended path for this layer.

(7) The recommended path is obtained by step (6) on the current floor, in order to make indoor floors recommended path connected together to form a complete interior recommended path, just need to connect the end of the current floor recommended path to the nearest stair node, find the stair node closest to the tail node r_m of the path in the set S_t, making it the last node of the current floor. At the same time, the next floor is planned as the starting point.

(8) Steps (1) to (7) is the complete single-floor path plan method, starting from the bottom floor, each floor can follow the process (1) to (7) and finally calculate the recommended path forward.

(9) Step (8) completes the planning of the interior tour route, while the design of the return route is more concise compared with the flow of the forward route. When the last floor tour is completed, the last exhibition hall is taken as the starting point of the return route.

3 Simulation Experiments

Multi-floor path planning in this paper takes the space structure of The National Museum of China as the experimental environment and uses the plane structure of the first floor, second floor, third floor and the underground exhibition hall in the interior of the National Museum as the multi-level structure of the path planning. Due to space limitations, the planar structure will not be displayed, please refer to the published planar structure of the museum.

Key positions such as exhibition hall, staircase, entrance and exit are marked in detail in the plane structure drawing. Next, it is necessary to construct an indoor road network. Firstly, the node of the indoor path is selected and each exhibition hall is set as a single path node. The number of the node is taken as the location of cultural relics, indicating the location of the path node where cultural relics sit. The corridor between the exhibition halls serves as the path node connecting each exhibition hall, and the corridor in the public area is represented by several limited nodes. The nodes connecting the upper and lower floors, such as elevators, stairs, and escalators, are also marked as key path points and appear simultaneously in the upper and lower floors. Finally, the adjacent nodes are connected to form an indoor road network structure. According to the marked location of indoor nodes and the adjacence of indoor nodes, the indoor path topology diagram is constructed, which is shown in Fig. 1.

Figure 1 is partial indoor path topology diagram. The nodes in the topology diagram represent the entrances, exits, exhibition halls, staircases and elevators and the key nodes of the path inside the museum, and the connecting lines between the path nodes represent the walking distance between the nodes. It is assumed that the indoor path topology is a directed graph because indoor walking routes are simpler than outdoor walking routes, people walk in different directions, and we adopt the strategy of stratification for multi-floor path planning, so it is assumed that the positive and negative weights of edges in the directed graph are the same. According to the calculation method provided in Sects. 2.2 and 2.3, the indoor path simulation experiment was carried out.

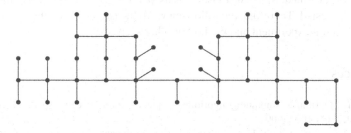

Fig. 1. Indoor path topology

The following is the display diagram of marking the experimental results of indoor path planning. The indoor nodes are connected and marked according to the path node data output from the experimental results. The recommended routes calculated by the path planning algorithm at each layer can be intuitively seen, and the results show that the strategy of stratification is realized. In Fig. 2, the endpoint of the first floor and the

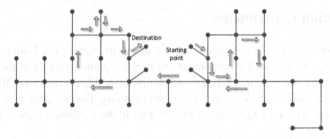

Fig. 2. Simulation results of route

starting point of the second floor are the same stair node, indicating that the planned route of the first floor can be entered into the next floor through the stair node after the end of the planned route of the first floor, realizing the connection of the planned route of the upper and lower floors.

After comparing with the results calculated by the recommendation algorithm, the exhibition hall nodes in the planning route calculated in this experiment contain most of the cultural relics in the recommendation list. The method in Sect. 2.1 is then used to calculate the recommended nodes of the path, and the calculated results all appear in the planned path. It can be seen that the method of recommending path in this chapter is effective, and the way a route is traveled is also reasonable.

4   Conclusion

This paper discusses the problem of indoor multi-floor path planning. Through the recommendation of cultural relics of the indoor path recommended node, adopt hierarchical planning strategy to solve the problem of complex interior space structure, combining indoor path planning of Dijkstra algorithm, finally give a smart recommendation feature several floors of the path planning scheme, and the national library of China for application scenario is verified. Since the museum is used as the application scene, other scenes have not been tested. Therefore, the following work focuses on the further improvement of the application experiment and method of other scenes.

References

1. Zhang, H.: Optimized smoothing algorithm for plane parameter cubic spline curve. J. Eng. Graph. **02**, 105–108 (2009)
2. Wu, J.: Research on multi-UAV collaborative track planning and effectiveness evaluation method. Nanchang Aeronaut. Univ. **05**, 45–47 (2012)
3. Li, Y.: Design and implementation of coordinate transformation system. China Univ. Geosci. (Beijing) **03**, 14–18 (2010)
4. Liu, Z.: Analysis and control of the singular inflection point of parametric curve. Nanchang Aeronaut. Univ. **01**, 10–12 (2012)
5. Dai, S.: Research on adaptive cubic spline interpolation approximation algorithm. Dalian Univ. Technol. **03**, 22–25 (2008)

6. Kang, Y., Zhou, J.: General aviation development status, trends and countermeasures analysis. Mod. Navig. **05**, 360–367 (2012)
7. Bai, Q.: Geometric continuity of general cubic parameter spline curve and its interpolation method. Northeast Normal Univ. **02**, 14–16 (2006)
8. Fang, Z.F., He, Q.S., Xiang, B.F., Xiao, H.P., Du, Y.X.: A finite element cable model and its applications based on the cubicspline curve. China Ocean Eng. **27**(5), 683–692 (2013)
9. Wu, W.C., Wang, T.H., Chiu, C.T.: Edge curve scaling and smoothing with cubic spline interpolation for image up-scaling. J. Signal Process. Syst. **78**(1), 95–113 (2015)

E-commerce Review Classification Based on SVM

Qiaohong Zu[1], Yang Zhou[1(✉)], and Wei Zhu[2]

[1] School of Logistics Engineering, Wuhan University of Technology, Wuhan 430063, China
1935044138@qq.com
[2] Hongyun Honghe Group Kunming Cigarette Factory, Kunming 650000, China

Abstract. In order to classify the massive historical review information of e-commerce platforms, efficiently extract review information and visualize it, this paper establishes an SVM-based e-commerce review classification model (using the combination of word frequency and information gain for feature selection), using the SVM classification model effectively classifies the review text, and uses J2EE as the developed technical framework to realize the B/S mode of the e-commerce review information system, combined with the JFreeChart plug-in to realize the visual display of review data classification, and provide conciseness for merchants and consumers Intuitive reference. This paper compares the two algorithm classification models of random forest and SVM. By comparing the results of classification experiments, it is verified that SVM can solve the small sample data classification problem in this paper more efficiently and accurately.

Keywords: E-commerce comments · SVM · Comments classification · Classification model · Visualization

1 Introduction

With the development of mobile e-commerce, a huge amount of commodity review data is generated. Although the review system divides reviews into good and bad reviews based on customer review scores automatically, merchants and customers cannot obtain clear substantive information related to commodities from a large number of complex historical review contents. In order to get useful information from the data quickly so that data utilization can be fully improved, an automated technology is needed to categorize and visualize reviews and provide useful references for merchants and customers.

2 Introduction of SVM Algorithm and Information Gain Method

2.1 SVM Algorithm

The basic idea of SVM is to obtain a hyperplane with the maximum geometric spacing that can be used to divide data, and then use this hyperplane to classify data or do regression analysis. Figure 1 is a hyperplane on a two-dimensional plane.

© Springer Nature Switzerland AG 2021
Q. Zu et al. (Eds.): HCC 2020, LNCS 12634, pp. 246–257, 2021.
https://doi.org/10.1007/978-3-030-70626-5_26

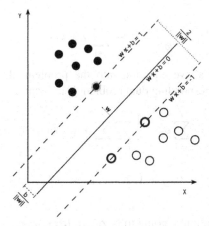

Fig. 1. Two-dimensional hyperplane

The x and ω are vectors in the formula, b is a real number, let $f(x) = \omega \cdot x + b$, when $\omega \cdot x + b = 0$, x is a point on the hyperplane, the point $f(x) < 0$, its corresponding y is equal to negative 1, and $f(x) > 0$ corresponds to the data point $y = 1$. The maximum spacing of the classifications is $2/\|\omega\|$.

Typical support vector machine target attributes are of two categories. Given a training sample set (x_i, y_i), $i = 1, 2, \ldots l, y \in R^n, y \in \{\pm 1\}$, l represents the number of training sample sets, and the hyperplane is denoted as: $(\vec{\omega} \cdot \vec{x}) + b = 0$.

Where, the constraints met are:

$$y_i[(\vec{\omega} \cdot \vec{x_i}) + b] \geq 1, \quad i = 1, 2, \ldots, 1 \tag{1}$$

According to the calculation formula of classification interval, the classification interval is b/‖ω‖. When the maximum classification interval 2/‖ω‖ is obtained in other words, the optimal classification hyperplane can be obtained. In order to facilitate calculation, the following transformation can be carried out:

$$\min \theta(\vec{\omega}) = \frac{1}{2}\|\vec{\omega}\|^2 = \frac{1}{2}\left(\vec{\omega} \cdot (\vec{\omega})'\right) \tag{2}$$

Lagrange function is introduced and set as $L(\omega, b, \alpha)$, and is expressed as:

$$L(\vec{\omega}, b, \alpha) = \frac{1}{2}\|\vec{\omega}\|^2 - \sum_{i=1}^{l} \alpha_i(y_i((\vec{\omega} \cdot \vec{x_i}) + b) - 1) \tag{3}$$

Where, α is the Lagrange multiplier. Usually, the saddle point of the function is the optimal solution of the problem. In this case, the solution of the function and the partial derivative of b are set to 0, and the following equation can be obtained:

$$\frac{\partial L}{\partial \vec{\omega}} = 0 \rightarrow \vec{\omega} = \sum_{i=1}^{l} \alpha_i y_i x_i \tag{4}$$

$$\frac{\partial L}{\partial b} = 0 \rightarrow \sum_{i=1}^{l} \alpha_i y_i = 0 \tag{5}$$

By substituting the above equation into the problem, the QP problem can be transformed into the corresponding dual problem [1]. Then:

$$\maxQ(a) = \sum_{j=1}^{l} \alpha_j - \frac{1}{2} \sum_{i=1}^{l} \sum_{j=1}^{l} \alpha_i \alpha_j y_i y_j (x_i \cdot x_j)$$

$$st. \sum_{j=1}^{l} \alpha_j y_j = 0, \ j = 1, 2, \ldots, l, \ \alpha_j \geq 0, \ i = 1, 2, \ldots, l \tag{6}$$

And the optimal solution is going to be $\alpha* = (\alpha_1*, \alpha_2*, \ldots, \alpha_l*)^T$. According to the above formula we can get $\vec{\omega}*, b* = 1 - y_i(\vec{\omega}*x_i), \alpha_i > 0$, and we can see from the dual expression that only some of these points satisfy $\alpha_i > 0$, they will become support vectors. The final decision function is:

$$f(x) = \text{sgn}(\vec{\omega} * \cdot \vec{x} + b*) = \text{sgn}\{\sum_{i=1}^{l} \alpha_i * y_i(\vec{x} \cdot \vec{x_i}) + b*\}, x \in R^n \tag{7}$$

If the N-dimensional space of the sample data is not linearly separable, then a hyperplane cannot be found to separate it in the N-dimensional space. SVM maps the input variables to a higher-dimensional eigenvector space, and constructs a classification hyperplane in the new space. Mapping x from the input space to the feature space H [2]:

$$\vec{x} \rightarrow \Phi(x) \tag{8}$$

Support vector machines avoid solving problem directly in high dimensional space by kernel function mechanism.

Put the space transformation into, and the final decision function is:

$$f(x) = \text{sgn}\{\sum_{i=1}^{l} \alpha_i * y_i(\Phi(\vec{x}) \cdot \Phi(\vec{x_i})) + b*\}, x \in R^n \tag{9}$$

Only the inner product of the feature space needs to be calculated. SVM uses kernel function to calculate the inner product [3]:

$$k(\vec{x} \cdot \vec{x_i}) = \Phi(\vec{x}) \cdot \Phi(\vec{x_i}) \tag{10}$$

After plugging in the kernel function, the final decision function is:

$$f(x) = \text{sgn}\{\sum_{i=1}^{l} \alpha_i * y_i k(\vec{x} \cdot \vec{x_i}) + b*\}, x \in R^n \tag{11}$$

This function is the support vector machine, or the optimal hyperplane classification rule. The inner product of high dimensional space is solved by kernel function, and the classification problem of nonlinear separable space is solved perfectly.

The SVM methods above require data to meet the constraint conditions: $y_i[(\vec{\omega} \cdot \vec{x_i}) + b] \geq 1, i = 1, 2, \ldots, l$. If there are some outliers in the data, the penalty factor and relaxation variable are introduced. The relaxation variable reduces the restriction condition of the data and allows certain deviation of the data, and the restriction condition becomes: $y_i[(\vec{\omega} \cdot \vec{x_i}) + b] \geq 1 - \zeta_i, i = 1, 2, \ldots, l$.

ζ_i represents the relaxation variable of sample i and allows certain deviation of the sample data. After the relaxation variable is introduced into the sample constraint, penalty factor C is introduced into the optimized objective function. Based on the optimization of the new objective function, the relevant model is finally obtained, and the new optimized objective function is as follows:

$$\min \theta(\vec{\omega}) = \frac{1}{2} \| \vec{\omega} \|^2 + C \sum_{i=1}^{l} \zeta_i \tag{12}$$

2.2 Information Gain Method

Information gain is the concept in information theory, information entropy is also one of the concepts of it. Information entropy is referring to the amount of information in current data, information gain refers to the difference of information entropy change after introducing the feature term. Information gain takes the amount of information of a phrase as the basis for selecting feature items, the more information the feature has, the more information gain will be, which means the more important the feature is. Information entropy is defined as follows:

$$Entropy(S) = - \sum_{i=1}^{P} P(c_i) \log_2 P(c_i) \tag{13}$$

Where, c_i represents the ith category of comments, $P(c_i)$ represents the probability of the ith category of data. After the introduction of the feature term, the definition of information entropy is as follows, where t_k represents a feature term [4]:

$$Entropy(S|t_k) = - \sum_{i=1}^{P} P(c_i|t_k) \log_2 P(c_i|t_k) \tag{14}$$

As can be seen from the above two formulas, the probability of each category is calculated before the introduction of the feature term, then the information entropy is obtained, and the new comment frequency is calculated after the introduction of the feature term. Therefore, according to the definition of information gain, which is the difference of information entropy after the introduction of feature terms, it is defined as follows:

$$Gain(t_k) = - \sum_{i=1}^{P} P(c_i) \log_2 P(c_i) - P(t_k)(- \sum_{i=1}^{P} P(c_i|t_k) \log_2 P(c_i|t_k))$$

$$- P(\overline{t_k})(- \sum_{i=1}^{P} P(c_i|\overline{t_k}) \log_2 P(c_i|\overline{t_k})) \tag{15}$$

Where, $P(c_i|\overline{t_k})$ is the probability value of appearing in the comment c_i without appearing the feature item, $P(t_k)$ is the probability value of the feature item appearing, and $P(\overline{t_k})$ is the probability value of the feature item not appearing [5].

The information gain can obtain the information of features by calculating the conditional probability of feature items. The information of features reflects the importance of features, which can be used as the criteria for feature selection [6]. The information of all feature items can be sorted, and the feature items with large information can be obtained as features to establish the classification model of comments.

3   The Realization of e-commerce Review Information System

3.1   System Structure and Function Design

System Structure. In this paper, the python language is used to crawl the comment content of each large shopping website to obtain the original data, and the text data is classified and stored in the database. The client can manage and analyze the information through the corresponding client device. Logically, the system can be divided into data acquisition layer, data transmission layer and system application layer.

- Data acquisition layer: Obtain text comment information, and then to remove and re-clean the information to obtain pure text.
- Data transmission layer: The information transmission layer is responsible for the data transmission between the data acquisition system and the database system. The data transmission layer is mainly responsible for data transmission through Internet, wireless network, mobile network and other networks.
- System application layer: At the top end of the system architecture, the system application layer is mainly responsible for the processing, analysis and presentation of comment data. The system application layer is the core part of the comment data visualization system.

System Function Modules. The main functional modules involved in the system are shown in Fig. 2, including analysis platform management, basic information management and system management. This system adopts the basic architecture pattern of J2EE, namely the B/S pattern of the three-tier architecture of presentation layer, business logic layer and data access layer.

Analysis platform mainly displays and analysis the data to realize visualization, including two modules: comment analysis and sales analysis. The comment analysis module mainly consists of comment type analysis and emotion analysis, which can realize the visual display of user comments. The sales analysis module mainly consists of two modules, namely, e-commerce sales analysis and sales comment analysis, to realize the visual display of different sales of users.

Basic information management module is mainly for the maintenance and management of e-commerce information, brand information and commodity information. Users can add, delete, change and check such information to achieve basic maintenance.

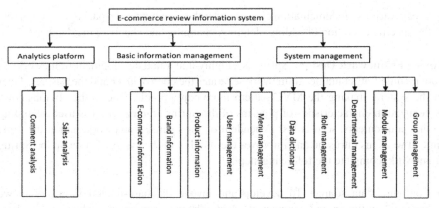

Fig. 2. System function module diagram

3.2 Comprehensive System Framework Design of J2EE

This system adopts J2EE as the technical architecture of development, combines with Struts, Spring and Hibernate technical framework, uses MySQL database management system, realizes the successful interaction of internal information, combines basic processing business with management, and realizes the Information system based on B/S mode.

J2EE Architecture. A typical J2EE application is usually divided into three layers, the presentation layer, the business logic layer, and the data layer, which run in different locations, namely the client, server, and backend database server [7]. In the process of actual development projects, the three-tier architecture can be extended according to the needs of the project. One of the commonly used extensions is to add the control mediation layer and data persistence layer to the three-tier architecture [8]. Different application components are distributed in different tier-dependent J2EE applications according to their logical functions. As shown below:

- Application components running in J2EE clients;
- WEB layer components running on J2EE server side, namely Servlet and JSP components;
- Enterprise business logic layer component running on J2EE server side -- EJB component;
- Enterprise Information (EIS) system-layer components running on the EIS server.

Struts Framework Design. Struts framework mainly carries on the JSP, Servlet, JavaBean as well as some custom tags and the information resource integration, has realized a good MVC classic pattern. Struts framework mainly completes the following steps:

- Complete the basic information configuration of Struts, and authorize the control file;
- Realize the development of each JSP page;
- Verify login in LoginFilter.java under cn.whut.filter package;

- Create various implementation classes under the cn.whut.action package, to complete the specific operation of adding, deleting, changing and checking the database.

Spring Framework Design. Spring is a highly integrated development framework. Its core module IOC container framework separates the control layer and the business layer of the system, making the MVC pattern more configurable. Due to the obvious componentization characteristics of plug-ins of the Spring framework, it is used according to the requirements, so it is less dependent on the overall framework, and at the same time, it weakens intrusion and improves the stability of the system. To set up the Spring framework, complete the following steps:

- Configure data source. To develop and invoke database-related information, we have to configure and connect it. Add <bean> component whose id is dataSource in beans.xml file. Configure the attributes driverClass, jdbcUrl, user, password, initialPoolSize, minPoolSize, maxPoolSize, maxIdleTime, acquireIncrem-ent, idleCo-nnectionTestPe-riod.
- Configure sessionFactory, which is also done in beans.xml file. By adding <bean> component with id of sessionFactory, and by adding mapping file in mappingResources inside < bean> component, database connection objects are provided for DAO layer.
- Configure the transaction, use the annotation-based method to configure the transaction and inject it into the configured sessionFactory.
- Configure DAO, configure DAO class based on HibernateTemplate in <bean> component with id userDaoTarget, and inject it into sessionFactory.
- Configure DAO transaction. In order to be restricted by the Spring layer when accessing DAO, it is required to configure DAO transaction component, inject it into DAO transaction object and add the desired transaction management strategy together.

Hibernate Framework Design. Before building control layer of Struts, we need to complete DAO layer based on Hibernate. First, create the persistent class and its corresponding mapping file, namely, create related *.Java classes under the cn.whut.bean package, design the field names and types in the class and construct simple constructors, all mapping files are placed directly in the cn.whut.bean package, design a database field name, length, node, etc. inside the mapping file, different mapping file corresponding to different database table. When the system adds, deletes, modifies and queries data, it needs to complete the relevant operation through HibernateTemplate. Therefore, the following steps are required: (1) Define interface classes in the cn.whut.service package; (2) Write the relevant implementation classes in the cn.whut.service.impl package; (3) Configure and authorize DAO in the mapping file.

4   Experiment of Comment Classification Based on SVM

4.1   Data Preprocessing

Before the model training of classification, the original comment data should be preprocessed. The original comment data should go through the following key steps:

- Data cleaning: The original comment was mixed with many advertisements or impurities. This article mainly used to build the filter library of the comments to achieve the cleaning of the comments.
- Comment participle: This paper selects ICTCLAS participle system of Chinese Academy of Sciences for comment participle. The system has a word-segmentation accuracy of more than 97% and a recall rate of 90%.
- Eliminate the high-frequency words and stop words: such as "of", "I", "in" and other words, without substantive effect on the classification of comments.
- Feature selection: Feature selection method is required to select important feature vocabulary to form feature subset. In this paper, the method of combining lexical frequency and information gain is used to select features.
- Weight vector representation: This paper mainly adopts the weighted word frequency TF-IDF algorithm. This vector space model converts the original comment data into a fixed-dimension weight vector, which is used for classification of the model.

4.2 Data Acquisition

In this experiment, more than 5,000 comments were collected from Tmall, jingdong, amazon, yihaodian and suning, five domestic e-commerce websites. According to statistics, 7 categories with large data volume were selected as the labels of SVM classification model. Table 1 shows the number of valid training data for each category.

Table 1. Quantity of effective training data

Category	Logistics	Quality	Price	Service	Logistics quality	Logistics price	Quality of price
Training data quantity	1304	1012	1526	1207	561	602	649

4.3 Establishment of SVM-Based Comment Classification Model

One-versus-rest (OVR) method is a kind of SVM multi-classification method widely used at present. Its basic idea is that you have k classes and you build k classifiers, one of the class will be marked as positive class, while the other is negative. Then training data set is trained. At the time of classification, the test samples were classified by k classifiers, and marked respectively, if the final mark has only one negative or positive marking, the test sample is classified correctly, otherwise, it is misclassified. In this paper, k is 7, therefore, 7 SVM classifiers are established. Table 2 is a series of parameters of 7 SVM models using OVR method.

- Number is the model number, Alpha is the non-zero value α in the SVM model, Bias is the b value in the SVM model, SupportVectors are the Support vectors, namely, non-zero points in the α, KernalFunction is the kernel, TrainScale is the size of the training sample.

Table 2. Parameters of SVM model

Number	Alpha	Bias	Support Vectors	Kernal Function	Train Scale
S1	M129*1	5.8464	M129*256	rbf_kernel	2000
S2	M112*1	6.1717	M112*256	rbf_kernel	1600
S3	M66*1	7.3693	M66*256	rbf_kernel	1800
S4	M478*1	0.4535	M478*256	rbf_kernel	1800
S5	M90*1	1.0579	M90*256	rbf_kernel	1300
S6	M76*1	1.0579	M76*256	rbf_kernel	1300
S7	M80*1	1.0579	M80*256	rbf_kernel	1300

Note:

- Mm*n represents a matrix with m rows and n columns;
- Rbf_kernel is a Gaussian kernel function.

Finally, the relevant parameters of Alpha, Bias and KernalFunction are substituted into the following expressions to obtain the corresponding classification hyperplane:

$$f(x) = \mathrm{sgn}\{\sum_{i=1}^{l} \alpha_i y_i k(\overrightarrow{x} \cdot \overrightarrow{x_i}) + b\}, x \in R^n \qquad (16)$$

The specific implementation of this SVM model is based on the LIBSVM library developed by Lin Zhiren team. The SVM learning library provides a simple and complete SVM model training, classification and prediction interface.

This paper selects RBF as kernel function of SVM, the optimal parameters of C and g are selected by cross validation, this paper uses 10-fold cross-validation, training data can be divided into 10 portions, one as a test data set. Ten results of the model will eventually be averaged. Figure 3 shows model accuracy under cross-validation of various parameter combinations.

As shown in Fig. 3, the optimal parameters C and g are selected by the experimental method in this paper. C represents the relaxation variable penalty factor of the SVM model, and g represents the parameter of the Gaussian kernel function. In this experiment, Gaussian kernel function is used to set up the comment classification model. The Gaussian kernel function is mainly convenient to calculate the inner product of space vectors in high dimensional space. In this experiment, X coordinate represents log2g, Y coordinate represents log2C, and Z coordinate represents the accuracy of classification. Then, the space of C and g parameters is traversed to find the best combination of C and g parameters.

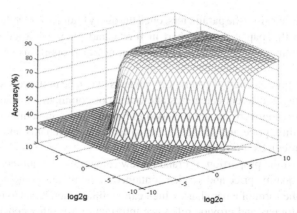

Fig. 3. Model accuracy under cross-validation of various parameter combinations

5 Experimental Results Analysis and Comment Information System Presentation

5.1 Analysis of Experimental Results

The precision rate and recall rate are mainly used to evaluate the quality of classifier results. Precision mainly refers to the accuracy of classification, which is an index to measure the classification accuracy. Recall rate measures the recall rate of the classifier. F1-Measure is to combine accuracy and recall rate into one index, which is used as the overall classification comment index. Its basic definition is as follows:

$$\text{Precision rate} = \frac{\text{Number of text correctly classified}}{\text{Number of texts actually classified}} \qquad (17)$$

$$\text{Recall rate} = \frac{\text{Number of text correctly classified}}{\text{Expected number of texts}} \qquad (18)$$

$$\text{F1} = \frac{2 \times \text{precision rate} \times \text{recall rate}}{\text{precision rate} + \text{recall rate}} \qquad (19)$$

The results of the review classification model experiment are analyzed as follows:

- Average accuracy rate: SVM (87.88%) > random forest (80.47%);
- Average recall rate: SVM (66.33%) > random forest (60.16%);
- Average F1 value: SVM (74.87%) > random forest (68.77%);

By comparison, SVM multi-classification model is superior to the random forest classification model. Therefore, SVM is selected as the classification model for comments in this paper.

5.2 Verification of Main Function Modules of the Comment Information System

The verification of the system example is as follows: check the parameters such as time, category, brand, e-commerce enterprise and comment type according to the required query data. After selecting the parameters, click the query button to enter the data analysis display page. In this paper, the pie chart in JFreeChart plug-in is used to display the proportion of comment types, while the bar chart shows the proportion of brand comment categories.

From the proportion chart of commodity reviews, it can be intuitively seen that in the selected time range, different categories and brands of commodities sold on the major e-commerce shopping websites are included in their reviews, including the proportion of logistics, quality, price and service, so as to facilitate merchants and consumers to understand the previous information points.

In the chart of the proportion of brand review categories, the number of comments about logistics, quality, price and service contained in the review content of each brand is displayed in the form of a bar chart, which can facilitate merchants to understand the feedback of customers and provide reference information for other consumers quickly and concisely. The display results are shown in Fig. 4.

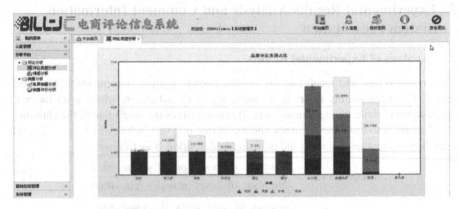

Fig. 4. Proportion of brand review categories

6   Conclusion

This paper uses the random forest, and SVM classification algorithm in the classification experiment, and the accuracy, the recall rate and the F1 of experimental results are compared, it turns out that the SVM algorithm is more suitable for the classification of small sample data in this paper. And then we analyses the comments information system operation mode, and display the classified comment data in the visualization system. Through many tests of the system, the system meets the requirements of the classification and visual display in this paper.

References

1. Yuanhai, S., Liming, L., Lingwei, H., Naiyang, D.: Key issues and prospects of support vector machines. Sci. China Math. **50**(09), 1233–1248 (2020)

2. Yuexuan, A., Ding Shifei, H., Jipu. : A review of twin support vector machines. Comput. Sci. **45**(11), 29–36 (2018)
3. Fangyuan, L., Shuihua, W., Yudong, Z.: Overview of support vector machine models and applications . Comput. Syst. Appl. **27**(04), 1–9 (2018)
4. Wang, H., Ge, J., Zhang, D., Liu, G.: Sensor selection for target tracking based on single dimension information gain. J. Eng. **2019**(20), 6562–6565 2019
5. Tiejun, C., Shasha, L., Guang, H., Fuchuan, J.: Research on dimensionality reduction method of fault influencing factors in SFT based on information gain. J. Saf. Environ. **18**(05), 1686–1691 (2018)
6. Lidong, W.: Text feature selection method based on information gain. Comput. Knowl. Technol. **13**(25), 242–244+254 (2017)
7. Jiang, Q., Qiyan, J.: Design and Implementation of Company Financial Management System based on J2EE Technology. J. Phys. Conf. Ser. **1578**(1), 012145 (2020)
8. Jin, Y.: Design and implementation of J2EE-based bank account information query system. Comput. Inf. Technol. **27**(06), 26–29 (2019)

Experimental Research of Galfenol Composite Cantilever

Qinghua Cao[✉], Xiaoxing Zhang, Nannan Wu, Zhifeng Zhou, and Jinming Zang

Jiangxi Province Key Laboratory of Precision Drive and Control, Nanchang Institute of
Technology, Nanchang, China
154280397@qq.com

Abstract. In order to research the output characteristics of Galfenol composite
cantilever beam, an experimental platform is built for the experiment of beam. The
influence of external magnetic field, load and substrate thickness on the strain of
Galfenol surface and substrate surface is deeply analyzed, and the magnetostrictive mechanism of composite cantilever beam is discussed. The conclusion that
Galfenol alloy is subjected to the coupling effect of tension and bending in the
cantilever beam can provide reference for further exploring the application of the
cantilever beam.

Keywords: Magnetostrictive · Galfenol · Composite cantilever beam ·
Experiment

1 Introduction

There are few literatures about the experiment of Galfenol alloy in the flexible cantilever
structure. Because there are many factors affecting the Galfenol composite cantilever,
the relevant static and dynamic experimental research can deeply understand its output
characteristics, and provide reference for the composite cantilever as actuator, sensor
and energy collector in the future.

2 Construction of Experimental Platform

The working mechanism of the Galfenol composite cantilever beam built on the experimental platform is to rely on the magnetostrictive deformation of the Galfenol sheet to
drive the substrate to produce bending, and the output deflection and force of the composite cantilever beam can be controlled by controlling the input current. Because the
expansion and contraction of Galfenol alloy is micron, in order to better study the performance of composite cantilever driven by Galfenol sheet, DC power supply is used The
bias coil of the experimental prototype is supplied with current. A sinusoidal waveform
signal is generated by a function signal generator and amplified by a power amplifier.
The amplified signal is connected with the ad port of DSpace. The cantilever prototype
is driven by the excitation current. The laser micrometer detects the displacement signal
of the cantilever beam and connects with the ad port of DSpace The displacement signal

Q. Zu et al. (Eds.): HCC 2020, LNCS 12634, pp. 258–268, 2021.
https://doi.org/10.1007/978-3-030-70626-5_27

Computer

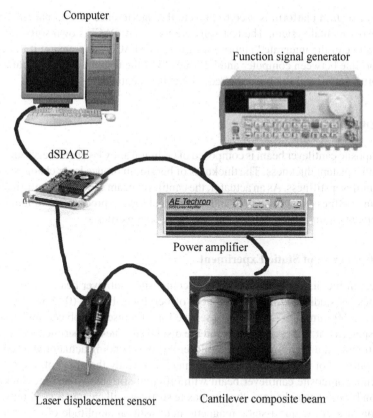

Function signal generator

dSPACE

Power amplifier

Laser displacement sensor   Cantilever composite beam

Fig. 1. Schematic diagram of experimental device

is collected and processed and input to the computer to obtain the required experimental data. The experimental device is shown in Fig. 1.

The whole experimental platform uses computer as its data acquisition system. After the test, the experimental data is exported. The laser micrometer is fixed by magnetic table holder to facilitate its measurement. The port of the laser micrometer is connected with the screw micrometer, and the connection is fixed on the fixed end of the magnetic base. During the debugging of the instrument and equipment, it can be adjusted by adjusting The magnetic gauge base is used to approximate the distance between the laser micrometer and the free end of the cantilever beam, so that the measurement range is close to the test distance of the laser micrometer. Then, the measurement distance is precisely adjusted by adjusting the spiral micrometer until the laser is green, so the distance between the cantilever beam and the laser micrometer is in the measurable range.

Due to the small expansion and contraction of Galfenol composite cantilever beam, it has high requirements for the test environment. The micro interference in the environment will have a great impact on the cantilever beam and laser micrometer. Therefore, the vibration isolation platform is used to prevent the interference of external environment on the experimental system. The test software used is dSPACE's own software, which can directly call the integrated algorithm signal of dSPACE and transmit it through the serial port line between computer and DSpace. After the test, the experimental data can be imported into Excel to obtain the required experimental data.

3 Experimental Research

The composite cantilever beam is composed of Galfenol alloy layer and beryllium bronze layer with constant thickness. The thickness of beryllium bronze layer changes with different cantilever stiffness. As an actuator, the cantilever beam can provide information for optimizing stiffness, magnetic field and mechanical load to produce maximum bending displacement Optimization of structure provides design criteria.

3.1 Introduction of Static Experiment

The effect of the displacement of Galfenol composite cantilever beam on the composite stiffness is studied. Different thickness of beryllium bronze (0.45 mm, 0.90 mm, 1.80 mm, 3.60 mm and 7.20 mm) is selected as the research object, and the maximum displacement values at the free end are displayed. These experiments can provide information for the performance of cantilever beam as bending actuator and sensor.

The purpose of no-load actuator experiment is to study the effect of magnetic field in Galfenol composite cantilever beam with different stiffness. When the thickness of beryllium bronze layer changes, the composite stiffness will change accordingly. Under the action of a cyclic quasi-static magnetic field with an amplitude of 65 ka/M and a frequency of 0.01 Hz, the strain of the composite cantilever beam was measured by a resistance strain gauge attached to the free surface of the Galfenol and beryllium bronze layers. Demagnetization is required before the test, and the test results are repeated more than 4 times to ensure the consistency of the experimental data.

The purpose of pre loading actuator experiment is to study the effect of magnetic field and mechanical preloading in composite cantilever beam with different stiffness. When different loads are hung on the free end, in order to obtain obvious bending moment, Galfenol layer is pasted near the fixed end, and a longer beryllium green copper layer is used. In order to avoid bending of cantilever beam due to gravity, the thickness of beryllium bronze layer is 1.80, 3.60 and 7.20 mm. Different precision weights are suspended on the free end of the composite cantilever beam. After the cantilever beam is stable, demagnetization is carried out by quasi-static magnetic field. The strain on the free surface of Galfenol alloy and beryllium bronze layer near the fixed end is measured by strain gauge. The same as the previous no-load experiment, the pre strain caused by mechanical strain is also recorded before the strain data generated by quasi-static cyclic magnetic field To be recorded. Three cantilever beams are tested according to different thickness. The same load will produce different stress profiles along the thickness of these composite cantilever beams. Therefore, it is not accurate to directly compare the characteristics of different cantilever beams. In order to avoid this problem, slightly different loads are selected for different cantilever beams.

Under the combined action of different magnetic fields and bending loads, the hall sensor captures the signal which is proportional to the magnetic induction intensity in Galfenol alloy. Two different methods are used to test the sensor characteristics to determine the consistency of the experimental data. The first method is that when different loads are suspended at the free end of the composite cantilever beam, the magnetic field is obtained by the hall sensor under quasi-static cycle The second method is to use a constant driving current to generate a bias magnetic field. Under the action of the bias magnetic field, the mechanical load changes quasi statically, and the response of Hall sensor is recorded at the same time. The first method is similar to measuring B-H curve, and the second method is similar to measuring B - σ curve.

3.2   Effect of Substrate Thickness Without Load

This section describes the experimental results of the composite cantilever without load. The external magnetic field and beryllium bronze thickness of the composite cantilever are used as its control parameters. The strain effect of different beryllium bronze thickness on Galfenol alloy surface and beryllium bronze surface is analyzed in depth.

This section describes the experimental results of the composite cantilever without load. The external magnetic field and beryllium bronze thickness of the composite cantilever are used as its control parameters. The strain effect of different beryllium bronze thickness on Galfenol alloy surface and beryllium bronze surface is analyzed in depth. The effect of thickness variation of beryllium bronze on the surface strain of Galfenol and beryllium bronze is shown in Fig. 2 and Fig. 3, respectively. For each composite cantilever beam, Fig. 2 shows that the strain on the surface of Galfenol alloy increases monotonously until it is saturated in magnetic field. Its behavior is similar to that of Galfenol material. With the decrease of the thickness of beryllium bronze layer, the slope and saturation value of the strain curve gradually increase. This behavior can be explained as a thinner beryllium bronze with a certain thickness of Galfenol alloy The layer provides less binding to the Galfenol alloy layer, so the Galfenol can expand more freely. Except for the thickness of 7.20 mm, the maximum strain on the surface of Galfenol alloy is higher than that of saturated magnetostriction. This is because the strain on the surface of Galfenol alloy is formed under the combined action of tension and bending.

However, the strain behavior of beryllium bronze will increase with the increase of strain amplitude in magnetic field, as shown in Fig. 2. When the thickness of beryllium bronze is 3.60 mm and 1.80 mm, the surface strain of beryllium bronze layer decreases monotonously with the increase of magnetic field, which clearly shows the bending behavior of these cantilever beams. When the thickness of beryllium bronze is 0.90 mm, the surface strain of beryllium bronze layer decreases with the increase of magnetic field until 32 ka/M, When the magnetic field exceeds this value, it increases with the increase of its size, and even changes the sign when the magnetic field reaches 45 kA/m. This behavior shows that the deformation of beryllium bronze layer is bending behavior at low magnetic field, and the bending effect is reduced when the magnetic field is higher than 45 kA/m When the thickness of beryllium bronze is 0.45 mm, the strain on the surface of beryllium bronze is positive under the action of most of the magnetic fields. This is mainly caused by the tensile deformation, which results in the surface strain of the

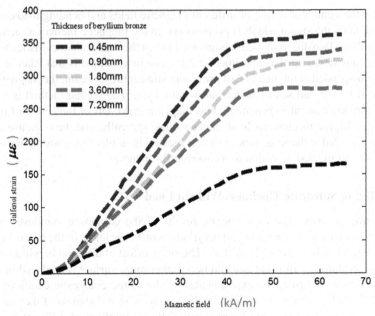

Fig. 2. Strain of Galfenol with different thickness of beryllium bronze

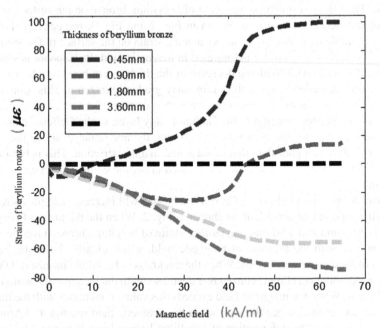

Fig. 3. Strain of beryllium bronze with different thickness

surface. The negative value of saturation strain on the surface of beryllium bronze layer decreases with the decrease of thickness from 3.60 to 1.80, and the saturation strain is positive at the thickness of 0.90 and 0.45 mm, and increases with the decrease of thickness. For beryllium bronze with different thickness, there is no monotonic change in the saturation strain value. When the thickness of beryllium bronze layer is large, the saturation strain almost remains unchanged, but the thickness of beryllium bronze layer is small. This behavior is related to the contribution of beryllium bronze layer to the overall stiffness of the composite structure.

In order to further elaborate the tensile and bending behaviors of the composite cantilever, the strain in the middle plane can be calculated according to the magneto mechanical coupling theory of the composite cantilever beam in reference 3 and the experiment, as shown in Fig. 4.

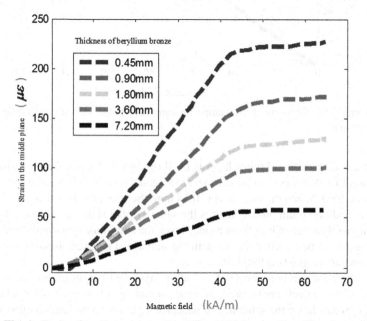

Fig. 4. Strain of medium plane under different thickness of beryllium bronze

Figure 4 shows that the medium plane strain is a function of magnetic field. When the thickness of beryllium bronze is constant, the medium plane strain monotonically increases with the increase of magnetic field, and the maximum value can reach 220 $\mu\varepsilon$. When the magnetic field is constant, the medium plane strain monotonically increases with the decrease of the thickness of beryllium bronze. As the thickness of beryllium bronze layer decreases, its contribution to the composite stiffness also decreases, and the structure of composite cantilever beam tends to be more free, With the decrease of the thickness of beryllium bronze, the medium plane strain gradually approaches the free strain of Galfenol alloy layer.

The normalized displacement of the composite cantilever beam with different thickness of beryllium bronze layer is a function of magnetic field, as shown in Fig. 5. When

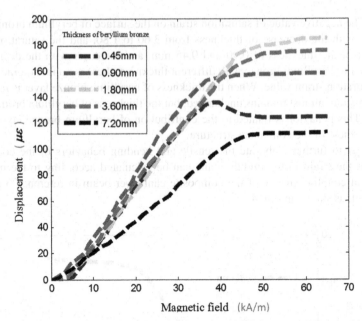

Fig. 5. Normalized displacement of composite cantilever beam with different thickness of beryllium bronze

the thickness of beryllium bronze layer is fixed (except 0.45 mm), the normalized displacement of Galfenol composite cantilever increases monotonously with the increase of magnetic field. When the thickness is 0.45 mm, the strain on the surface of Galfenol alloy layer and beryllium bronze layer is the same symbol If the strain rate of Galfenol alloy is higher than that of beryllium bronze layer, the amplitude of normalized displacement of cantilever beam will decrease with the increase of magnetic field, and this effect is more obvious in thin beryllium bronze layer.

Since the adhesive between Galfenol alloy layer and beryllium bronze layer and any other related defects are ignored, as an additional layer between Galfenol layer and beryllium bronze layer, the influence of adhesive layer on normalized displacement of cantilever beam can be estimated as a function of elastic modulus. The thickness of bonding layer is shown in Fig. 6, since the exact value of Young's modulus of adhesive layer is unknown. The elastic modulus range of typical amino tetrafunctional epoxy resin and nitrile rubber can be determined by simulation experiments. Because the thickness of adhesive layer will change in different cantilever beams, the thickness range of adhesive layer can also be simulated. The results show that the normalized free end displacement will increase with the increase of adhesive layer thickness and elastic modulus of adhesive layer. This is a reasonable explanation for the error between the experimental data and the magneto mechanical coupling model of the Galfenol composite cantilever.

3.3 Influence of Bending Moment Load Without Magnetic Field

Figure 7 and Fig. 8 show the relationship between surface strain and stress of various cantilever beams without magnetic field. In all stress-strain curves, only beryllium bronze

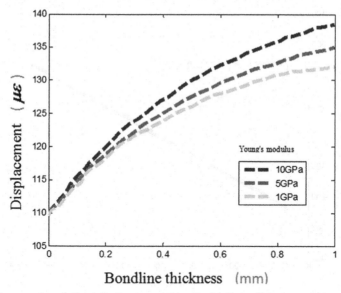

Fig. 6. Influence of normalized displacement with adhesive layer thickness and Young's modulus

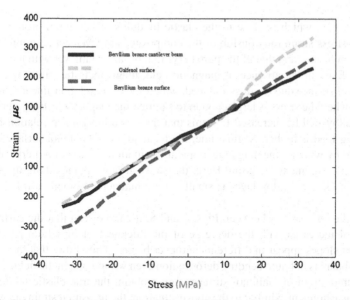

Fig. 7. Relationship between strain and stress with 7.2 mm beryllium bronze thickness

cantilever beam has very good linearity and shows that its slope range (elastic modulus) is 128 ± 5 gpa, which is close to the elastic modulus of beryllium bronze.

Figures 7 and 8 show the relationship between surface strain and stress of various cantilever beams without magnetic field. In all stress-stain curves, only beryllium bronze cantilever beam has very good linearity and shows that its slope range (elastic modulus)

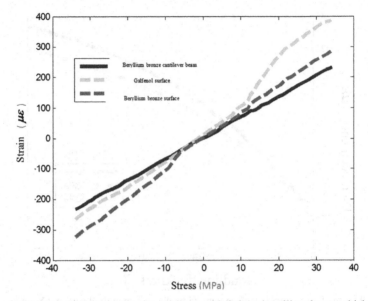

Fig. 8. Relationship between strain and stress with 3.6mm beryllium bronze thickness

is 128 ± 5 gpa, which is close to the elastic modulus of beryllium bronze. Figure 8 shows the stress-strain relationship of the composite cantilever with 3.6 mm thickness of beryllium bronze and that of the pure beryllium bronze cantilever with the same total thickness. Even in the absence of magnetic field, magnetoelastic strain will occur on the Galfenol surface due to stress-induced, which shows a high non-linear relationship. However, when the stress is large enough to saturate the magnetoelasticity, the strain of Galfenol alloy will be increased It seems that there is a local linear relationship when the bending load is higher. At this time, the local stress of Galfenol surface is mainly determined by whether the magnetic moment orientation is parallel or perpendicular to the strain measurement direction. Here, the stress at other points along the thickness direction of Galfenol alloy layer is small, so the magnetic moment still has different orientation.

From the comparison between Figs. 7 and 8, it can be seen that the surface strain of Galfenol increases with the decrease of the thickness of beryllium bronze layer, so different slopes appear in different cantilever beams. This shows that the composite cantilever beam is not pure bending deformation even without magnetic field. Therefore, the measured strain of Galfenol surface cannot obtain the true elastic modulus from bending measurement. Similar to the above situation, the measured strain on the surface of beryllium bronze layer in the composite cantilever beam is also different from that in the pure beryllium bronze cantilever beam. Its strain nonlinearity is mainly caused by the Galfenol nonlinear elastic modulus (∇E effect) which affects the composite stiffness of the cantilever beam. In this case, the elastic modulus of Galfenol alloy changes significantly along its thickness due to the change of stress and stress-induced magnetoelastic strain, which results in the cantilever beam becoming a highly nonlinear elastic structure.

4 Conclusion

The static performance of the Galfenol composite cantilever beam was studied by setting up an experimental platform. The effects of external magnetic field, bending load and substrate thickness on the surface strain of Galfenol and substrate were analyzed. Under the no load magnetic field, due to the combined effect of tension and bending on Galfenol surface, it was found that the maximum strain on the surface of Galfenol was larger than that of saturated magnetostriction When the magnetic field is high, the bending behavior dominates; when the magnetic field is high, the tensile behavior dominates the deformation; under the bending load without magnetic field, the surface strain of Galfenol will increase with the decrease of the substrate thickness. Because the nonlinear elastic modulus (effect) of Galfenol affects the composite stiffness of cantilever beam, the strain of the substrate also shows nonlinearity, so the whole complex is complex Under the combined action of magnetic field and bending load, the induced Galfenol tensile stress is gradually increased with the increase of load, which will hinder the driving strain and should be avoided as far as possible. When the load direction is changed and gradually increased, the induced medium pressure stress of Galfenol layer will gradually increase, which will increase the free stress in Galfenol alloy layer. It can improve its driving performance and can be used.

Acknowledgements. The authors would like to acknowledge the financialsupport by Youth Science Fund of Jiangxi province office of education of China (Grant No. GJJ161102) and Key Laboratory Open Fund of of precision drive and control (Grant No. KFKT-201610).

References

1. Ueno, T., Higuchi, T.: Miniature magnetostrictive linear actuator based on smooth impact drive mechanism. Int. J. Appl. Electromagnet Mech. **28**, 135–141 (2012)
2. Shu, L., Dapino, M., Evans, P., Chen, D., Lu, Q.: Optimization and dynamic modeling of galfenol unimorphs. J. Intell. Mater. Syst. Struct. **2011**(8), 781–793 (2011)
3. Smith, R.C., Dapino, M.J., Seelecke, S.: Free energy model for hysteresis in magnetostrictive transducers. J. Appl. Phys. **93**(1), 458–466 (2003)
4. Datta, S.: Quasi-static Chatacterization and Modeling of the Bending Behavior of Single Crystal Galfenol for Magnetostictive Sensors and Actuator. University of Maryland, USA (2009)
5. Atulasimha, J., Flatau, A.B.: A review of magnetostrictive iron–gallium alloys. Smart Mater. Struct. **20**,1–15 (2011)
6. Guruswamy, S., Srisukhum bowornchai, N., Clark, A.E., Restorff, J.B., Wun-Fogle, M.: Strong, ductile, and low-field-magnetostrictive alloys based on Fe-Ga.Scripta Materiali **43**,239–244 (2013)
7. Ueno, T., Higuchi, T.: Miniature magnetostrictive linear actuator based on smooth impact drive mechanism. Int. J. Appl. Electromagnet. Mech. **28**,135–141 (2012)
8. Shu, L., Dapino, M., Evans, P., Chen, D., Lu, Q.: Optimization and dynamic modeling of galfenol unimorphs. J. Intell. Mater. Syst. Struct. J. (8), 781–793 (2011)
9. Shu, L., Headings, L.M., Dapino, M.J., Chen, D., Lu, Q.: Nonlinear model for Galfenol cantilevered unimorphs considering full magnetoelastic coupling. J. Intell. Mater. Syst. Struct. **25**(2),187–203 (2014)

10. Ghodsi, M., Modabberifar, M., Ueno, T.: Quality factor, static and dynamic responses of miniature galfenol actuator at wide range of temperature. Int. J. Phys. Sci. **6**(36), 8143–8150 (2011)
11. Sauer, S., Marschner, U., Adolphi, B., Clasbrummel, B., Fischer, W.-J.: Passive wireless resonant galfenol sensor for osteosynthesis plate bending measurement. IEEE Sensors J. **12**(5), 734–739 (2012)
12. Marana, M.A.: Development of a Bio-Inspired Magnetostrictive Flow and Tactile Sensor. University of Maryland, USA (2012)
13. Davino, D., Giustiniani, A., Visone, C., Adly, A.A.: Energy Harvesting tests with galfenol atvariable magneto-mechanical conditions. IEEE Trans. Mag. **48**(11), 3096–3099 (2012)
14. Yoo, J.-H., Flatau, A.B.: A bending-mode galfenol electric power harvester. J. Intell. Mater. Syst. Struct. **23**(6), 647–654 (2012)
15. Suryarghya Chakrabarti, B.S.: Modeling of 3D Magnetostrictive Systems with Application to Galfenol and Terfenol-D Transducers. Ohio State University, USA (2009)

A Capacitive Flexible Tactile Sensor

Dandan Yuan, Haoxin Shu[✉], Yulong Bao, Bin Xu[✉], and Huan Wang

Jiangxi Province Key Laboratory of Precision Drive and Control, Nanchang Institute of
Technology, Nanchang 330099, China
xubin84115@163.com

Abstract. In this paper, a capacitive flexible tactile sensor was designed to measure the pressure of objects based on MEMS technology. This sensor is a structure of a 4×4 array, with metal Ag as the capacitive electrode, which forms the tactile sensing unit of the sensor. The structure of capacitive flexible tactile sensor was designed and an experimental platform was established to test the performance. The tests show that when the thickness of the intermediate layer is 2 mm and the density is medium, the sensor's sensitivity is the best while the time of both the response and the rebound is fast.

Keywords: Tactile sensors · Flexible · MEMS · Capacitive · Pressure testing

1 Introduction

With the rapid development of society, a research hotspot field of robotics attracts more and more attention. The flexible tactile sensor is self-evidently more and more important to a robot in terms of sensing the pressure of external objects Many scholars at home and abroad have conducted researches and achieved some results. HU Xiao-Hui et al. [1] developed a flexible capacitive tactile array sensor with the microneedle structure in the robotics field. This device has better flexibility because PDMS microneedle layer is sandwiched between the upper and lower electrodes. M.-Y. Cheng et al. [2] designed a capacitive touch array sensor which is composed of a PDMS structure and a flexible printing plate with electrodes. It effectively reduces the complexity of the capacitor structure of each sensing element. In 2019, Jie Qiu and others prepared a flexible capacitive tactile sensor with fast response, low detection limit, and high sensitivity [3]. However, because the response and rebound time of a flexible tactile sensor is also an important performance metric, a capacitive flexible tactile sensor with this structure is proposed in this paper. According to the working principle, tactile sensors can be divided into piezoresistive type [4], piezoelectric type [1], capacitive type [5, 6], optical sensor [7], magnetic sensor [8], etc. Capacitive tactile sensors have received wide attention and have been extensively used benefiting from their good linear response, wide dynamic range, and favorable durability [10]. This study has made three kinds of intermediate medium layers with different thickness and density. The best one is obtained by the sensitivity test.

2 Structure Design

The overall three-dimensional structure of the new capacitive flexible tactile sensor is shown in Fig. 1. The sensor is a structure of a 5-layer device, including a contact layer, an upper electrode plate and an upper electrode, an intermediate dielectric layer, a lower electrode plate and a lower electrode.

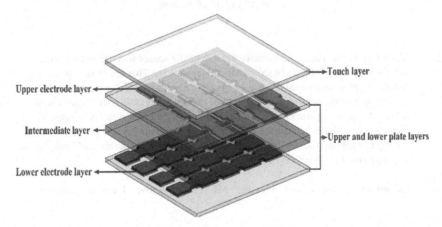

Fig. 1. Structure diagram of capacitive flexible tactile sensors.

The designed capacitive flexible tactile sensor uses PDMS film as the contact layer. PDMS is a kind of polymer organic silicide. It is able to withstand multiple touch and press, effectively protects the capacitive flexible tactile sensor from damage, and improves the durability of multiple touch and press. The upper and lower plates are made of PI material. It is a flexible polymer material that has good heat resistance, it has good corrosion resistance, excellent electrical insulation, good chemical stability and so on. The upper and lower electrodes are made of metal Ag, an excellent sensor electrode material which has low activity, good thermal conductivity and electrical conductivity. Ag is soft, ductile and not easily corroded by chemicals. The intermediate medium layer is made of PU material (polyurethane filter sponge) whose main body is a mesh structure. So it has excellent air permeability, good flexibility, and is durable, environmentally friendly and low-priced.

The overall size of the upper and lower electrode layers is a square of 24 mm × 24 mm, and the electrodes are made into a 4 × 4 sensor array. The upper and lower electrodes are staggered horizontally and vertically, and each intersecting place constitutes a sensing unit, so there are 16 independent sensing units in total. The structure of the electrode layer is shown in Fig. 2.

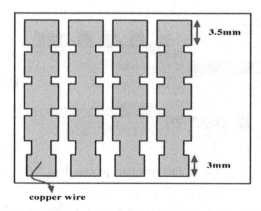

Fig. 2. Electrode layer structure diagram.

3 Preparation Process

The fabrication process of a capacitive flexible tactile sensor is shown in Fig. 3. It shows the process flow of the four layers using PI flexible substrate MEMS technology.

a) Silicon wafer is used as the bottom plate with a PI tape pasted on it. Then, it is cleaned in sequence with acetone, anhydrous ethanol and deionized water. After the cleaning, it is put on the heating platform for drying.

b) A layer of positive photoresist is uniformly sprayed on the surface of PI film. The negative film is fixed on the disc by vacuum pump. When the negative film is rotating at a high speed, the photoresist is evenly distributed on the negative film by the glue throwing machine.

c) The photoetching process is carried out by a photoetching machine. The photoetching machine exposes the photoresist through an ultraviolet light source and puts the exposed film into the developer. The photoresist in the exposed area is etched away and the photoetching pattern is fully displayed. Then it is rinsed in deionized water and then taken out for natural air drying.

d) The sample is sputtered with a high vacuum triple target magnetron coater. The vacuum is pumped to 10^{-4} Pa, the flow rate controlled to 80, and the power controlled to 100 W. After sputtered with Cr target for three minutes and then with Ag target for 30 min, the sample is cooled and taken out.

e) The purpose of stripping process is to remove the remaining photoresist, so that the electrode pattern can be displayed. After the remaining photoresist and Ag attached to the photoresist are removed with acetone, the Ag electrode remained.

4 Experiment and Analysis

The experimental platform of the capacitive tactile sensor is shown in Fig. 4(a). It consists of the LCR digital bridge tester and digital display push-pull tester. The capacitive

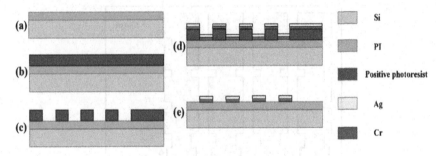

Fig. 3. Electrode layer process flow.

flexible tactile sensor is fixed on a circular plate so that its circular contact is just above the contact element. A certain pressure is applied through moving the push-pull gauge vertically downward by hand cranking. The physical picture of the device is shown in Fig. 4(b).

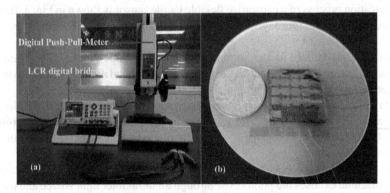

Fig. 4. (a) Experimental Platform (b) Fabricated flexible tactile sensors.

In this paper, we have made three kinds of PU materials for the intermediate dielectric layer. They are ones of 2 mm thick and medium density, 3 mm thick and medium density, and 3 mm thick and high density respectively. Because a single different parameter need to be set for comparison, we compare the experimental results of different intermediate dielectric layers with medium density and high density with the thickness of 3 mm PU. Figure 5(a) shows the test results of two capacitive flexible tactile sensors. The results show that for the same thickness of 3 mm, the initial value of high density is 0.53 pf, the sensitivity 12.58%/N, the initial value of medium density 1.45 pf, and the sensitivity 15.98%/n. In comparison with the experimental results of the same medium density, 2 mm thick and 3 mm thick, as shown in Fig. 5(b), for the same medium density, the initial value of 2 mm thickness is 1.4 pF and the sensitivity is 22.62%/N. Therefore, through the comparison of three different types of capacitive tactile sensors, the sensitivity of the sensor is the highest when the thickness of the intermediate dielectric layer is 2 mm and the density is medium density.

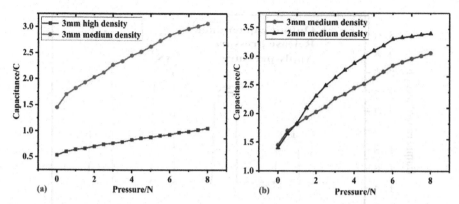

Fig. 5. (a) 3 mm medium and high-density capacitance value comparison (b) 2 mm and 3 mm medium-density thickness capacitance comparison.

When the applied pressure reaches 6N and the intermediate dielectric layer is 2 mm thick, which indicates that the capacitive sensor of this model has reached the maximum range. The sensitivity of the capacitive sensor can be expressed as:

$$K = \frac{\Delta C/C_0}{\Delta N} \tag{1}$$

Through date analysis, when the applied pressure is 6N, the sensor with 3 mm high density has the lowest sensitivity of 12.58%/N, and the sensor with 2 mm medium density has the highest sensitivity of 22.62%/N.

The response time and rebound time of the capacitive touch sensor to the pressure signal are also very important in application, which can reflect the performance of the capacitive touch sensor. In this paper, response time and rebound time of a 2 mm medium-density capacitive tactile sensor are tested. The pressure of 1N, 2N, 3N, 4N is applied in sequence to a tactile unit for 2 s, and the time for releasing the pressure at intervals is 3 s and a set of data is taken every 0.1 s. The experimental results are shown in Fig. 6. It can be seen from the figure that the response time of the capacitive touch sensor is 0.1 s, and the rebound time is 0.3 s. Therefore, the response time of the capacitive touch sensor is faster.

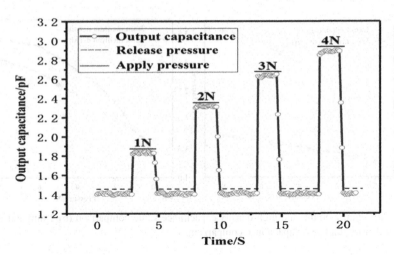

Fig. 6. Capacitive tactile sensor response time and rebound time.

5 Conclusion

In this paper, a capacitive flexible tactile sensor is fabricated by MEMS technology. Three sensors with different intermediate dielectric layers has been designed for comparison. It studies the relationship between the sensitivity and the thickness and density parameters of capacitive flexible tactile sensors, as well as the range, response time and rebound time. After many experimental tests, it can be seen that the comprehensive performance of the intermediate medium layer with thickness of 2 mm and medium density is the best.

Acknowledgments. This work was supported in part by Science Foundation of the department of education of Jiangxi province (GJJ180938, GJJ170987, GJJ180936). Open project of Shanxi Key Laboratory of Intelligent Robot (SKLIRKF2017004).

References

1. XiaoHui, H., et al.: A flexible capacitive tactile sensor array with micro structure for robotic application. Sci. China Inf. Sci. **57**(12), 1–6 (2014). https://doi.org/10.1007/s11432-014-5191-8
2. Qiu, J., Guo, X., Chu, R., et al.: Rapid-response, low detection limit, and high-sensitivity capacitive flexible tactile sensor based on three-dimensional porous dielectric layer for wearable electronic skin. ACS Appl. Mater. Interfaces. **11**(43), 40716–40725 (2019)
3. Fu, Y.M., Chou, M.C., Cheng, Y.T., et al.: An inkjet printed piezoresistive back-to-back graphene tactile sensor for endosurgical palpation applications. In: 2017 IEEE 30th International Conference on Micro Electro Mechanical Systems (MEMS), pp. 612–615. IEEE (2017)
4. Liu, W., Yu, P., Gu, C., et al.: Fingertip piezoelectric tactile sensor array for roughness encoding under varying scanning velocity. IEEE Sens. J. **17**(21), 6867–6879 (2017)

5. Sun, C.T., Lin, Y.C., Hsieh, C.J., et al.: A linear-response CMOS-MEMS capacitive tactile sensor. In: SENSORS, 2012 IEEE, pp. 1–4. IEEE (2012)
6. Jiang, Q., Xiang, L.: Design and experimental research on small-structures of tactile sensor array unit based on fiber Bragg grating. IEEE Sens. J. **17**(7), 2048–2054 (2017)
7. Volkova, T.I., Böhm, V., Naletova, V.A., et al.: A ferrofluid based artificial tactile sensor with magnetic field control. J. Magn. Magn. Mater. **431**, 277–280 (2017)
8. Zou, L., Ge, C., Wang, Z.J., et al.: Novel tactile sensor technology and smart tactile sensing systems: a review. Sensors **17**(11), 2653 (2017)

Multi-objective Optimization of E-commerce Logistics Warehouse Layout Based on Genetic Algorithm

Qiaohong Zu and Yanchang Liu[✉]

School of Logistics Engineering, Wuhan University of Technology, Wuhan 430063, China
zuqiaohong@foxmail.com, 410323118@qq.com

Abstract. With the support of Internet technology and infrastructure construction, fast and convenient online shopping has become an important part of urban consumption. E-commerce has a wide variety of products, large demand, short consumption cycles, and consumers have relatively high time requirements for online shopping services. A good warehouse layout strategy and inventory allocation strategy can help companies shorten inventory and pick-up time, reduce labor costs, and optimize the quality of delivery services. Therefore, studying the layout and distribution of e-commerce logistics warehouses is of inestimable value for improving the market competitiveness of e-commerce enterprises. This paper takes a E-commerce company's storage center as the object, considers goods' turnover rate and shelves' stability as principles to construct a multi-objective optimization mathematical model. By setting up random goal weights to improve traditional genetic algorithm, and to use MATLAB software platform to optimize the solution with mixed multitargets genetic algorithm on basis of the background of a specific warehouse position distribution. The result shows that this model is practical and effective. It can realize the reasonable distribution of the layout problem and reduce handling loss, as well as improve warehouse's space utilization.

Keywords: Warehouse layout · Genetic algorithm · Multi-objectives optimization · Slotting allocation · MATLAB

1 Introduction

Modern logistics companies not only pay attention to the reputation and reputation of the company to attract more customers while increasing their storage capacity, but also pay attention to how to ensure an appropriate storage distribution plan based on reasonable management. By optimizing the location of the inventory system, it will minimize unnecessary movement of warehouse operations to reduce operating costs and maximize storage efficiency, ultimately increasing the company's profitability.

2 Description of Goods Slotting Problem

According to its objective, there are five different optimization goals. With the goal of improving space utilization in ware housing slotting optimization; Aim to reduce equipment costs of warehousing slotting optimization; With the goal of maintaining minimum

Q. Zu et al. (Eds.): HCC 2020, LNCS 12634, pp. 276–282, 2021.
https://doi.org/10.1007/978-3-030-70626-5_29

inventory warehousing slotting optimization; With the goal of improving efficiency of warehousing slotting optimization. This paper combines current realities of the most storage, taking improving the efficiency of warehousing as optimization target [1].

2.1 Slotting Optimization Principles

The formulation of reasonable distribution principles is essential to improve work efficiency and achieve efficient management. Optimization principles can be inventory turnover rate, type of goods, shelf stability. This article establishes optimization goals based on inventory turnover and shelf stability [2].

Turnover Rate. Turnover ratio, also known as turnover times, is a comprehensive index to measure and evaluate the management status of each link of the company's inventory purchase, production, and sales recovery [3].

Weight Distribution. The load capacity of any kind of shelf is limited, so the goods on the shelf must be reasonably distributed within the shelf load range [4]. Carrying out cargo space allocation with certain distribution rules can ensure the uniform carrying capacity of the shelves and improve the efficiency of storing and storing goods.

2.2 The Basis Assumption of the Problem

This article assumes a warehouse model of a third-party logistics company. Supposing there is a row with b columns and c layers of one shelf in reservoir storage. We mark the row as row1 which is nearest to the entrance and exit. Similarly, we get column 1 and layer 1. Then we set a coordinate plane, for example, the coordinate of x row y column z layer is point (x, y, z), the shipping area is point $(0, 0, 0)$.

It also supposes that goods in warehouse can be stored in k different types (different size, weight, etc.) Each cargo can only store one type of goods. Two or more different types of goods cannot be stored in the same cargo space. Finally, assuming that the turnover rate of the kind k goods can be expressed as S_k, and the quality is M_k. In order to facilitate the optimization processing, this paper sets each cargos length, width and height as the same value L_0. Then this paper is based on these assumptions: Types of goods are known; different kinds of goods shape the same size and mass distribution is uniform. Each goods' turnover rate and mass are already known. Each position can store only one kind of goods. Single way to bring goods out of storage area. The length, width, height of each cargo and the roadway width are fixed value L_0. Time-consuming of accessing to goods is neglected and only considering the selection time [5].

3   Mathematical Model of Slotting Optimization

3.1 Target Function and Constraint Conditions

Out of storage efficiency analysis. In order to improve the storage efficiency, it is necessary to shorten the time of goods out of storage, obviously the key is to shorten the picking path. Assuming the coordinates of a location are (x, y, z)(row, column, height), coordinates shipments as the origin $(0, 0, 0)$, set Vx, Vy, Vz as average speed along x, y, z direction respectively. Goods located at (x, y, z) has a turnover rate of S_k. The time spending on taking goods to shipment area is:

$$(xVx + yVy + zVz) \cdot L_0 \cdot S_k \tag{1}$$

To shorten the time, we get the follow target function:

$$minf_1(x, y, z) = \sum_{x=1}^{a} \sum_{y=1}^{b} \sum_{y=1}^{c} Z \cdot L_0 \cdot M_{xyz} \tag{2}$$

In formula 2, S_k means the turnover rate of kind k goods, L_0 means the length of each goods cell.

Shelf Stability Analysis. The center of gravity of an object is related to stability, the lower the center of gravity [6], the stability is better. Suppose position (x, y, z) has stored goods with the mass M_{xyz}, in order to make the center of gravity to be the lowest and ensure the shelves to be stable, we should minimize the product of M_{xyz} and layer z, therefore we get the follow function:

$$minf_2(x, y, z) = \frac{\sum_{x=1}^{a} \sum_{z=1}^{c} Z \cdot L_0 \cdot M_{xyz}}{\sum_{x=1}^{a} \sum_{z=1}^{c} M_{xyz}} \tag{3}$$

In formula 3, M_{xyz} means the mass of the goods stored at (x, y, z), L_0 means the length of each goods cell.

We also need to consider the stability of the level of a separate shelf. The weight at both ends of the shelf should remain balanced. The location of the center should be in the horizontal *direction/2b* (Fig. 1).

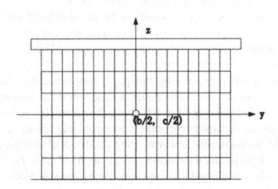

Fig. 1. Center map of a single row

Like principles in vertical direction, we get functions:

$$minf_3(x, y, z) = \frac{\sum_{x=1}^{a} \sum_{z=1}^{c} (y - \frac{1}{2}b)^2 \cdot L_0 \cdot M_{xyz}}{\sum_{x=1}^{a} \sum_{z=1}^{c} M_{xyz}} \tag{4}$$

Constraint Conditions. According to hypothetical general warehouse model, there are a row of shelves in reservoir storage with b columns and c layers for each shelf [7]. We use x, y and z to represent row, column and layer respectively. Then, we get the space restrain conditions of slotting optimization:

$$1 \le x \le a, 1 \le y \le b, 1 \le z \le c \tag{5}$$

4 Solution of Layout Optimization Model Based on Genetic Algorithm

4.1 Fitness Function Determination and the Operator Designation

Fitness Function. Genetic algorithm uses fitness to measure how close of each individual could be to the optimal solution. Our fitness function is:

$$
\begin{cases}
minf_1(x, y, z) = \dfrac{1}{\sum_{x=1}^{a} \sum_{y=1}^{b} \sum_{z=1}^{c} \left(\frac{x}{V_x} + \frac{y}{V_y} + \frac{z}{V_z} \right) \cdot L_0 \cdot P_i + 1} \\
minf_2(x, y, z) = \dfrac{1}{\frac{\sum_{x=1}^{a} \sum_{z=1}^{c} Z \cdot L_0 \cdot M_{xyz}}{\sum_{x=1}^{a} \sum_{z=1}^{c} M_{xyz}} + 1} \\
minf_3(x, y, z) = \dfrac{1}{\frac{\sum_{x=1}^{a} \sum_{z=1}^{c} (y - \frac{1}{2}b)^2 \cdot L_0 \cdot M_{xyz}}{\sum_{x=1}^{a} \sum_{z=1}^{c} M_{xyz}} + 1}
\end{cases} \tag{6}
$$

Selecting Operators. When we choose $N/2$ pairs of male parents to do crossover operation, we can get $N/2$ weight vectors by formula (6), then to figure out the fitness of each individual by formula (4). Through the linear transformation we can get each individual's choice probability as follows:

$$P_5 = \frac{f(X) - f_{min}}{\sum_{R=1}^{N} (f(X) - f_{min})} \tag{7}$$

Designation of Crossover Operator. This paper uses the binary intersection operator. Parent individual $x_i^{(1,t)}$ and $x_i^{(2,t)}$ generate offspring $x_i^{(1,t+1)}$ and $x_i^{(2,t+1)}$. The process is: First choose a random number $u_i \in [0, 1)$, then figure out β_i according to formula (7):

$$\beta_i = \begin{cases} (2u_i)^{\left(\frac{1}{\eta\mu+1}\right)} - 1 & u < 0.5 \\ 1 - [2(1 - u_i)]^{\frac{1}{\eta\mu+1}} & u \ge 0.5 \end{cases} \tag{8}$$

Parent crossover to produce offspring:

$$
\begin{aligned}
x_i^{(1,t+1)} &= 0.5\left((1 + \beta_i)x_i^{(1,t)} + (1 - \beta_i)x_i^{(2,t)}\right) \\
x_i^{(2,t+1)} &= 0.5\left((1 - \beta_i)x_i^{(1,t)} + (1 + \beta_i)x_i^{(2,t)}\right)
\end{aligned} \tag{9}
$$

Assuming the crossover probability is P_c, then $100P_c\%$ of the individuals in the population is used to cross, the remained $100(1 - P_c)\%$ access to the new offspring populations directly.

Designation of Mutation Operator. We use the polynomial mutation operator in this paper. Parent individual $x_i^{(1,t)}$ generates offspring $x_i^{(1,t+1)}$. The process is: First choose a random number $u_i \in [0, 1)$, then figure out β_i

$$\beta_i = \begin{cases} (2u_i)^{\left(\frac{1}{\eta\mu+1}\right)} - 1 & u < 0.5 \\ 1 - [2(1 - u_i)]^{\frac{1}{\eta\mu+1}} & u \geq 0.5 \end{cases} \tag{10}$$

Parent crossover to produce offspring:

$$x_i^{(1,t+1)} = x_i^{(1,t)} + \beta_i \tag{11}$$

5 Case Simulation and Analysis

5.1 The Basic Parameter Settings and Data Entry

We set the optimized shelves with 4 rows, 8 columns and 4 layers. Average movement speed along the x, y, z directions are set as $Vx = Vy = 1\,\text{m/s}$, $Vz = 0.5\,\text{m/s}$, the length of cell L_0 equals to 1-m, initial population are 40 and the maximum iterations are 100. We select a third-party logistics storage center's warehouse in Class A cargo area as data samples. Entering the initial data into MATLAB according to Table 1. The initial cargo space coordinates of the 15 kinds of goods are: (1,4,2), (2,4,1), (3,2,1), (1,3,4), (3,1,4), (1,1,1), (2,4,2), (1,1,4), (3,2,2), (4,4,4), (2,2,3), (1,4,3), (3,3,4), (4,3,4), (4,4,2).

Table 1. Data of the goods to be optimized

Goods number	1	2	3	4	5	6	7	8	9	10	11	12	13	14	15
Turnover rate (%)	17	37	31	11	46	29	32	41	69	22	58	34	10	58	24
Mass (%)	18	13	2	43	21	3	21	1	11	12	12	9	2	18	41

5.2 Analysis of Simulation Results and Model Validation

Analysis of Simulation Results. In order to ensure the accuracy of the simulation, we treat the turnover rate data and mass data as equal, for example, if turnover rate is 0.62, we use 62. Operating the solving program in MATLAB and its tracing result shows in Fig. 2:

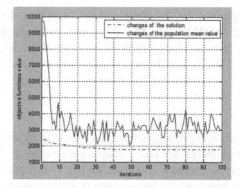

Fig. 2. Tracing of solution and populations means

Convert the solved chromosome into cargo space coordinates: (2,3,1), (3,2,1), (2,1,2), (1,2,2), (1,2,1), (1,3,1), (2,4,1), (2,1,1), (1,1,1), (1,1,2), (1,4,1), (2,2,1), (4,1,1), (3,1,1), (3,3,1). Compare the data before optimization with the new coordinates. We get Table 2:

Table 2. Three objective function values

Function value	Before	After	Reduce	Reduce rate (%)
First	59.51	32.02	27.49	46.19391699
Second	2.9765	1.2156	1.7609	59.16008735
Third	0.7851	0.4365	0.3486	44.40198701

From the table above, we could find that: our three optimization goals have been greatly improved. They have been reduced 46.19%, 9.16%, 4.40% respectively.

Model Validation. We use MATLAB to simulate cargo space allocation status with three-dimensional simulation diagram (Fig. 3).

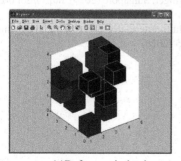
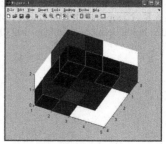

    (a)Before optimization   (b)After optimization

Fig. 3. D simulation of position distribution based on turnover rate

Obviously, after optimization, distribution of the cargo is as close as possible to the warehouse entrances and shelf low-rise, the layout of the cargo space allocation becomes order and reasonable. We use black to represent the goods of which the $Sk \geq 0.04$, similarly, $0.2 \leq Sk \leq 0.04$, $Sk < 0.2$, we can mark them blue and red respectively.

6 Conclusion

This paper uses multi-objective hybrid genetic algorithm theory and MATLAB software to optimize, emulate and analyze the layout problem at an ordinary warehouse. It finally improves the current layout problem. Although this paper takes goods' turnover rate and shelves' gravity as optimization principles, in practical, we may choose other principles such as goods relevance, "first in first out" and so on. Under these circumstances, we can add a target function and don't need to change the solving method. So, this proposed optimization algorithm has extensive adaptability.

References

1. Yue, L., Guan, Z., He, C., et al.: Slotting optimization of automated storage and retrieval system (AS/RS) for efficient delivery of parts in an assembly shop using genetic algorithm: a case Study. In: IOP Conference Series-Materials Science and Engineering, vol. 215 (2017)
2. Liu, L.Q., Yang, W.W.: Study on the warehousing slotting optimization based on improved adaptive genetic algorithm. Comput. Era **2018**(08), 57–60 (2018)
3. Bao, S., Zhang, M., Cai, Z.: The slotting optimization of tobacco automated stereoscopic warehouse based on fault tree and field test. In: 2017 2nd IEEE International Conference on Intelligent Transportation Engineering (ICITE), pp. 346–351. IEEE (2017)
4. Sutaria, U., Shah, V., Shah, K., Shah, C., Khatawate, V.: Optimization of brake rotor slotting using finite element analysis. In: Vasudevan, H., Kottur, V.K.N., Raina, A.A. (eds.) Proceedings of International Conference on Intelligent Manufacturing and Automation. LNME, pp. 803–813. Springer, Singapore (2020). https://doi.org/10.1007/978-981-15-4485-9_78
5. Li, Z., Xu, K., Tang, K.: A new trochoidal pattern for slotting operation. Int. J. Adv. Manuf. Technol. **102**(5–8), 1153–1163 (2019)
6. Zhang, W., Zhu, J., Yuan, R.: Optimization of automated warehouse storage location assignment problem based on improved genetic algorithm. In: Zhang, J., Dresner, M., Zhang, R., Hua, G., Shang, X. (eds.) LISS2019, pp. 297–311. Springer, Singapore (2020). https://doi.org/10.1007/978-981-15-5682-1_22
7. Lang, M.A.K., Cleophas, C., Ehmke, J.F.: Multi-criteria decision making in dynamic slotting for attended home deliveries. Omega, 102305 (2020)

Magnetic Circuit Design of Galfenol Composite Cantilever

Qinghua Cao[1](✉), Jinming Zang[1], Nannan Wu[2], and Zhifeng Zhou[2]

[1] Jiangxi Province Key Laboratory of Precision Drive and Control, Nanchang, China
154280397@qq.com
[2] Nanchang Institute of Technology, Nanchang, China

Abstract. The magnetic circuit design is one of the most important factors in the design of giant magnetostrictive materials. In order to improve the uniformity of magnetic field, the composite cantilever beam is placed in the center of the hollow coil, the analytical solution of the driving magnetic field is solved, the basic parameters of the coil are determined, and the magnetic leakage reason is analyzed by finite element method. At the same time, the magnetic circuit design method is optimized by using finite element method. After optimization, the uniformity of magnetic induction intensity is improved, which provides reference for the magnetic circuit design of giant magnetostrictive material devices.

Keywords: Magnetostrictive · Galfenol composite cantilever · Magnetic circuit design · Finite element analysis

1 Introduction

The design of magnetic circuit is regarded as one of the most important factors in the design of devices based on giant magnetostrictive materials. In order to improve the uniformity of magnetic field, the magnetic circuit design of hollow solenoid coil is carried out with magnetostrictive material. The equivalent magnetic circuit method in electrical theory is used to solve the problem of magnetic circuit. In the same time, combining with the finite element simulation analysis, the magnetic circuit structure design criterion of uniform magnetic field is further established.

The magnetic circuit structure of Galfenol composite cantilever is shown in Fig. 1, which is mainly composed of Galfenol composite cantilever, driving coil, fixed end, framework, shell and end cover. The Galfenol sheet works under the driving magnetic field generated by the electrified coil, so the structure and size parameters of the driving coil play a key role in the magnetic field uniformity and Galfenol operation, and also affect the energy transfer efficiency of the whole system. The driving magnetic field of Galfenol alloy sheet is produced by cylindrical coil. The main factors to be considered in the design are the design of magnetic circuit, the selection of materials for each part of driving coil and the determination of coil geometric parameters.

The maximum magnetic field intensity generated by cylindrical coil is within the inner diameter of coil. Galfenol alloy sheet can be placed in this area to improve energy

© Springer Nature Switzerland AG 2021
Q. Zu et al. (Eds.): HCC 2020, LNCS 12634, pp. 283–294, 2021.
https://doi.org/10.1007/978-3-030-70626-5_30

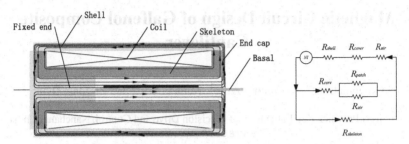

Fig. 1. Schematic diagram of driving structure and its equivalent magnetic circuit

conversion rate. The main components of driving structure include coil, coil framework, fixed end (placed in the framework to fix one end of cantilever beam), shell and end cover. In order to make the composite cantilever beam have better magneto mechanical conversion efficiency, the main components of driving structure are coil, coil framework, fixed end (placed in the framework to fix one end of cantilever beam), shell and end cover, The following principles should be followed in the design of magnetic circuit: the selection of each component of the magnetic circuit of composite cantilever beam should be as reasonable as possible to make the magnetic flux in the magnetic circuit as large as possible, and the magnetic line of force of the driving sheet through Galfenol should be as much as possible; the magnetic leakage should be reduced to maximize the magnetic flux of the driving sheet.

The equivalent magnetic resistance R of the whole magnetic circuit is as follows:

$$\frac{1}{R} = \frac{1}{R_{shell} + R_{cover} + R_{air}} + \frac{1}{R_{core} + \frac{1}{\frac{1}{R_{patch}} + \frac{1}{R_{air}}}} + \frac{1}{R_{skeleton}} \tag{1}$$

The total magnetic flux of equivalent magnetic circuit can be expressed as follows:

$$\varphi = \frac{NI}{R} \tag{2}$$

In the process of magnetic circuit design of Galfenol composite cantilever beam, the determined factors that can not be changed are Galfenol material and size, In order to increase the total magnetic flux of equivalent circuit From Eq. (2), the case magnetoresistance is decreased, the fixed end magnetoresistance and so on, the skeleton magnetic resistance is increased, so 45 steel is used as the shell, end cover and fixed end material, and the coil framework is made of non-magnetic material copper; at the same time, the 45 steel copper framework can not only increase the total magnetic flux of the circuit, but also reduce the magnetic resistance of the fixed end and the branch of the valve plate and increase the magnetic resistance of the parallel skeleton branch, so as to increase the magnetic field passing through the cantilever beam and improve the utilization of the magnetic field. Since the relative permeability of Galfenol alloy material is 60–100, the distribution of magnetic field in the air domain can be reduced by reducing the volume of air domain around the cantilever beam, resulting in the increase of magnetic flux through the cantilever beam. Thus, the driving structure and the materials of each part of the cantilever beam have been shown as Table 1.

Table 1. The materials of driving structure

Name	Materials
Shell	Steel
End cap	Steel
Skeleton	Copper
Basal	Beryllium bronze
Driving source	Galfenol

2 Analysis of Magnetic Field

Firstly, the axisymmetric solenoid coil is used to provide magnetic field for Galfenol material. In reality, the hollow solenoid coil should be wound close to each other, so each turn of excitation coil has complete and independent helicity. Because the conductor is enameled wire, the coil helicity and current density inhomogeneity of magnetized hollow solenoid are ignored regardless of the shape of the wire cross section.

For the axisymmetric solenoid coil, the cylindrical coordinate system can be used to analyze and solve various physical quantities. At this time, there is no tangential component in the magnetic field, and its radial component is an odd function of the spatial coordinate Z, while the axial component is its even function. For a constant magnetic field, $\nabla \cdot B = 0$. However, for the magnetic vector potential, $B = \nabla \cdot A$, so the analytical solution of magnetic field B can be obtained by solving A. First of all, the single turn coil can be analyzed. Since the radius of the wire of each single turn coil is far less than that of the solenoid coil, it can be treated as an ideal situation in which the current and coil radius are known and the wire radius is infinitesimal.

Fig. 2. Schematic diagram of ring coil

Suppose that the radius of the ring is R, the current is I, and its plane is parallel to the rectangular coordinate plane xoy. As shown in Fig. 2, the center of the ring is located at. According to Biot Savart law, for any point in the magnetic field, the vector magnetic potential generated by the field source is as follows:

$$A(P) = \frac{\mu_0}{4\pi} \oint_l \frac{dl(Q)}{a} \tag{3}$$

Where, l is the loop circuit, dl is the vector differential of tangent direction at any point $Q(x',y',z')$ on each single turn wire of solenoid coil, and its direction is the same as that of the current at that point a, which is vector of the point Q to point P.

$$dl = dx'i + dy'j + dz'k \qquad (4)$$

$$R = \left[(x - x')^2 + (y - y')^2 + (z - z')^2 \right]^{\frac{1}{2}} \qquad (5)$$

The vector magnetic potential of point P can be expressed as follows:

$$A(x, y, z) = \frac{\mu_0 I}{4\pi} \oint_l \frac{dx'i + dy'j + dz'k}{\left[(x - x')^2 + (y - y')^2 + (z - z')^2 \right]^{\frac{1}{2}}} \qquad (6)$$

The Eq. (6) can be expressed in cylindrical coordinates as follows:

$$A(\rho, \phi, z) = \frac{\mu_0 I R}{4\pi} \int_0^{2\pi} \frac{[\sin(\phi - \phi')e_\rho + \cos(\phi - \phi')e_\phi]d\phi'}{\left[\rho^2 + R^2 - 2\rho R \cos(\phi - \phi') + (z - r)^2\right]^{\frac{1}{2}}} \qquad (7)$$

Where e_ρ is the radial unit vector of point P and e_ϕ is the tangential unit vector of point P.

Let $\theta = \phi - \phi'$, then Eq. (7) can be simplified as follows:

$$A(\rho, \phi, z) = \frac{\mu_0 I R}{4\pi} \int_{-\pi}^{\pi} \frac{\sin\theta d\theta}{\left[\rho^2 + R^2 - 2\rho R \cos\theta + (z - r)^2\right]^{\frac{1}{2}}} e_\rho$$
$$+ \frac{\mu_0 I R}{4\pi} \int_0^{2\pi} \frac{\cos\theta d\theta}{\left[\rho^2 + R^2 - 2\rho R \cos\theta + (z - r)^2\right]^{\frac{1}{2}}} e_\phi \qquad (8)$$

The integrand in the first integral on the right of Eq. (8) is an odd function of θ, so its value is zero. The results are as follows:

$$A(\rho, \phi, z) = \frac{\mu_0 I R}{4\pi} \int_0^{2\pi} \frac{\cos\theta d\theta}{\left[\rho^2 + R^2 - 2\rho R \cos\theta + (z - r)^2\right]^{\frac{1}{2}}} e_\phi \qquad (9)$$

It is well known that the vector magnetic potential has only component A_φ in tangential direction from Eq. (9), and other components are zero, and A_φ has no relationship with coordinate tangential component φ. At the same time, the above conclusion can be satisfied not only for circular coil, but also for axisymmetric solenoid coil. Let the magnetic induction intensity r from the center of the circle be B_r, which can be expressed as follows:

$$B_r = \frac{\mu_0 I R^2}{2(r^2 + R^2)^{\frac{3}{2}}} \qquad (10)$$

If the magnetic induction intensity from the center of the circle is r, it can be expressed as: therefore, if the inner radius, the outer radius, the length and the current of the axisymmetric solenoid coil are set as, l and I, respectively, the magnetic induction intensity at the distance from the axis r on the axis can be obtained by integrating the solenoid coil on the basis of the ring coil B. When the Galfenol composite cantilever is placed in an axisymmetric solenoid, the permeability of Galfenol material is greater than that of air, which leads to the change of magnetic induction. Ignoring the reaction of Galfenol eddy current and demagnetizing field on the driving magnetic field, and assuming the relative permeability of Galfenol material as constant, the magnetic induction intensity of coil axis after adding Galfenol composite cantilever beam can be expressed as follows:

$$B = \frac{1}{2}\mu\mu_0 nI\{\left(r + \frac{L}{2}\right) \ln \frac{R_2 + \left[R_2^2 + (r + L/2)^2\right]^{1/2}}{R_1 + \left[R_1^2 + (r + L/2)^2\right]^{1/2}} - \left(r - \frac{L}{2}\right) \ln \frac{R_2 + \left[R_2^2 + (r - L/2)^2\right]^{1/2}}{R_1 + \left[R_1^2 + (r - L/2)^2\right]^{1/2}}\} \tag{11}$$

Equation (11) only preliminarily estimates the solenoid driving coil [12], and the experimental results show that the magnetic field of the solenoid coil at the Galfenol composite cantilever beam is small, which is caused by The magnetic route of Galfenol composite cantilever beam is composed of different conductive materials (Galfenol material, air and high permeability material). Equation (3) caused by biotsavart's law is not fully applicable. Therefore, it is necessary to modify the parameters of solenoid coil after using Eq. (11) for initial calculation.

3 Basic Parameters of the Coil

The purpose of determining the coil parameters is to drive the Galfenol alloy to the maximum deformation. The magnetic field intensity on the axis line of the hollow cylinder coil is the largest, so the Galfenol alloy is driven by the exciting magnetic field The position of the alloy should be located at the axis line as far as possible. From the previous analysis, reducing the air area around the Galfenol alloy sheet can increase its magnetic field strength, so the inner diameter of the framework should not be too large. However, under the magnetic field driving, the cantilever beam will bend and deform, enough bending deformation range is reserved, and The inner diameter of coil frame is selected as 20 mm. According to Eq. (11), relations Key information is as follows: the magnetic field intensity reaches the maximum at the center of the coil; the magnetic field intensity and uniformity decrease significantly on both sides of the coil. Therefore, a more reliable way to improve the efficiency of magnetic field utilization is to place the Galfenol alloy sheet at the center of the axis of the hollow cylinder coil, and increase the length of the hollow cylinder coil to ensure the uniformity of magnetic induction intensity at the center line.

When designing the structure size of the hollow cylindrical coil, it is necessary to leave space for the skeleton, and the position of its bonding Galfenol alloy sheet is in the center of the hollow cylindrical coil. Finally, the appropriate coil height is selected according to the material technology, the required magnetic field performance and the materials provided by the market.

Driven by the magnetic field generated by electrifying, Galfenol alloy sheet will bend, at the same time, Joule heat will also be generated. As time goes on, the accuracy of cantilever beam will be affected, so it is very important to determine the coil diameter.

In order to ensure the Galfenol chip to work under the condition of long time and high precision, the heat generation power W of the coil is as follows:

$$W = I^2 R \tag{12}$$

It can be seen from Eq. (12) that the heat can be controlled by reducing the input current I or reducing the resistance of the coil, and the control effect of the current on the thermal power is more obvious, but the magnetic flux ϕ of the magnetic circuit generated by the low current is also reduced. Under the condition that the magnetic potential energy in the magnetic circuit is constant, the coil turns N should be increased when the input current is reduced, so the copper wire with smaller wire diameter is used The resistivity of enameled wire with different wire diameter is shown in Table 2.

Based on the above considerations, the wire diameter ($d = 1$ mm) is selected as the winding scheme of the coil, and the number of turns N of the coil can be calculated as follows:

$$N = \frac{(R_2 - R_1) \cdot L}{d} \tag{13}$$

The number of turns of the coil is 2025.

Table 2. Resistivity of enameled wire with different wire diameter

Diameter d (mm)	Resistance (Ω/m)			Diameter d (mm)	Resistance (Ω/m)		
	Minimum	Nominal	Maximum		Minimum	Nominal	Maximum
0.70	0.04317	0.04442	0.04571	0.95	0.02342	0.02412	0.02484
0.75	0.03756	0.03869	0.03987	1.00	0.02116	0.02176	0.02240
0.80	0.03305	0.03401	0.03500	1.10	0.01748	0.01799	0.01851
0.85	0.02924	0.03012	0.03104	1.20	0.01467	0.01511	0.01558
0.90	0.02612	0.02687	0.02765	1.30	0.01250	0.01288	0.01327

4   Finite Element Analysis of Magnetic Circuit

The methods to improve the magnetic field efficiency of Galfenol thin plate is summed up, and preliminarily determines the material and geometric size of each part of the cantilever structure. However, there are many factors influencing the actual model. Different geometric dimensions affect the magnetic flux and magnetic leakage of the magnetic circuit. Single magnetic theory is difficult to meet the requirements of complex design. Large finite element analysis software provides a convenient and effective calculation method, which is more and more used in the calculation of magnetostrictive devices.

According to the geometric model size determined previously, a two-dimensional axisymmetric finite element model is established, as shown in Fig. 3(a), and the properties of the selected material are assigned to the corresponding region of the model. After the current density of the area where the coil is located is loaded, the mesh is divided as shown in Fig. 3(b), and then the solution is carried out.

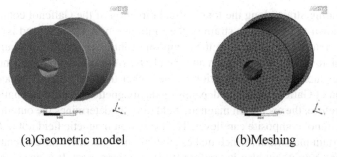

| (a)Geometric model | (b)Meshing |

Fig. 3. Magnetic field simulation process of Galfenol composite cantilever beam

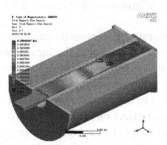

Fig. 4. Distribution of magnetic lines of force   **Fig. 5.** Magnetic induction intensity.

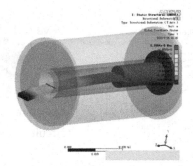

Fig. 6. Strain diagram of cantilever beam driven by current

It can be seen from Figs. 4 and 5 that the magnetic flux leakage at the opening of the end cover is relatively large, and the influence of the fixed end on the magnetic line of force is relatively large. However, the end cover of the driving structure needs to be opened to facilitate the cantilever beam test. It can be seen from Fig. 4 that the position with the maximum magnetic induction intensity appears at the side near the fixed end of the cantilever. As the magnetic field intensity far away from the fixed end decreases rapidly, the magnetic induction intensity of the free end of the cantilever beam is very small, and the magnetic field intensity of the Galfenol sheet basically meets the best driving range of the Galfenol material.

The bending strain along the longitudinal direction of the Galfenol composite cantilever is shown in Fig. 6. The strain of the cantilever near the fixed end is the largest. Since the strain of the Galfenol in the composite cantilever is solved by superposition of the magnetostrictive deformation and the elastic strain caused by the bending of the beam, the greater the magnetostriction is, the higher the total strain is. Since the magnetostriction of Galfenol material depends on the magnetic induction strength, the same is true Therefore, the excitation magnetic field directly determines the output characteristics of Galfenol composite cantilever. The excitation magnetic field not only needs to satisfy a certain magnitude, which makes the Galfenol cantilever have a relatively large displacement output, but also its uniformity is very important. It is necessary to avoid damage or aging of Galfenol alloy due to uneven internal excitation magnetic field in long-term high-strength work, and Galfenol is in uniform driving magnetic field, which is conducive to accurate control of displacement.

In recent years, it has become a trend to separate the composite cantilever beam from the coil for magnetic circuit design, which is conducive to the measurement of the structure and easy to be miniaturized. According to the above basic viewpoints, reference 59 adopts the magnetic circuit structure as shown in Fig. 7(a), and suryarghya in reference 117 Chakrabarti adopts the magnetic circuit structure shown in 7(b). The former adopts the laminated form, which has good effect on restraining eddy current loss, while the latter uses block material, which has large error at high frequency. In this paper, combined with the above two magnetic circuit design methods, another composite cantilever beam structure is designed, as shown in Fig. 8.

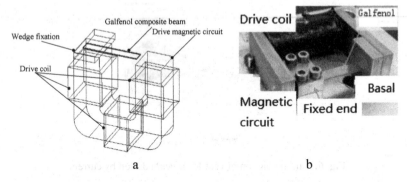

Fig. 7. Magnetic circuit structure of Galfenol composite cantilever beam

Figure 9 shows the grid division diagram, and Fig. 10 shows the magnetic induction intensity distribution map of the structure. It can be seen from the figure that the distribution of magnetic induction intensity is not ideal due to the large gap in the middle, so the magnetic circuit needs to be further optimized.

A ferrite rod with high permeability is added to the right side of the Galfenol composite cantilever beam to make the magnetic line of the whole magnetic circuit structure pass through the Galfenol sheet of the composite cantilever beam as far as possible. Figure 11 shows the geometric model of the optimized driving structure, and Fig. 12 and

Fig. 8. Schematic diagram of magnetic drive structure designed

Fig. 9. Meshing diagram of driving structure with cantilever beam separated from coil

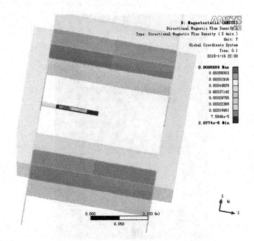

Fig. 10. Distribution of magnetic induction intensity of driving structure

Fig. 11. Geometry model of optimized driving structure

Fig. 13 show the distribution of magnetic field intensity and magnetic induction intensity distribution after optimization, and the uniformity of magnetic induction intensity
is significantly improved after optimization.

Fig. 12. Distribution of optimized magnetic lines of force

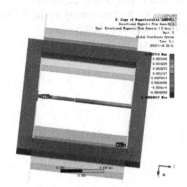

Fig. 13. Optimized magnetic induction distribution

5 Conclusion

In the design process of the magnetic circuit of Galfenol composite cantilever prototype, firstly, the composite cantilever is placed in the center of the air core coil, the analytical solution of the driving magnetic field is solved, the basic parameters of the coil are determined, and the driving structure is analyzed by finite element method. Then, combined with the magnetic circuit design in the literature, the composite cantilever beam and the coil are separated to design a new magnetic circuit The advantages and disadvantages of the magnetic circuit are analyzed. At the same time, the magnetic circuit design method is optimized by using finite element method, and the uniformity of magnetic induction intensity is obviously improved after optimization.

Acknowledgements. The authors would like to acknowledge the financial support by Youth Science Fund of Jiangxi province office of education of China (Grant No. GJJ161102) and Key Laboratory Open Fund of precision drive and control (Grant No. KFKT-201610).

References

1. Tu, J., Liu, Z., Li, Z.: Magnetic circuit optimization design and finite element analysis of giant magnetostrictive actuator. J. Chongqing Univ. **3**(20) (2020)

2. Hailong, C.A.O., Shijian, Z.H.U., Jingjun, L.O.U., et al.: Magnetic circuit design and simulation analysis of giant magnetostrictive actuator. Ship Sci. Technol. **37**(6), 109–113 (2015)
3. Hongbo, Y.A.N., Hong, G.A.O., Hongbo, H.A.O., et al.: Design and simulation of exciting coil in rare earth giant magnetostrictive actuator. Mech. Sci. Technol. Aerosp. Eng. **38**(10), 1569–1575 (2019)
4. Xiaohui, G.A.O., Yongguang, L.I.U., Zhongcai, P.E.I.: Optimization and design for magnetic circuit in giant magnetostrictive actuator. J. Harbin Inst. Technol. **48**(9), 145–150 (2016)
5. Atulasimha, J., Flatau, A.B.: A review of magnetostrictive iron–gallium alloys. Smart Mater. Struct. **20**, 1–15 (2011)
6. Ueno, T., Higuchi, T.: Miniature magnetostrictive linear actuator based on smooth impact drive mechanism. Int. J. Appl. Electromagn. Mech. **28**, 135–141 (2012)
7. Shu, L., Dapino, M., Evans, P., Chen, D., Lu, Q.: Optimization and dynamic modeling of galfenol unimorphs. J. Intell. Mater. Syst. Struct. **22**(8), 781–793 (2011)
8. Smith, R.C., Dapino, M.J., Seelecke, S.: Free energy model for hysteresis in magnetostrictive transducers. J. Appl. Phys. **93**(1), 458–466 (2003)

A MEMS-Based Piezoelectric Pump with a Low Frequency and High Flow

Dehui Liu[1](✉), Zhen Wang[2], Liang Huang[2], Shaojie Wu[2], and Dandan Yuan[2]

[1] GongQing Institute of Science and Technology, Jiujiang, China
dehuiliu@163.com
[2] Jiangxi Province Key Laboratory of Precision Drive and Control, Nanchang Institute of Technology, Nanchang 330099, China

Abstract. This paper mainly presents a piezoelectric pump based on a single piezoelectric vibrator under MEMS technology. The piezoelectric vibrator of the piezoelectric pump is made by MEMS technology, and the pump body is completed by 3D printing. Driven by a low-frequency AC voltage of 220 V 50 Hz, the center displacement of the piezoelectric vibrator of the pump can reach 136 μm with high output characteristic. Finally, the piezoelectric pump was used to perform a cooling test on the heating sheet PTC under the condition of 26 °C. It was found that the temperature of the PTC could be reduced by about 20 °C under the condition of the maximum output flow of 159 ml/min. Therefore, the piezoelectric pump has important practical significance in many aspects such as micro flow control and computer CPU cooling.

Keywords: Piezoelectric pump · MEMS · Piezoelectric vibrator

1 Introduction

With the gradual deepening of piezoelectric research, the use of piezoelectric elements as piezoelectric driving mechanisms for micro-pumps has developed rapidly [1–5]. When the inverse piezoelectric effect of the piezoelectric element is used, the piezoelectric material can realize the conversion from electrical energy to mechanical energy without a transmission mechanism [6]. In addition, the driving mechanism has a fast response speed and is easy to drive a small mechanism. Therefore, the application research of piezoelectric elements in micro and small machinery and precision machinery has become a current research hotspot.

In this paper, it proposes a piezoelectric pump driven by a single piezoelectric vibrator based on MEMS technology. The pump combines PZT and beryllium bronze with conductive adhesive in a concentric circle to form a piezoelectric vibrator. The piezoelectric pump has a good output by inputting alternating current with a suitable frequency and has a certain practical value.

© Springer Nature Switzerland AG 2021
Q. Zu et al. (Eds.): HCC 2020, LNCS 12634, pp. 295–301, 2021.
https://doi.org/10.1007/978-3-030-70626-5_31

2 Structure and Principle

The structure diagram of the piezoelectric pump designed in this paper is shown in Fig. 1(a). It can be seen from the figure that the piezoelectric pump is mainly composed of a pump body, a pump cover, a piezoelectric vibrator, an inlet and outlet valve, a fastening screw, and a diaphragm. The pump cover and the pump body are clamped with fastening screws to play a role of fixing the piezoelectric vibrator. At the same time, in order to prevent the water in the pump cavity from leaking to the surface of the piezoelectric vibrator when the piezoelectric pump is running, the bottom surface of the piezoelectric vibrator is added a waterproof diaphragm plays a sealing role. The inlet valve and outlet valve in Fig. 1(a) communicate with the outside water inlet and outlet, and its function is to form a periodic enclosed space together with the piezoelectric vibrator to complete the pumping process.

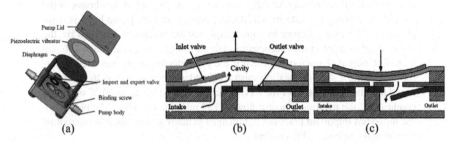

Fig. 1. Piezoelectric pump principle structure diagram (a) structure diagram (b) piezoelectric pump suction (c) piezoelectric pump drainage.

As shown in Fig. 1(b)(c), it is a schematic diagram of the device structure. The working principle of the device is that under the action of alternating voltage, the piezoelectric vibrator is bent up and down to form a cavity volume and pressure changes in conjunction with the opening and closing of the one-way valve, thereby realizing one-way flow of fluid in the whole process. When the piezoelectric vibrator is bent upward and deformed, the volume of the cavity increases and the pressure is lower than the atmospheric pressure, the inlet valve is pushed open, and the liquid is sucked into the cavity, as shown in Fig. 1(b). When the piezoelectric vibrator bends and deforms downward, the piezoelectric vibrator presses the liquid to open the outlet valve, the inlet valve is in a closed state and the liquid is discharged from the outlet, as shown in Fig. 1(c).

3 Simulation Analysis

The piezoelectric pump can work normally depends on normally bending and deforming of the piezoelectric vibrator. Therefore, this paper uses ANSYS software to simulate and analyze the natural frequency and vibration mode of the piezoelectric vibrator, and explore the effective working frequency of the piezoelectric vibrator. The modeling parameters of the piezoelectric vibrator are shown in Table 1. Figure 2(a)(b) shows the first few modes of the piezoelectric vibrator, It can be clearly seen from the figure that

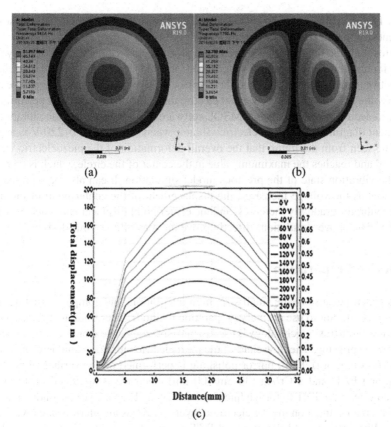

Fig. 2. Piezoelectric vibrator simulation results (a) first-order mode (b) second-order mode (c) The relationship between voltage and center displacement of piezoelectric vibrator.

Table 1. Material parameters of piezoelectric vibrator.

Material name	Diameter/mm	Thickness/mm	Density/kg.m	Elastic Modulus/GPa	Poisson's ratio
PZT	25	0.1	7600	63	0.32
Beryllium Bronze	35	0.1	8920	118	0.35

the deflection of the first-order mode shape gradually and uniformly increases from the edge of the piezoelectric vibrator to the center of the circle. The other vibration modes are irregularly and unevenly distributed. According to the working principle of the piezoelectric pump, in order to maximize the output flow, the cavity generated by the disturbance should also be the largest. Therefore, only in the first-order vibration mode, the piezoelectric vibrator can produce the largest volume change in the pump cavity. Therefore, the frequency of the selected AC voltage is smaller than the first-order frequency of the piezoelectric vibrator, which is 842 Hz.

After studying the vibration shape of the piezoelectric vibrator, the frequency range of the input alternating current is determined. In order to further explore the relationship between different voltages and the displacement of the center of the piezoelectric vibrator, the coupling analysis of piezoelectric and solid mechanics was performed on the piezoelectric vibrator using COMSOL software. Figure 2(c)shows the vibration displacement diagram of the piezoelectric vibrator when a voltage of 0–220 V is applied. It can be seen from the figure that the overall deformation of the piezoelectric vibrator is arched and reaches the maximum value at the center of the circle, which is consistent with the vibration state of the previous model simulation. It can also be seen from the figure that as the voltage increases, the displacement of the center point of the piezoelectric vibrator gradually increases, and the center point displacement reaches 181 μm at 220 V. The above simulations directly prove the feasibility of this design.

4 Device Preparation

In this paper, because of the small size, high integration, and complex structure of the device, it is difficult to use ordinary preparation methods. Therefore, in order to prepare better devices, MEMS technology is used to make piezoelectric vibrators, and the shell is made by 3D printing. The flow chart of the piezoelectric vibrator manufacturing process is shown in Fig. 3(a). First, mirror polishing is performed on the surface of beryllium bronze and PZT, and a Cr/Ag (30 nm/400 nm) electrode is sputtered on the mirror polished side of the PZT to form a fine electrode layer. Then a layer of conductive silver paste is screen printed on the Ag electrode layer of PZT as an intermediate layer. When the beryllium bronze and the prepared PZT are bonded together in a vacuum drying oven, the bonding pressure is 0.1 MPa, and the temperature rises from 60 °C to 175 °C, with a 15 °C gradient every 10 min, and heat preservation at 175 °C more than 1 h. After bonding, the thickness of the PZT is reduced to about 100 μm by mechanical grinding and polishing is performed. Then the Cr/Ag layer as the upper electrode was sputtered onto the thin PZT layer, and finally the device was processed for lead wires [7]. As shown in Fig. 3(b), it is the actual picture of the piezoelectric vibrator after fabrication.

The aforementioned shell is made by 3D printing. 3D printing adopts a layer-by-layer stacking method, so devices with more complex structures can be printed quickly and easily. The main steps of printing the shell are as following. First, the pump model is designed in Solidworks 3D software. Then it is saved in stl format and imported into Creality 3D software. The stl file is converted into G code and imported into the 3D printer. Then the printer is started to run, the printing material PLA is melted and extruded from the printer nozzle according to the system requirements, and deposited in the designated position, after a period of time, the shell is printed and formed [8]. The printed physical map is shown in Fig. 3(c). After the shell is printed and formed, it is assembled according to the working principle described above, and the physical picture after assembly is shown in Fig. 3(d).

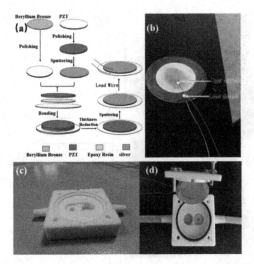

Fig. 3. Production of piezoelectric pump (a) Production flow chart (b) Physical map (c) Pump body (d) Assembly exploded view.

5   Experimental Test and Analysis

After the device is assembled, the prerequisite to ensure that the device can work normally is that the piezoelectric vibrator can have enough displacement at the external AC voltage input. Therefore, this article first built a test platform to test the vibration displacement of the piezoelectric vibrator, as shown in Fig. 4(a). Under a series of AC voltage input with a frequency of 50 Hz, the voltage displacement graph shown in Fig. 4(b)is obtained. It can be seen from the figure that the maximum displacement of the center of the piezoelectric vibrator has a linear relationship with the input voltage, which increases with the increase of the voltage, which is consistent with the previous simulation experiment. And the maximum center displacement is 136 μm under the input voltage of 220 V, which shows that the device has certain actual working ability.

In order to prove the practicability of the piezoelectric pump, this paper carried out an experiment using the piezoelectric pump to pump water to cool the PTC heating element. Figure 4(c) shows the experimental graph, and the experimental data are shown in Fig. 4(d). In general, the temperature of the PTC heating element is reduced by about 20 °C on average and the maximum output flow reached 159 ml/min. This proves the practicability and large flow output characteristics of the piezoelectric pump.

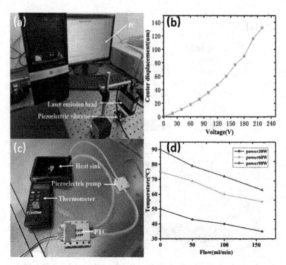

Fig. 4. Test platform and results (a) Center displacement test platform (b) Curve of between voltage and center displacement(c) Cooling PTC experiment (d) Relationship between flow output and equilibrium temperature.

6   Conclusion

This article mainly describes a piezoelectric pump based on a single piezoelectric vibrator under MEMS technology. The piezoelectric vibrator in the piezoelectric pump is made by bonding PZT and beryllium bronze substrate in a concentric circle combined with MEMS technology, the whole shell is made in one piece using 3D printing, and then assembled manually. Under the excitation of 220 V 50 Hz low-frequency AC voltage, the measured maximum displacement of the center of the piezoelectric vibrator is 136 µm. Finally, the device was tested for the practicality of cooling PTC by pumping water at a room temperature of 26 °C, It was found that the device can reduce the temperature of the PTC by about 20 °C and the maximum output flow reached 159 ml/min under the same conditions.

Acknowledgments. This work was supported by the Science and Technology Project of Jiangxi Provincial Education Department (GJJ181425, GJJ180938).

References

1. Xu, J.W., Liu, Y.B., Shao, W.W., et al.: Optimization of a right-angle piezoelectric cantilever using auxiliary beams with different stiffness levels for vibration energy harvesting. Smart Mater. Struct. **21**(6), 065017 (2012)
2. Dong, J., Liu, C., Chen, Q., et al.: Design and experimental research of piezoelectric pump based on macro fiber composite, p. 112123. Phys., Sens. Actuators A (2020)
3. Lu, S., Yu, M., Qian, C., et al.: A quintuple-bimorph tenfold-chamber piezoelectric pump used in water-cooling system of electronic chip. IEEE Access **8**, 186691–186698 (2020)

4. Liu, H., Lee, C., Kobayashi, T., et al.: Piezoelectric MEMS-based wideband energy harvesting systems using a frequency-up-conversion cantilever stopper. Sens. Actuators, A **186**, 242–248 (2012)

5. Zhang, Q., Huang, J., Li, K., et al.: A flexible valve based piezoelectric pump for high viscosity liquid transportation (2020)

6. Zhao, X., Zhao, D., Wang, J., et al.: Research on inlet and outlet structure optimization to improve the performance of piezoelectric pump. Micromachines **11**(8), 735 (2020)

7. Tang, G., Liu, J.Q., Yang, B., et al.: Piezoelectric MEMS low-level vibration energy harvester with PMN-PT single crystal cantilever. Electron. Lett. **48**(13), 784–786 (2012)

8. Huang, L., Tang, G., Hu, M., et al.: Study on double piezoelectric layers driven pump based on MEMS and 3D printing. In: International Conference on Human Centered Computing, pp. 571–579. Springer, Cham (2018). https://doi.org/10.1007/978-3-030-15127-0_57

Micro-video Learning Resource Portrait and Its Application

Jinjiao Lin[1], Yanze Zhao[1], Tianqi Gao[1], Chunfang Liu[1], and Haitao Pu[2(✉)]

[1] Shandong University of Finance and Economics, Jinan, China
[2] Shandong University of Science and Technology, Jinan, China
pht@sdust.edu.cn

Abstract. The emergence of a large number of online learning platforms changes the learners' demands and learning styles, thus the society puts forward higher requirements for the personalization, intelligentization and adaptability of learning resource platforms. For large-scale, multi-source and fragmented micro-video learning resources and personalized education problems, based on micro-video online learning resources data, the paper studies the accurate, comprehensive and usable micro-video learning resources portrait method. And through the application of deep learning technology, it studies the theory and method of micro-video learning resource data analysis and personalized learning resource recommendation. It explores and forms the basic theories and methods of data-driven micro-video learning resources analysis to support the research of personalized education theories and methods.

Keywords: Micro-video · Learning resources · Resource portrait · Personalized recommendation

1 Introduction

Micro-video learning resources have the characteristics of multi-source, multi-dimensional and fragmentation. It can meet learners' ubiquitous, mobile and personalized learning characteristics and requirements in the age of intelligence. Especially because of the COVID-19 in 2020, micro-video learning resources online have attracted unprecedented attention. Massive micro-video learning resources promotes the teaching from "curriculum" to "knowledge point", and at the same time, the knowledge transfer has changed from the linear structure to the networked structure, and the traditional teaching methods and the recommendation of learning resources cannot fully meet the learning needs of learners. In addition, people's learning is based on knowledge points and its logical relationships, and learners' previous knowledge and experience will greatly affect the learning effect [1]. So it has great research significance to organize the existing micro-videos to explore the accurate, comprehensive and usable micro-video learning resources portrait method and personalized learning resource recommendation.

© Springer Nature Switzerland AG 2021
Q. Zu et al. (Eds.): HCC 2020, LNCS 12634, pp. 302–307, 2021.
https://doi.org/10.1007/978-3-030-70626-5_32

2 Related Work

2.1 Learning Resource Portrait

In China, the study of resource portrait and its application is the research focus for both pedagogy and computer science researchers. Professor Yu Shengquan proposed the framework of international standards for learning meta-level from the perspective of basic education [2]. Professor Yu Ping and Zhu Zhiting put forward the content shareability standard of open education resources [3]. Professor Yang Jiumin studied various interaction designs in videos from the perspective of learning effects of video resources [4]. These studies focus on video learning resources portrait and its applications in the foundation education. There is a lack of research on fine-grained and fragmented micro-video learning resources in higher education.

2.2 Micro-video Learning Resource Portrait

Micro-video learning resource portrait refers to the use of consistent concepts, relationships and properties to describe micro-video learning resources under certain technical specifications. Jiang et al. [5] proposed a multi-modal LDA model to mine the content portrait of video learning resources. Minxin et al. [6] used the existing classification relationships in text mining and domain ontology to find candidate keywords that can represent semantic relationships. Yang et al. [7] proposed an attention mechanism based on relation representation to extract the directed relation information among elementary mathematical knowledge points. These existing researches focus on text, they only extract the low-level features, and They don't extract the relationship between multi-source network knowledge.

2.3 Personalized Learning Resource Recommendation

At present, the existing personalized learning resources recommended method which can be roughly divided into the following types: based on collaborative filtering (CFB) [8], based on the content (CB) [9], based on sequence mining (SMB) [10], mixing method. These researches didn't fully consider the semantic part of learning resources and paid little attention to the logical structure and the systematization of learning resources.

Therefore, based on unsolved problems in the above studies, this paper explores the portrait and application of micro-video learning resources, and proposes a method to carry out learning resources portrait and personalized recommendation.

3 Portrait and Application Analysis of Micro-video Learning Resources

The main system framework of this paper is shown in Fig. 1, which mainly includes the micro-video learning resources portrait of and the personalized recommendation.

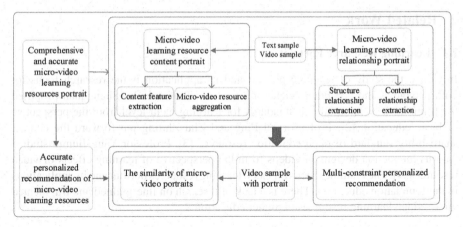

Fig. 1. System framework

3.1 Micro-video Learning Resource Portrait

Micro-video learning resource portrait mainly includes micro-video learning resource content portrait and micro-video learning resource relationship portrait.

1. Micro-video learning resource content portrait

Micro-video learning resource content portrait mainly includes the concept and properties of micro-video learning resources. When the content layer of micro-video learning resource is depicted, it is necessary to restore the source properties of the learning resource and label these properties. Specific as follows:

Firstly, we should extract the content feature of micro-video learning resource. Because of online learning resource covers all disciplines and fields and their content creators have different levels of knowledge, the same knowledge exist many different expressions, and it is not reality to determine the features of micro-video learning resources artificially. Therefore, we should study how to combine text, image and audio to mine the content features of micro-video learning resources. These features not only include low-level features such as keywords, but also contain a high-level feature, such as discipline, knowledge domain, knowledge unit, knowledge level, etc.

Secondly, we should aggregate micro-video learning resources. Different from basic education, which has standardized subject knowledge system, the knowledge system of higher education is open, the knowledge points are named according to their respective cultivation characteristics in higher education. Therefore, it is necessary to work out the domain knowledge point label system based on the above content features.

2. Micro-video learning resource relationship portrait.

It contains structural relationship and content relationship.

Firstly, we need to extract the structure relationship. The logical relationship between knowledge points may be different for different fields. A knowledge point may belong to a number of knowledge fields, and each knowledge field corresponds to a number of micro-videos. So the extraction of the micro-video learning resource relationship is a multi-dimensional problem. Therefore, we need to study how to combine text, image and voice data to mine relationship features. These features should not only include low-level features such as hierarchical relationship and association relationship, but also include high-level features such as co-reference and preorder.

Secondly, we need to extract the content relationship. Micro-video learning resources are based on the knowledge point granularity, it includes concept, principle, test questions and other types of content relations. Therefore, we need to study how to carry out transfer learning based on small sample data such as expert knowledge to accurately predict content relations.

3.2 Personalized Recommendation of Micro-video Learning Resources

It is implemented based on the above portrait and learner needs.

Firstly, the similarity of micro-video portraits is the basis of the recommendation algorithm. It has multi-dimensional characteristics, and the dimensions are not the same. Therefore, we need to study the measurement of the similarity of micro-video portraits.

Secondly, personalized recommend is based on micro-video portraits, and it is necessary to fully consider students' personalized learning needs and other constraints, such as the learner's professional background, previous knowledge, field experience, learning needs, learning objectives, and so on, so we need to study personalized micro-video recommendation under multiple constraints.

4 The Implementation of Micro-video Learning Resource Portrait and Application

Based on the problems that need to be solved, combined with the application analysis of current artificial intelligence and other technologies, this paper proposes the method of micro-video learning resource portrait and personalized recommendation system.

4.1 Micro-Video Learning Resource Portrait

The purpose of this paper is to study the iterative discovery method of the concepts of content layer and hidden properties in multiple fusion of text, image and audio. In this method, subjects, fields, knowledge level and relationships are taken as semantic annotation factors. This technology is an important technology to solve the problem of feature extraction of data-driven micro-video learning resources, and it is the basis of personalized guidance. According to the technical characteristics of deep learning, we think that a Convolutional Neural Networks (CNNs) data processing model can be adopted to solve this problem. As shown in Fig. 2, during the construction of a federation classifier for implicit properties, the system extracts the content features of multivariate learning resources data (such as text, image, audio, etc.), and combined with multivariate data fusion, the system extracts the common features of multivariate data as the important features of the classifier.

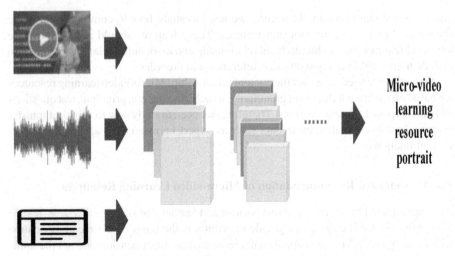

Fig. 2. Convolutional neural networks model

4.2   Personalized Recommendation System

This paper designed a micro-video learning resource portrait similarity method based on small sample. According to cognitive load theory, it provides appropriate methods to support the selection of micro-video learning resources. The collaborative filtering method can also be used to achieve the recommendation of micro-video learning resources, so it is necessary to calculate the portrait similarity of micro-video learning resources. In the definition of computational portrait similarity, we not only consider the content feature and structure feature, but also consider the timing factor of micro-video learning resources, and we use the latest change part of micro-video increment to calculate the result similarity. Different from other research similarities, in the field of education, whether the similarity of learning resources is accurate or not requires expert knowledge for final verification. Therefore, the sample micro-video data set needs to be reviewed online by corresponding experts and labeled as similar or not. Then, these labeled data are used as training sets to make accurate similarity prediction for micro-video learning resources. Multi-constraint personalized micro-video learning resource recommendation needs to consider the matching degree of students' personalized needs and micro-video learning resource portrait. According to the principle of homogeneity, we can match students who have similar personalized needs with micro-video learning resources which have similar portraits. Graph Convolutional Neural Network (GCN) is a neural network of learning graph structure, whose learning goal is to obtain the hidden state of graph perception of each node. We can take micro-video learning resources as nodes, and take their portrait as its characteristic value, then we can input this feature graph into graph convolution network for training and obtain corresponding similarity results.

5   Conclusion

This paper takes into account the disciplinary logic, domain and knowledge level of micro-video learning resource data, and proposes to use deep learning method to integrate multiple data such as text, image and audio to depict micro-video learning resources accurately. And it proposes a personalized recommendation method to calculate the similarity of micro-video learning resources by GCN. This paper explores and forms the basic theories and methods of data-driven micro-video learning resources analysis. In this paper, artificial intelligence technology is integrated into education, it provides a feasible way for micro-video learning resource portrait and its application.

Acknowledgment. Youth Innovative on Science and Technology Project of Shandong Province (2019RWF013), Postgraduate Education Reform Research Project of Shandong University of Finance and Economics (SCJY1911), Teaching Reform Research Project of Shandong University of Finance and Economics in 2020 (jy202011), Teaching Reform Research Project of Shandong Province (M2018X169, M2020283).

References

1. Zhi, C., Guangyuang, S., Xiuyun, P.: The knowledge point is the cognitive unit of the person. Psychol. Sci. **25**(3), 369–370 (2002)
2. Shengquan, Y., Qi, W., Yuntao, Y., et al.: Research on international standards proposal learning cell. China Educ. Technol. (11) 64–70 (2018)
3. Ping, Y., Zhiting, Z.: The study of content standards for open educational resources. Open Educ. Res. **20**(1), 111–120 (2014)
4. Yang Jiumin, X., Ke, H.J., et al.: The interaction effects of the type of cues and learners' prior knowledge on learning in video lectures. Modern Distance Educ. Res. **32**(01), 93–101 (2020)
5. Bian, J., Huang, M.L.: Semantic topic discovery for lecture video. In: Bi, Y., Bhatia, R., Kapoor, S. (eds.) Intelligent Systems and Applications, IntelliSys 2019, Advances in Intelligent Systems and Computing, vol. 1037, pp. 457–466. Springer, Cham (2020). https://doi.org/10.1007/978-3-030-29516-5_36
6. Shen, M., Liu, D.R., Huang, Y.S.: Extracting semantic relations to enrich domain ontologies. J. Intell. Inf. Syst. **39**(3), 749–761 (2012)
7. Dongming, Y., Dawei, Y., Hang, G., et al.: Research on knowledge point relationship extraction for elementary mathematics. J. East China Normal Univ. (Social Sciences) **2019**(5), 53–65 (2019)
8. Li, J., Chang, C., Yang, Z., Fu, H., Tang, Y.: Probability matrix factorization algorithm for course recommendation system fusing the influence of nearest neighbor users based on cloud model. In: Tang, Y., Zu, Q., Rodríguez García, J.G. (eds.) HCC 2018. LNCS, vol. 11354, pp. 488–496. Springer, Cham (2019). https://doi.org/10.1007/978-3-030-15127-0_49
9. Apaza, R.G., Cervantes, E.V., Quispe, L.C., et al.: Online courses recommendation based on LDA. In: Proceedings of the 1st Symposium on Information Management and Big Data, Cusco, Peru, 8–10 September, pp. 42–48 (2014)
10. Zhang, H., Huang, T., Lv, Z., Liu, S., Zhou, Z.: MCRS: a course recommendation system for MOOCs. Multimedia Tools Appl. **77**(6), 7051–7069 (2017). https://doi.org/10.1007/s11042-017-4620-2

Research on Downscaling and Correction of TRMM Data in Central China

Hanbo Zhang, Shiqing Dou$^{(\boxtimes)}$, Yong Xu, and Nan Zhang

School of Surveying and Mapping Geographic Information, Guilin University of Technology,
Guilin 541006, Guangxi, China
2120191109@glut.edu.cn

Abstract. The Central China has abundant rainfall, and rainfall is unevenly distributed in space and time, which can easily cause floods and soil erosion. It is of great significance to obtain precipitation information accurately and quickly. At present, remote sensing precipitation data has been widely used, but its spatial resolution and data accuracy still cannot meet actual application requirements. Therefore, this paper fully considers the applicability of TRMM 3B43 data from 2001 to 2019 in the Central China, based on a geographically weighted regression model, combined with NDVI, EVI, elevation, slope and aspect data. Different combinations were selected to downscale $TRMM_{EVI}$ data, and perform GDA and GRA corrections on the optimized TRMM data, and finally perform accuracy evaluation and result analysis on annual, quarterly, and monthly scales. The research results showed that: (1) The accuracy of the $TRMM_{EVI}$ data is better than the $TRMM_{NDVI}$ data when the spatial resolution is increased from 0.25° to 1 km, (2) The GDA correction result is more satisfactory than the GRA correction result and the data stability is better, so it is more suitable for TRMM data correction in the Central China. (3) The R^2 of $TRMM_{NDVI}^{GDA}$ data and the measured data of the site has high accuracy on the annual (0.91–0.986), quarter (0.704–0.88), and monthly (0.625–0.89) scales, and its detailed characteristics are better than TRMM data. (4) The better the downscaling and correction effect will be in the months with greater precipitation. Through downscaling and correction of TRMM data, it can better reflect the real precipitation information in the Central China, and provide reliable data support for agricultural production, optimal allocation of water resources, and flood prevention and disaster reduction.

Keywords: TRMM3B43 · GWR · Scale down · GDA · Central China

Precipitation is an important component of global surface material exchange, ecological succession, hydrological cycle and other processes [1, 2]. It is of great significance to monitor and forecast precipitation quickly and accurately.

At present, traditional regional precipitation is mostly obtained by spatial interpolation method based on observations from meteorological stations, but when using this method, it is prone to problems such as uneven point density and distribution, and the accuracy of interpolation results is difficult to guarantee the rapid development of satellite remote sensing technology provides a new method for large-area simultaneous precipitation prediction [3]. It has the characteristics of high temporal and spatial resolution, wide

© Springer Nature Switzerland AG 2021
Q. Zu et al. (Eds.): HCC 2020, LNCS 12634, pp. 308–318, 2021.
https://doi.org/10.1007/978-3-030-70626-5_33

coverage, and is not restricted by topographical conditions. The use of satellite remote sensing to detect precipitation has now become an important source for obtaining spatial precipitation data [4].

In recent years, a variety of remote sensing precipitation products have been produced successively on the global and regional scales. Among them, TRMM data products can provide relatively high spatial and temporal resolution precipitation data. However, in practical application, the spatial resolution of 0.25° is a little rough, which does not meet the requirements of regional spatial accuracy, and there is an overestimation when compared with the measured data at the site. Therefore, it is very necessary to conduct downscaling and correction research on TRMM precipitation data.

Scholars at home and abroad have obtained TRMM data by various methods. Li Qiong [5] used three downscaling methods: stepwise regression, BP neural network, and GWR model to downscale TRMM data in the headwater area. The results of the study showed that the GWR model has the best downscaling effect; Duan [6] realized the downscaling of precipitation data in the Tana Lake basin and the Caspian Sea through the nonlinear relationship between TRMM and NDVI, and obtained monthly precipitation data with a high accuracy of 1 km resolution through GDA correction; The current domestic downscaling research on TRMM precipitation products is mainly achieved by establishing a global regression model of single element or multiple elements and TRMM data, and most of them use NDVI as the main influencing factor.

This study uses the GWR model to establish the functional relationship between the TRMM data in Central China and the NDVI, EVI, elevation, slope, and aspect data of the same period to achieve the downscaling of the TRMM data. Obtain 1 km spatial downscaling precipitation data (see Table 1), and using GDA and GRA two correction methods to correct the downscaling data.

Table 1. Downscale data information.

Downscaling data type	Dependent variable	Control variable	Independent variable
NDVI_downscale TRMM data ($TRMM_{NDVI}$)	TRMM 3B43	Elevation, slope and aspect data	NDVI
EVI_downscale TRMM data ($TRMM_{EVI}$)	TRMM 3B43	Elevation, slope and aspect data	EVI

1 Data and Methods

1.1 Research Area

Central China (see Fig. 1) is composed of the three provinces of Hunan, Hubei and Henan. It is located between $24°38'–36°24'$N and $108°21'–116°39'$E. It is mainly composed of plains, hills, basins, Lake composition, with a total area of 560,000 km^2. The geographical location of Central China is special, belonging to the middle and lower reaches of the Yellow River and the middle reaches of the Yangtze River, and the river system is developed. During the same period of rain and heat in Central China, the precipitation in the year has obvious seasonal characteristics, mainly in summer (June to August) [7], which is prone to flood disasters and soil erosion.

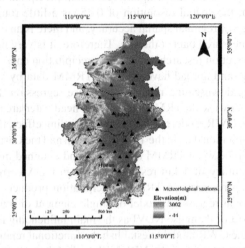

Fig. 1. The spatial distribution of meteorological stations in Central China.

1.2 Data Sources

This study used TRMM 3B43 (TRMM) data (Version 7), MOD13A3 data, DEM data and actual weather station data from 2001 to 2019. Extract NDVI and EVI data from MOD13A3 data, and use DEM data to calculate slope and aspect data.

1.3 Research Methods

Geographically Weighted Regression Models. Geographically Weighted Regression (GWR) is a local parameter estimation method used to quantify spatial heterogeneity, which is an extension of the ordinary linear regression model. In the case of considering the spatial weight of adjacent points, the regression model is established by estimating the parameters of the dependent variable and the independent variable at each position. The basic formula is as follows:

$$y_i = \beta_0(u_i, v_i) + \sum_{t=1}^{n} \beta_t(u_i, v_i)x_{it} + \varepsilon(u_i, v_i), i = 1, 2, \cdots, m \qquad (1)$$

In the formula, y_i is the precipitation of the dependent variable at the i sample point; x_{it} is the observation value of the i sample point of the t independent variable; (u_i, v_i) represents the latitude and longitude coordinates of the i sample point; $\beta_0(u_i, v_i)$ is the constant regression parameter of the i sample point; $\beta_t(u_i, v_i)$ is the linear regression parameter of the t influence factor on the i sample point, $\varepsilon(u_i, v_i)$ is the model in the i sample point. The residual value calculated for each sample point, m is the number of points.

Data Calibration. Through downscaling, the resolution of TRMM data is greatly improved, but the data still has large errors compared with the measured data, so it is necessary to correct the downscaling data. Respectively use geographic ratio analysis (GRA) and geographic difference analysis (GDA) to correct the downscaling data to obtain higher precision precipitation data.

Accuracy Indicators. In this study, the measured data from the weather station is taken as the "true value", and three indicators of R2, BIAS and RMSE are introduced to analyze the accuracy of the downscaling results. R2 assesses the degree of linear correlation between the measured precipitation data of the meteorological station and the downscaled precipitation data; BIAS reflects the degree of deviation between the measured precipitation data of the meteorological station and the downscaled precipitation data; RMSE evaluates the overall level of error. The formula is as follows:

$$R^2 = \frac{\sum_{i=1}^{n}(Y_i - \overline{y})^2}{\sum_{i=1}^{n}(y_i - \overline{y})^2}, i = 1, 2, \cdots, n \tag{2}$$

$$BIAS = \frac{\sum_{i=1}^{n} x_i}{\sum_{i=1}^{n} y_i} - 1, i = 1, 2, \cdots, n \tag{3}$$

$$RMSE = \sqrt{\frac{\sum_{i=1}^{n}(x_i - y_i)^2}{n}}, i = 1, 2, \cdots, n \tag{4}$$

In the formula, x_i represents the true value of the TRMM data of the data, y_i represents the estimate of the measured data, $\overline{y}$ represents the average size of the estimate, Y_i represents the fitted value of linear correlation, and n represents the number of points in the study area.

2 Results and Analysis

2.1 TRMM Data Suitability Analysis

Year Scale Applicability Analysis. Taking the annual measured precipitation of 56 stations in Central China from 2001 to 2019 as the independent variable, and the corresponding annual TRMM data precipitation as the dependent variable, a univariate linear regression analysis was performed (see Fig. 2). On the whole, the TRMM data has a high degree of fit with the site measured data on an annual scale, with R^2 reaching 0.630, BIAS being 0.072, and RMES being 250.09 mm, which is somewhat overestimated.

It can be seen from the distribution map of the correlation coefficient (R) of meteorological stations in Central China (see Fig. 3). The correlation coefficient between the measured data of 56 sites and the TRMM data is between 0.54 and 0.97, of which 85% of the sites are higher than 0.75, and all the sites have passed the 95% significance test, which can indicate that there is a good linear correlation between the two data.

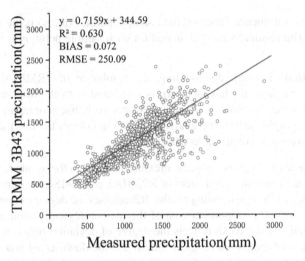

Fig. 2. Scatter plot of TRMM annual data and measured annual data.

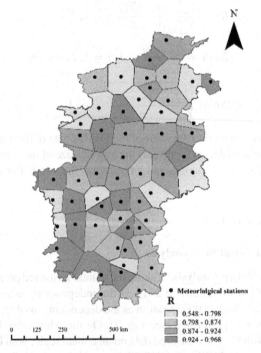

Fig. 3. The spatial distribution of the correlation coefficient based on meteorological station.

Monthly Scale Applicability Analysis. Analyze TRMM monthly data through measured data. It can be seen from Table 2 that the precipitation data of TRMM has a good correlation with the precipitation measured at the station on the monthly scale. R^2 of each month is greater than 0.63, and R^2 of June to September is lower than that of other months, in terms of the BIAS indicators, the lowest in June (BIAS = 0.0301), the highest in January (BIAS = 0.0983) and RMSE ranged between 12.11–62.41 mm. This indicates that the precipitation of TRMM data is slightly larger than the measured data at the station on the monthly scale.

Table 2. Verification results of the applicability of TRMM monthly data.

Month	R^2	BIAS	RMSE/mm	month	R^2	BIAS	RMSE/mm
January	0.847	0.0735	12.687	July	0.715	0.0301	62.414
February	0.854	0.0707	16.939	August	0.637	0.0662	49.031
March	0.839	0.0983	24.349	September	0.731	0.0722	36.705
April	0.791	0.0487	35.796	October	0.813	0.0491	23.060
May	0.742	0.0491	46.267	November	0.828	0.0638	21.182
June	0.724	0.0625	54.891	December	0.875	0.0637	12.115

Quarter-Scale Applicability Analysis. The precipitation data of 56 stations in the study area from 2001 to 2019 were divided into four seasons, and linear fit-ting was carried out with TRMM data of the same period. The results (see Table 3) showed that the fitting effect was the best in winter ($R^2 = 0.865$), the second in spring ($R^2 = 0.818$) and the worst in summer ($R^2 = 0.710$).

Table 3. TRMM quarterly data applicability verification results.

Season	R^2	BIAS	RMSE/mm	Season	R^2	BIAS	RMSE/mm
Spring	0.818	0.0609	35.897	Autumn	0.788	0.0625	27.855
Summer	0.710	0.0519	55.716	Winter	0.865	0.0697	14.079

2.2 Analysis of Downscaling and Calibration Results

Annual Downscaled Corrected Precipitation Data Results. The GWR model was used to obtain the annual TRMM downscaling data ($TRMM_{Year-EVI}/TRMM_{Year-NDVI}$), and the multi-year TRMM and downscaling TRMM data were averaged to obtain the multi-year average data. It can be seen from Fig. 4 that the spatial distribution of data precipitation tends to be consistent before and after the downscaling processing, show-ing a decrease from southeast to northwest, and the image details are enhanced. The $TRMM_{Year-EVI}$ data (b) is closer to the TRMM data (a) than the $TRMM_{Year-NDVI}$ data (c). According to the accuracy analysis of the two kinds of downscale data by means of the measured data (Fig. 5), the $TRMM_{Year-EVI}$ data R^2 is 0.847, BIAS −0.044 and RMSE 173.55 mm. The overall accuracy is superior to $TRMM_{Year-NDVI}$ data, which can better reflect the precipitation characteristics in central China.

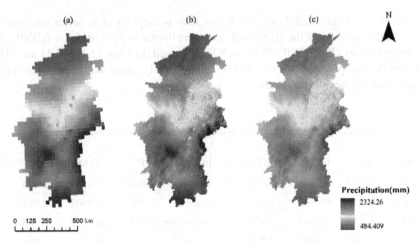

Fig. 4. Average annual downscaling result chart.

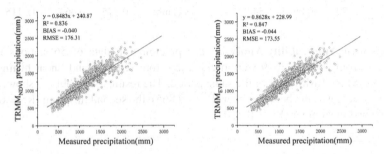

Fig. 5. Scatter plot of downscaled data and measured seasonal data.

After the scale reduction, the spatial resolution of the scale reduction data is greatly improved, but there is still a large error in precipitation compared with the measured data. On the basis of the optimal [TRMM]_(Year-EVI) data, GDA and GRA correction were performed respectively to obtain GDA correction TRMM data and GRA correction TRMM data. It can be seen from Fig. 6 that the precipitation distribution of the two correction data tends to be consistent, mainly distributed in southeastern Hubei, central and southern Hunan and parts of Henan. The 56 stations in the research area were divided into experimental points (36) and verification points (20), and the accuracy of verification points was analyzed through the measured data. It can be seen from Table 4 that the BIAS and RMSE accuracy of [TRMM]_(Year-NDVI)^GRA data are improved, and the coefficient of determination R^2 is overall better than TRMM data, but lower than [TRMM]_(Year-EVI) data; The [TRMM]_(Year-EVI)^GDA data shows consistent superiority in the three indicators (R^2: 0.956, BIAS: -0.026, RMSE: 86.233), effectively alleviating the overestimation problem in the TRMM data.

Monthly Precipitation Scale Corrected Precipitation Data Results. Data verification is performed on $TRMM_{Month}$ data and $TRMM_{Month-EVI}^{GDA}$ data through actual measured data. It can be seen from Fig. 7 that the verification index curves of the two data show

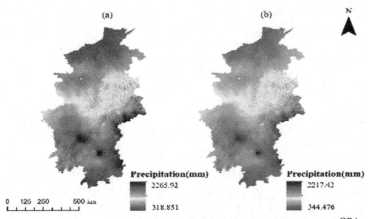

Fig. 6. Multi-year average downscaling correction result map (a is $TRMM_{Year-EVI}^{GDA}$, b is $TRMM_{Year-EVI}^{GRA}$).

Table 4. Calibration year data accuracy analysis table

Data	R^2	BIAS	RMSE/mm
$TRMM_{Year}$	0.630	0.072	250.09
$TRMM_{Year-EVI}$	0.847	−0.044	173.55
$TRMM_{Year-EVI}^{GRA}$	0.831	−0.043	111.921
$TRMM_{Year-EVI}^{GDA}$	0.956	−0.026	86.233

obvious regularity: both curves have the best R^2 and RMSE correction effect during the abundant water period (may-august); BIAS has a certain improvement of accuracy in other time periods except for the cold season (december-february) when the improvement is not obvious. There is a clear agreement between the validation indicator curves for the two data types: both curves have the same trend, and the $TRMM_{Month-EVI}^{GDA}$ data are more accurate than the $TRMM_{Month}$ data.

Results of Seasonal Precipitation Scale Corrected Precipitation Data. As shown in Fig. 8, the spatial distribution of $TRMM_{Season-EVI}^{GDA}$ data and $TRMM_{Season}$ data is consistent, and the spatial resolution of the former is much better than that of the latter. On the whole, the four-season precipitation is mainly distributed in the southeastern part of central China, with the least amount of precipitation in the northern part of Henan.

Accuracy analysis of the three seasonal data through measured data. It can be seen from Table 5: The accuracy of the three indicators in each season is in the order of $TRMM_{Season-EVI}^{GDA}$ > $TRMM_{Season-EVI}$ > $TRMM_{Season}$; From the perspective of the accuracy of the four seasons data, the R^2 of the summer data before and after correction is increased from 0.709 to 0.811, the absolute value of BIAS is reduced by 0.095, and

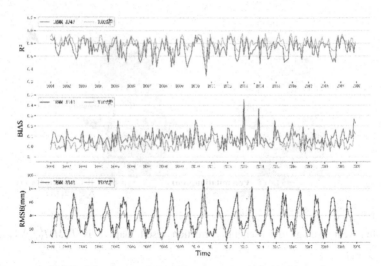

Fig. 7. The change chart of data accuracy index in November.

the RMSE is reduced from 55.715 mm to 42.154 mm, the correction effect is better than the other three seasons. From the above conclusions, it can be concluded that the more precipitation there is in the season, the better the precipitation and calibration results.

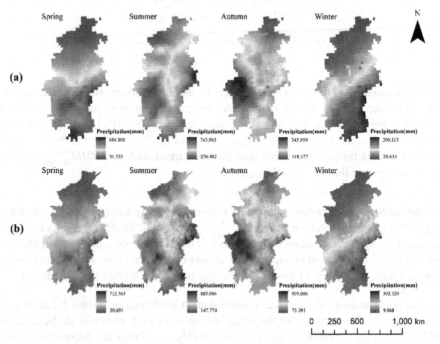

Fig. 8. Downscaling correction result chart of quarterly data (a is [TRMM]_Season data, b is TRMM_(Season-EVI)^GDA data).

Table 5. Quarterly data verification results.

Season	Data	R^2	BIAS	RMSE/mm
Spring	$TRMM_{Season}$	0.818	0.0614	35.897
	$TRMM_{Season-EVI}$	0.812	0.0515	34.787
	$TRMM_{Season-EVI}^{GDA}$	0.859	0.0086	28.257
Summer	$TRMM_{Season}$	0.709	0.0519	55.715
	$TRMM_{Season-EVI}$	0.706	0.0497	53.812
	$TRMM_{Season-EVI}^{GDA}$	0.811	−0.0129	42.154
Autumn	$TRMM_{Season}$	0.787	0.0626	27.855
	$TRMM_{Season-EVI}$	0.808	0.0619	27.245
	$TRMM_{Season-EVI}^{GDA}$	0.861	−0.0049	22.469
Winter	$TRMM_{Season}$	0.864	0.0697	14.219
	$TRMM_{Season-EVI}$	0.869	0.0702	14.081
	$TRMM_{Season-EVI}^{GDA}$	0.873	0.0601	13.268

3 Conclusion

The TRMM 3B43 data show good applicability in Central China. Accuracy inspection is carried out through the measured data: annual R^2 is 0.63, and seasonal and monthly R^2 generally reach above 0.7; BIAS performance is positive and is basically controlled within 0.1; RMSE is also within the allowable error range. Summer precipitation in Central China is abundant, and monsoon activity is unstable. Satellites are affected by monsoon and rainfall, and they are prone to deviations in precipitation estimation, which leads to a decrease in the accuracy of summer precipitation inspection. The reason for the poor estimation of winter precipitation is that the winter precipitation in Central China is dominated by snow, which affects the inversion of precipitation by satellite.

Through downscaling, the spatial resolution of TRMM data is increased from 0.25° to 1 km. The $TRMM_{EVI}$ data is more connected to the measured data than the $TRMM_{NDVI}$ data, the reason may be that the relationship between EVI and TRMM data in Central China is closer. The GDA and GRA calibrations were performed on the screened $TRMM_{EVI}$ data. Through the validation of the 20 validation sites in the study area, GDA calibration results are better than GRA calibration results. The accuracy of the corrected annual data is better than that of seasonal data and monthly data, and the correction results of summer and autumn are better than those of spring and winter. In the month with heavier precipitation, the correction effect is better.

This study can provide a reliable scientific basis for the downscaling study of precipitation products in Central China and other areas with complex terrain, the study of data correction, and the study of the temporal and spatial distribution characteristics of regional precipitation.

Acknowledgements. This research was supported by the national Natural Science Foundation of China (41801071; 42061059); the Natural Science Foundation of Guangxi (2020JJB150025); Research Project of Guilin University of Technology.

References

1. Jin, Y.F., Goulden, M.L.: Ecological consequences of variation in precipitation: separating short- versus long-term effects using satellite data. Glob. Ecol. Biogeogr. **23**(3/4), 358–370 (2014)
2. Bhattacharya, A., Adhikari, A., Maitra, A.: Multi-technique observations on precipitation and other related phenomena during cyclone Aila at a tropical location. Int. J. Remote Sens. **34**(6), 1965–1980 (2013). https://doi.org/10.1080/01431161.2012.730157
3. Michaelides, S., Levizzani, V., Anagnostou, E., Bauer, P., Kasparis, T., Lane, J.E.: Precipitation: measurement, remote sensing, climatology and modeling. Atmos. Res. **94**(4), 512–533 (2009)
4. Lv, Y., Dong, G.T., Yang, S.T., et al.: Spatio-temporal variation in NDVI in the Yarlung Zangbo River Basin and its relationship with precipitation and elevation. Resour. Sci. **36**(03), 603–611 (2014)
5. Li, Q., Wei, J.B., An, J., et al.: Correcting the TRMM 3B43 precipitation over the source region of the Yellow River based on topographical factors. J. Basic Sci. Eng. **26**(06), 6–22 (2018)
6. Duan, Z., Bastiaanssen, W.G.M.: First results from Version 7 TRMM 3B43 precipitation product in combination with a new downscaling–calibration procedure. Remote Sens. Environ. **131**, 1–3 (2013)
7. Huang, M.J., He, X.G., Lu, X.A., et al.: Analysis on the temporal and spatial characteristics of non-stationary SPI drought in Central China. Resour. Environ. Cent. China **29**(07), 1597–1611 (2020)

Research on the Knowledge Map of Combat Simulation Domain Based on Relational Database

Li Guo[1]([⊠]), Boao Xu[2]([⊠]), Hao Li[1], Dongmei Zhao[1], Shengxiao Zhang[1],
and Wenyuan Xu[1]

[1] China Shipbuilding Industry Systems Engineering Research Institute, Beijing, China
guolicssc@163.com
[2] 96901 PLA troops, Beijing, China
xuboao@126.com

Abstract. The analysis and evaluation of simulation data, especially the selection of assessment indexes and the establishment of index system, need the knowledge base or knowledge map of simulation application field as the expert knowledge support. The relational database of the existing combat simulation application system provides a reliable and easy to obtain data source for the construction of combat simulation domain knowledge map. This paper proposes an effective method to construct the knowledge map of simulation data based on relational database, which lays a foundation for the construction of combat simulation domain knowledge map and the analysis of simulation data based on knowledge map.

Keywords: Combat simulation system · Knowledge map · The analysis of simulation data · Relational data

1 Introduction

The war system composed of confrontation systems is a typical complex system, involving sea, air, shore, space, submarine, information and other fields. It has the characteristics of spatial distribution of components, diversity of missions and tasks, rapid change of war environment, various operational elements, complex command relationship and complex information interaction relationship, which makes the corresponding military combat simulation system gradually developing towards the direction of complex system. Complex combat simulation system has the characteristics of large-scale, distributed, interactive, dynamic etc. It consumes and produces a large amount of data, and these data are often distributed on multiple heterogeneous nodes in the network, with different formats and purposes. Therefore, the analysis and evaluation based on a large number of simulation data is a tedious and difficult work. Knowledge map aims to identify, discover and infer the complex relationship between objects and concepts from data. It is a computable model of the relationship between objects. The association between data can be enhanced through knowledge map and semantic technology, so that

Q. Zu et al. (Eds.): HCC 2020, LNCS 12634, pp. 319–325, 2021.
https://doi.org/10.1007/978-3-030-70626-5_34

users can conduct association mining and analysis of data in a more intuitive way. It provides new opportunities and challenges for the analysis and evaluation of complex combat simulation data [1].

Simulation data analysis and evaluation, especially the selection of assessment indexes and the establishment of index system, need knowledge base or knowledge map of simulation application domain as expert knowledge support, just as experts need knowledge of professional field as support. The construction of domain knowledge map needs deep application practice in professional field. The input and output data of existing simulation application system provide a reliable and easy to obtain data source for the construction of combat simulation domain knowledge map. Since most of the existing simulation application system data are stored in relational database, how to quickly and effectively build simulation data knowledge map based on relational database is a problem to be solved.

2   Overview of Entity Relationship in Combat Simulation Domain

The combat simulation system needs to simulate the whole battlefield environment, simulate the roles involved in combat activities, as well as various physical characteristics and behaviors of the roles, and describes the initial state of the elements of the combat simulation system through the combat simulation scenario [2]. Therefore, the entities constituting the combat simulation system mainly include combat scenario, battlefield environment, combat entity, component model describing physical characteristics of combat entity, behavior model describing individual behavior of combat entity, combat task, combat activity, force organization and operational area. As shown in Fig. 1, E-R diagram describing typical combat simulation application entities and their relationship, which is the basis of relational database design of combat simulation application system, is also the foundation of concept level architecture of combat simulation data knowledge map based on relational data.

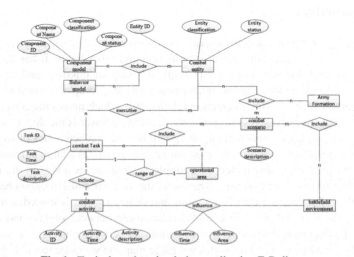

Fig. 1. Typical combat simulation application E-R diagram

3   Concept Architecture of Simulation Data Knowledge Map

The concept architecture of simulation data knowledge map refers to how to establish the concept architecture involving abstract concepts (entities), relationships and attributes in the knowledge map based on E-R architecture in relational database, and realize the mapping from E-R architecture to knowledge map architecture, including the extraction of concept elements, generation of relationship elements and generation of attribute elements [3, 4].

The concept in the knowledge map is the abstraction of data. The concept elements in the knowledge map of simulation data come from the entity, relationship and attribute in the E-R diagram of the relational data. The entity mapping into the knowledge map is called the entity concept, the attribute mapping into the knowledge map is called the attribute concept, and the relationship mapping into the knowledge map is called the relationship concept, as shown in Fig. 2.

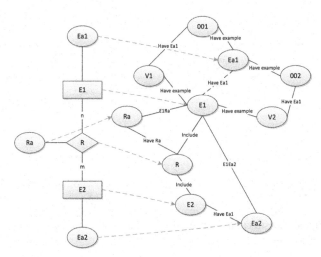

Fig. 2. Mapping from E-R framework of relational database to concept architecture of knowledge map

3.1   Extraction of Concept Elements

The concept in knowledge map, also known as type or category, refers to a class of people, things and events with the same description template [5]. The concept in the knowledge map of simulation data based on relational database is the abstraction of business objects in the simulation field, such as components, entities (combat equipment entities), formations, commanders, combat tasks, combat activities and other people and objects involved in simulation activities, as well as simulation events such as actions and tasks. The entity, relationship and attribute from E-R diagram of simulation application system database can be mapped to the concepts in knowledge map.

Taking the business scenario of combat entities performing combat tasks in Fig. 1 as an example, the entity concepts mapped to the knowledge map include combat entities and combat tasks, attribute concepts include entity identification, entity classification, entity status, task identification, task time, task description, execution location and execution time, and relationship concepts include execution, as shown in Fig. 3.

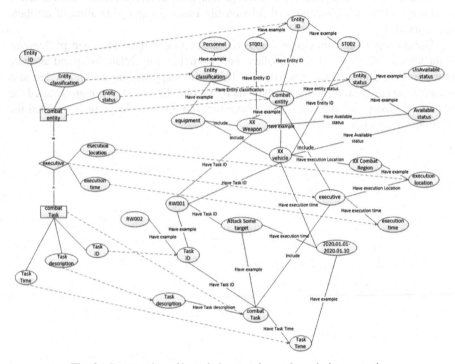

Fig. 3. Construction of knowledge map in combat mission execution

3.2 Generation of Relation Elements

The relationship in E-R diagram is regarded as the concept of relationship in the knowledge map. The concept has its own attributes and relationships. As shown in Fig. 2, R is a relationship in E-R diagram, and its attribute is Ra. If it is mapped to the knowledge map, R is a relational concept and its attribute Ra (attribute concept). Since the relationship in E-R diagram is regarded as a concept in the knowledge map, it is necessary to generate a new relationship to establish the relationship between concepts. The steps are as follows:

① the relationship in E-R diagram is usually an action, and if the action is regarded as a concept, it is an event concept. Therefore, the relationships between the event concept and the concepts of both ends of the action are inclusion, that is, R contains E1 and E2;

② entity E1 establishes a new relationship with the attribute Ra of R relationship in the original E-R diagram and the attribute of entity E2, and the relationship name is set as "E1Ra" and "E1Ea2"

In the combat mission execution scenario, the relationship "execution" in E-R diagram is an event concept in the knowledge map, which establishes the inclusion relationship with "combat entity" and "combat mission"; the attribute "execution time" of "execution" is an attribute concept. And the relationship of "combat entity execution task time" is established between "execution time" and "combat entity" in the transformation process. The relationship of "combat entity mission identification" is established between "combat entity" and "mission identification" of "combat mission".

3.3 Generation of Attribute Elements

The attribute in E-R diagram is regarded as the attribute concept in the knowledge map. As the attribute becomes the attribute concept after mapping to the knowledge map, it is no longer an attribute. Therefore, it is necessary to generate the relationship between the attribute and the entity concept to establish the relationship between them. As shown in Fig. 2, Ea1 is a concept in the concept architecture of knowledge map, and Ea1 is an attribute of entity E1. There is an attribute relationship between E1 and Ea1, which is implied in the E-R diagram, but it needs to be displayed in the knowledge map. Therefore, an attribute relationship must be generated. The steps are as follows:

Adding a word "Have" before the attribute name means "Have EA1".

As shown in Fig. 3, in the combat mission execution scenario, the relationship between "Combat entity" and "Entity status" is established with "Have entity status", and the relationship between "Combat task" and "Task ID" is established with "Have Task ID".

3.4 Extraction of Instance Elements

The instance is the specific concept in the knowledge map, such as the combat entity of a "XX weapon1", EA1 is "ST001". Among them, "XX weapon1" and "ST001" are both examples of concept "E1" and concept "Ea1". These instances are stored in the relational database, so instance element extraction is the process of mapping the data in the relational database to the concept it belongs to. The instance comes from the data in the relational database, because the entities, relationships and attributes in the E-R diagram have specific data corresponding in the database. After mapping to the knowledge map and forming concepts, these data are examples of concepts. The specific extraction steps are as follows: name the specific combat entity in each row, i.e., assume that the first line is "XX weapon1", Ea1 is "ST001", the second line is "XX weapon2", Ea1 is "ST002". The name of combat entity "XX weapon 1" and "XX weapon2" are taken as the instances of E1 respectively, The relationship between E1 and XX weapon1 and XX weapon2 is set as "with instance"; similarly, "ST001" and "ST 002" are taken as two instances of "Ea1" concept, and the relationship of "with instance" is established.

4 Construction of Knowledge Map of Simulation Data

After completing the concept architecture of knowledge map of simulation data, the mapping of entity, relationship and attribute in E-R diagram to concept, relationship and attribute of knowledge map is formed. Based on the inheritance of the relationship and attribute between the concept and the instance in the knowledge map, the simulation data knowledge map can be formed by using the data of the relational database.

In the knowledge map, the attributes of concepts are inherited, while for instances, the attributes of concepts are inherited, for example, "E1 - has Ea1-Ea1" means that there is "have Ea1" attribute between concept "E1" and concept "Ea1"; instances can inherit concept attributes, so instances "XX weapon1" and "XX weapon2" under concept "E1" also have "have Ea1" attribute, which correspond to "ST001" and "ST 002" instances under concept "Ea1"; according to this rule, the extracted concepts will be extracted. The knowledge map of simulation data is formed by association of attributes and relationships.

```
--Entity-Vehicle relationship data
:auto USING PERIODIC COMMIT
LOAD CSV WITH HEADERS FROM "file:///shiti_cheliang.csv" AS row
match(S:Entity{stID:row.stID}),(C:Vehicle{clID:row.clID})
merge(C)-[r:rel{scID:row.scID}]->(S)

--Entity-Task relationship data
:auto USING PERIODIC COMMIT
LOAD CSV WITH HEADERS FROM "file:///renwu_shiti.csv" AS row
match(R:Task{rwID:row.rwID}),(S:Entity{stID:row.stID})
merge(S)-[r:rel{rsID:row.rsID}]->(R)

--Entity-Weapon relationship data
:auto USING PERIODIC COMMIT
LOAD CSV WITH HEADERS FROM "file:///shiti_wuqi.csv" AS row
match(S:Entity{stID:row.stID}),(W:Weapon{wqID:row.wqID})
merge(W)-[r:rel{swID:row.swID}]->(S)

--Weapon Data
:auto USING PERIODIC COMMIT
LOAD CSV WITH HEADERS FROM "file:///wuqi.csv" AS row
CREATE (W:Weapon{wqID:row.wqID, wqMC:row.wqMC, wqLX:row.wqLX,wqDW:row.wqDW,wqSJ:row.wqSJ});

--Entity Data
:auto USING PERIODIC COMMIT
LOAD CSV WITH HEADERS FROM "file:///shiti.csv" AS row
CREATE (S:Entity{stID:row.stID, stMC:row.stMC, stFL:row.stFL,stZT:row.stZT,stBZ:row.stBZ});

--Vehicle Data
:auto USING PERIODIC COMMIT
LOAD CSV WITH HEADERS FROM "file:///cheliangs.csv" AS row
CREATE (C:Vehicle{clID:row.clID,clMC:row.clMC, clSJ:row.clSJ,clLX:row.clLX});

--Task Data
:auto USING PERIODIC COMMIT
LOAD CSV WITH HEADERS FROM "file:///renwu.csv" AS row
CREATE (R:Task{rwID:row.rwID, rwLB:row.rwLB, rwMC:row.rwMC,rwMS:row.rwMS,rwSJ:row.rwSJ});
```

Fig. 4. Transforming relational database data into neo4j graph database data

Based on the above rules, the relational database data is first transformed into the neo4j graph database data [4], as shown in Fig. 4, and finally the simulation data knowledge map as shown in Fig. 5 can be automatically generated.

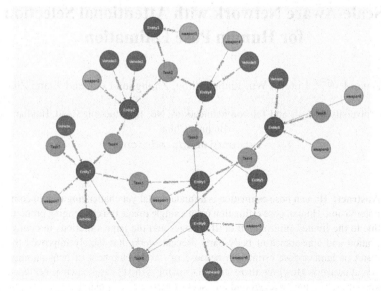

Fig. 5. The example of knowledge map of simulation data

5 Conclusions

As a kind of knowledge organization method in the field of combat simulation, the knowledge map of combat simulation data is more intuitive and easy to understand. According to the relationship between concepts and concept instances in the knowledge map, researchers can understand the rules of combat simulation data from the surface to the inside, from the shallow to the deep, so as to explore the relationship between data through the traceability of data. It provides the basis for simulation data analysis, evaluation index system establishment and index weight setting. Therefore, it is of great practical significance to construct a high-quality knowledge map of combat simulation data, which provides knowledge support for the application of combat simulation business system.

References

1. Xiao, Y., Xu, B., Lin, X., et al.: Concept and technology of knowledge map. Electronic Industry Press, Beijing (2020)
2. Kedi, H., Baohong, L., Jian, H., et al.: Overview of combat simulation technology. J. Syst. Simul. **9**, 1887–1895 (2004)
3. Qiao, L., Yang, L., Hong, D., et al.: Overview of knowledge mapping technology. Comput. Res. Dev. **53**(3), 582–600 (2016)
4. An automatic construction method of mobile data knowledge map based on relational database. China, invention patents, 201811320885.8, (2018)
5. Li, Y.: Research on Automatic Construction Method of Domain Specific Knowledge Map. Harbin Institute of technology, Harbin (2017)

Scale-Aware Network with Attentional Selection for Human Pose Estimation

Tianqi Lv[✉], Lingrui Wu, Junhua Zhou, Zhonghua Liao, and Xiang Zhai

Beijing University of Posts and Telecommunications, No. 10 Xitucheng Road, Haidian District, Beijing, China
lvtianqi@bupt.edu.cn

Abstract. Human pose estimation is a fundamental yet challenging task in computer vision. Human pose estimation from a single image is a challenging problem due to the limited information of 2D images and the large variations in configuration and appearance of body parts. Recent works has largely improved the result of human pose estimation because of the development of convolutional neural network. However, there still exists many difficult cases, such as occluded keypoints, complex background and scale variations of human body keypoints, which cannot be well dealt with. In this paper, we design a novel scale-aware network with attentional selection that extracts multi-scale semantic information and meaningful features. Specifically, we propose a Feature Pyramid Supervision Module (FPSM), which can improve the estimation accuracy of scale variations. Meanwhile, a Spatial and Channel Attention Module (SCAM) is designed for recalibrating the spatial and channel features. Based on the proposed algorithm, we achieve state-of-the-art result on LSP dataset and make competitive performance on MPII Human Pose dataset.

Keywords: Deep learning · Feature pyramid supervision · Spatial and channel attention · Human pose estimation

1 Introduction

Human pose estimation, the problem of localizing body parts from a RGB image, is a fundamental task in computer vision, and it serves as a key component to high-level vision tasks, e.g., action recognition and human re-identification. However, it is difficult to handle many difficult cases such as occluded keypoints, scale variations of human body parts and so on.

Deep convolutional Neural Networks (DCNNs) which can learn rich feature representations have achieved noticeable improvements on human pose estimation [1–3]. For example, a well-designed architecture, stacked hourglass network [4], is proposed to address human pose estimation. This work proves that the repeated bottom-up, top-down inference and intermediate supervision at the end of every stack are beneficial for the pixel location of keypoints. [5] designs an architecture using two same stacked hourglass networks based on generative adversarial networks. Its generator is a human pose

Q. Zu et al. (Eds.): HCC 2020, LNCS 12634, pp. 326–337, 2021.
https://doi.org/10.1007/978-3-030-70626-5_35

estimator and its discriminator back-propagates the adversarial loss to the generator. Although the stacked hourglass network and its variants [5–7] demonstrate significant performance in human pose estimation, it still needs further study on more accurate predictions.

Designing a model should consider to extract meaningful features and multi-scale semantic information which are beneficial for accurate estimation of body joints. As shown in Fig. 1(a), the estimated person is much smaller than the person in Fig. 1(b). The model ought to robustly deal with this problem. The sufficient contextual information of feature pyramid can enhance the robustness of DCNNs against scale variations. For another example, the woman in Fig. 1(b) with a twisted pose is not easy to be distinguished with background. Therefore, the model should focus on region of interest and adequately extract contextual representations.

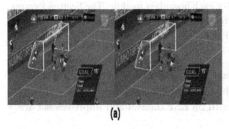 

(a)   (b)

Fig. 1. The estimations of difficult cases produced by our model. (a) The left picture is a person with small scale. The right picture is the estimation of the person. (b) The right picture is a woman with background of similar colors. The right picture is the estimation of the woman.

To enhance the ability of DCNNs and address the above problem, we aim at utilizing the information of feature pyramid and automatically inferring the contextual representation. Because hourglass network [4] is an effective architecture for human pose estimation, we use it as the basic model in experiments. A Feature Pyramid Supervision Module (FPSM) is proposed to extract multi-scale features which improve the ability of intermediate supervision. We concatenate the feature maps with same resolution got from each deconvolution layer of eight hourglass modules and add the low-resolution feature maps to adjacent high-resolution feature maps. The Feature Pyramid Supervision Module offers multi-scale features to learn and infer. Meanwhile, we design a Spatial and Channel Attention Module (SCAM) which can recalibrate the semantic information in spatial and channel dimensions. This module is used to automatically learn and infer the contextual representations, driving the model to focus on meaningful features. The structure of our model is shown in Fig. 2.

The model is evaluated on two human pose estimation benchmarks, i.e., MPII Human Pose dataset [8] and LSP dataset [9]. The experimental results prove that the Feature Pyramid Supervision Module and the Spatial and Channel Attention Module are effective for human pose estimation. In summary, our contributions are three-fold:

– We design an intermediate supervision method called Feature Pyramid Supervision Module, which can enhance the estimation ability of scale variations.

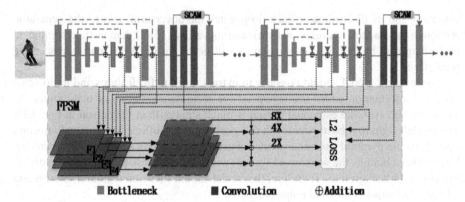

Fig. 2. Overview of our model. The basic structure is an 8-stack hourglass network. We place the Spatial and Channel Attention Module in the end of each hourglass module. The Feature Pyramid Supervision Module extracts scale feature maps from deconvolution layers and score maps from hourglass modules to calculate a Mean-Squared Error loss.

- We explore an attention mechanism, named Spatial and Channel Attention Module, to recalibrate the features in spatial and channel dimensions.
- Our algorithm achieves state-of-the-art result on LSP dataset and improves accuracy on MPII Human Pose dataset.

2   Related Work

2.1   Human Pose Estimation

Graph structures (e.g., mixture of body part, pictorial structures and loopy structures have been widely used to model articulated human poses. Recently, many methods on human pose estimation using deep convolutional neural networks have achieved state-of-the-art results for learning better feature representation [1–4, 10–12]. Among them, DeepPose [1], one of the earliest approaches based on DCNNs to human pose estimation, regresses the coordinates of body joints directly which results in high precision of joint detections. [4] proposes stacked hourglass network for human pose estimation. The hourglass architecture supports repeated bottom-up, top-down processing across different scales for large receptive field, which helps capture more correlations among human body joints. Many methods [5–7] design variants of stacked hourglass network for the well-designed architectures. [5] sets up two stacked hourglass networks, one as the generator and the other as the discriminator. [6] proposes a holistic framework to effectively address the drawbacks in the hourglass network. [7] uses an effective method with negligible additional parameters to solve the problem of using residual unit for hourglass.

2.2   Feature Pyramid Networks

[13] exploits the inherent pyramidal hierarchy of deep convolutional networks to construct feature pyramids. This architecture, named Feature Pyramid Network (FPN),

shows significant improvement as a generic feature extractor in several works. Mask R-CN, a semantic segmentation work, gains in both accuracy and speed using FPN and ResNet as backbone. FPN is adopted as the first stage in a human pose estimation algorithm named CPN. The pyramid feature representation based on FPN can provide sufficient context information.

2.3 Intermediate Supervision

Deep networks cause the characteristic difficulty of vanishing gradients during training. [2] provides a natural learning objective function that enforces intermediate supervision between conv-deconv pairs to prevent gradient vanish. Many pose estimation methods have demonstrated that intermediate supervision is critical to the final performance. [11] adopts a network that allows for repeated bottom-up, top-down inference across scales with intermediate supervision at the end of each stack. [12] utilizes the iterative prediction architecture which refines the predictions over successive stages with intermediate supervision at each stage. [6] introduces a new method of intermediate supervision to better capture the adjacency among the body keypoints.

2.4 Attention Mechanism

Attention mechanism is a method of the allocation of available computational resources to the most informative components of a signal [14]. Attention mechanism has indicated its utility in many tasks including sequence learning and Natural Language Processing (NLP). Meantime, attention mechanism is also increasingly utilized in the image vision field for using the high-level information to guide the feed-forward network. [15] introduces self-attention modules to find global context which can enlarge the receptive field and enhance the consistency of pixel-wise classification. [16] combines attention mechanism and spatial pyramid to extract dense features for pixel labeling. [17] proposes a Dual Attention Networks (DANet) to adaptively synthesize local features with their global dependencies.

3   Our Approach

3.1   Revisiting Stacked Hourglass Network

Hourglass network aims at capturing feature information across all scales of the RGB image and bringing these features together to obtain pixel-wise predictions. It first begins with the bottom-up processing that uses convolutional layers and max-pooling layers to a very low resolution and performs top-down processing by upsampling and combination of heatmaps across all scales. [4] expands the single hourglass with the other seven hourglass modules end to end. With the help of intermediate supervision which is placed at the end of each stack, the network architecture achieves significant improvement on two standard benchmarks.

3.2 Feature Pyramid Supervision Module

One challenge of localizing keypoints is scale variations of human body parts. The supervision of stacked hourglass network is only used at the output of each stack, which suffers from the influence of scale variations. To deal with this problem, we design a Feature Pyramid Supervision Module (FPSM) that can provide sufficient contextual information and integrate all levels of feature representations. It can extract rich feature information and can be built from a single input image scale. Figure 2 summarizes our Feature Pyramid Supervision Module.

We extract multi-scale feature maps from deconvolution layers of eight hourglass modules. Features with the same resolution are concatenated to obtain four high dimension feature maps (8, 16, 32, 64 pixels). We denote them as F1, ..., F4 respectively. As shown in Fig. 2, F1 undergoes a 1×1 convolutional layer to reduce the number of channels to 256 and is upsampled by a factor 2. Then we merge it with F2 by element-wise addition. This operation is iterated until F4 is dealt with. We attach two 1×1 convolutional layers on each merged map to reduce the channels. The first one fuses the information from difference feature maps followed by the batch normalization and ReLU. The second convolution layer converts high dimensional features into the number of body keypoints. We can get four predicted heatmaps after the processing. Finally, the predictions are upsampled to match the feature maps of ground-truth keypoints.

3.3 Spatial and Channel Attention Module

Contextual representations are essential for the inference of occluded joints, invisible joints and ambiguous background which is similar to body parts or limbs. Human understanding scenes effectively benefits from a mechanism of the human brain which we call visual attention. Motivated by this, we propose a Spatial and Channel Attention Module (SCAM) to automatically learn the contextual representations and lead the model to focus on meaningful features. This module is shown in Fig. 3. In this section, we first introduce the spatial and channel attention mechanism, and then describe Spatial and Channel Attention Module.

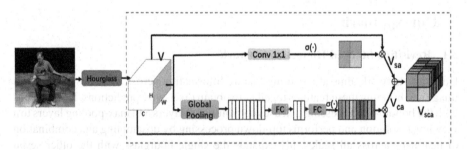

Fig. 3. Spatial and Channel Attention Module structure

Spatial Attention. The spatial attention provides the relative importance of spatial information of a feature map. We denote the feature map as $V = [v_{11}, \cdots, v_{h,w}, \cdots, v_{H,W}]$, where $v_{h,w} \in \mathbb{R}^{1 \times 1 \times C}$ representing the spatial location (h, w) with $w \in \{1, 2, \cdots, W\}$ and $h \in \{1, 2, \cdots, H\}$. To realize the spatial attention, the feature map first passes through a convolutional layer, generating a squeezed tensor $T \in \mathbb{R}^{H \times W}$ with single channel. We feed the tensor into a sigmoid layer $\sigma(\cdot)$ to bring the range of it to the interval $[0, 1]$. Finally, we multiply V by the tensor T:

$$V_{sa} = F_{sa}(T, V) = \left[\sigma(t_{1,1})v_{1,1}, \cdots, \sigma(t_{H,W})v_{H,W}\right] \tag{1}$$

Channel Attention. The channel attention focuses on recalibrating the channel-wise relationships to increase the representational power. Let $V = \{v_c\}_{c=1}^{C}$ denote the set of channels $V_c \in \mathbb{R}^{H \times W}$. We squeeze the global spatial information into a channel descriptor by using global average pooling. This operation is expressed as follows:

$$z_c = \frac{1}{H \times W} \sum_{i=1}^{H} \sum_{j=1}^{W} v_c(i, j) \tag{2}$$

z_c is the c-th element of vector $Z \in \mathbb{R}^{1 \times 1 \times C}$. We then employ a gating mechanism with a sigmoid activation to rescale the value:

$$S = F_{ca}(Z, W) = \sigma(W_2 \delta(W_1 Z)) \tag{3}$$

$w_1 \in \mathbb{R}^{C \times \frac{C}{2}}$ and $w_2 \in \mathbb{R}^{C \times \frac{C}{2}}$ are the weights of two fully-connected layers which are used to fully aggregate information. $\delta(\cdot)$ refers to the ReLU function and $\sigma(\cdot)$ is the sigmoid activation function. The vector S indicates the importance of feature map's channels. Finally, S is used to recalibrate V to:

$$V_{ca} = SV = [s_1 v_1, \cdots, s_c v_c] \tag{4}$$

Spatial and Channel Attention Module. The spatial and channel attention is a combination of the above two operations that recalibrate the semantic in formation in spatial and channel dimensions. We get the final output V_{sca} by element-wise addition of V_{sa} and V_{ca}. In our Spatial and Channel Attention Module (SCAM), the spatial and channel attention is integrated into some residual bottlenecks [18] following the final batch normalization. Feature maps after each hourglass are divided into two branches, one for predicting score maps used in intermediate supervision and another for the next hourglass. We insert the modified residual bottleneck into the second branch before integrating it into the next hourglass.

4 Experiments and Analysis

This model is evaluated on two public datasets, MPII Human Pose dataset and LSP dataset. We first introduce experimental setup in Sect. 4.1. Next, we carry out ablation experiments to reveal the effectiveness of individual proposed modules in Sect. 4.2. Finally, we compare the experimental results of the proposed model with state-of-the-art pose estimators.

4.1 Experimental Setup

Datasets and Evaluation Metrics. The MPII dataset includes images collected from YouTube videos with a wide range of real-world activities. It consists of around 25K images and 40K persons with full body annotations. Each person is annotated with 16 body joints. These data are divided into 28K images for training and 12K images for testing. With the high image resolution images in MPII dataset, the MPII dataset can be considered as a benchmark for human pose estimation. Our model used in LSP dataset is trained on the LSP dataset and the LSP-extended dataset. 11K samples are used for training and 1K for testing. The images are gathered from various sports such as skiing, baseball, gymnastics and so on. Each person is annotated with 14 body joints in LSP dataset. This dataset has a little noise in some annotations because the location of some occluded joints may not easy to confirm. The various poses of human and the noisy annotations make this dataset not easy to get a good result. Evaluation is conducted using the standard Percentage of Correct Keypoints (PCK) measurement [8] which shows the percentage of correct predictions that fall within a normalized distance of the ground truth. For the LSP evaluation, the distance is normalized by the torso size. The modified PCK measurement that the distance is normalized by the head size is denoted as PCKh [8] for the MPII.

Data Augmentation. Data Augmentation is beneficial for the robustness of the prediction network. The input images are cropped with the target human centered at the images and normalized to 256×256 pixels. As typical, we augment the training data with random flipping, rotation and scaling. The probability of flipping is 0.5, the rotation range is $[-30, 30]$ degrees and the scaling range is $[0.75, 1.25]$.

Implementation Details. All proposed models are trained using RMSProp optimizer [19] with an initial learning rate of $2.5e-4$. We drop the learning rate by a factor of 10 at the 90th and the 120th epochs. The network is trained using Pytorch and the training is on a workstation with two 11 GB NVIDIA 1080Ti GPUs. Mini-batch size is 12. A Mean-Squared Error (MSE) loss is used to compare the predicted heatmaps with the ground-truth heatmaps. The prediction of testing is conducted on six-scale image pyramids with flipping.

4.2 Ablation Study

To evaluate the performance of our model, we conduct ablative analysis on the MPII validation set. We use 8-stack hourglass network as our basic model.

Feature Pyramid Supervision Module. We first test the effectiveness of our proposed Feature Pyramid Supervision Module (FPSM). After using the module, our model obtains higher PCKh score (89.9) compared with the basic model (88.1). The results are shown in Table 1. It can be seen that the Feature Pyramid Supervision Module can refine the positions of keypoints because it integrates multi-level semantic features.

Table 1. Ablation study about our proposed modules on validation set of MPII Human Pose dataset

Methods	Head	Shoulder	Elbow	Wrist	Hip	Knee	Ankle	Total
[4]	–	–	–	–	–	–	–	88.1
FPSM	97.0	96.0	90.7	85.8	90.2	86.4	82.3	89.9
FPSM+SCAM	96.2	96.3	90.8	86.4	90.1	86.7	83.3	90.2

Spatial and Channel Attention Module. To figure out the effect of the Spatial and Channel Attention Module (SCAM), we conduct experiment by further adding this module to the current network. From Table 1, new structure can achieve a further 0.3% improvement that yields a 90.2% PCKh score. The Spatial and Channel Attention Module can recalibrate the spatial-wise and channel-wise relationships which is helpful for increase the accuracy of predictions.

4.3 Comparison with State-of-the-Art Methods

We further evaluate our proposed model by comparing against existing methods on two datasets, MPII Human Pose dataset and LSP dataset. Besides, we also show some qualitative results that are generated by our model.

Results on MPII. Table 2 shows the PCKh accuracy results at the threshold of 0.5 that the proposed network compares with state-of-the-art networks on the test set of MPII. Our model obtains 92.1% PCKh score which improves the hour-glass network with a margin by 1.2%. For keypoints that are hard to estimate, e.g. wrist and ankle, we achieve 1.3% and 1.9% improvements, respectively. Compared with other state-of-the-art models, our model gets a competitive result.

Results on LSP. We report the PCK scores at the threshold of 0.2 on LSP test set in Table 3. Following previous methods [7], we train our network by using two datasets, LSP dataset and MPII training set. Our method achieves 96.0% PCK score, which is the highest result by now. For the prediction of wrist, our model has 3.6% improvement.

Qualitative Examination. To provide visible set, Fig. 4 shows images of estimated pose generated by our method on MPII Human Pose dataset and LSP dataset. The 1st and the 2nd rows are some visible results of LSP test set. These persons in the images have various postures. The last three rows are the qualitative pose estimation evaluation on MPII validation set. The 3rd row shows some people with various twisted poses. The 4th and the 5th rows are the difficult cases: occluded keypoints, overlapping people and scale variations of human body. The multi-scale semantic information of feature pyramids can deal with the problem of scale variations. The meaningful feature that is chosen from contextual information can help to accurately localize body parts. Faced with difficult scenarios, our method can effectively improve performance of human pose estimation.

Table 2. Comparisons of PCKh@0.5 score on the MPII test set

Methods	Head	Shoulder	Elbow	Wrist	Hip	Knee	Ankle	Total
[10]	96.8	95.2	89.3	84.4	88.4	83.4	78.0	88.5
[2]	97.8	95.0	88.7	84.0	88.4	82.8	79.4	88.5
[3]	97.9	95.1	89.9	85.3	89.4	85.7	81.7	89.7
[4]	98.2	96.3	91.2	87.1	90.1	87.4	83.6	90.9
[20]	98.1	96.2	91.2	87.2	89.8	87.4	84.1	91.0
[11]	98.5	96.3	91.9	88.1	90.6	88.0	85.0	91.5
[5]	98.2	96.8	92.2	88.0	91.3	89.1	84.9	91.8
[7]	98.5	96.7	92.5	88.7	91.1	88.6	86.0	92.0
[6]	98.5	96.8	92.7	88.4	90.6	89.3	86.3	92.1
[21]	98.4	96.9	92.6	88.7	91.8	89.4	86.2	92.3
FPSM+SCAM	98.5	96.8	92.6	88.4	91.5	88.8	95.5	92.1

Table 3. Comparisons of PCK@0.2 score on the LSP test set

Methods	Head	Shoulder	Elbow	Wrist	Hip	Knee	Ankle	Total
[2]	97.8	92.5	87.0	83.9	91.5	90.8	89.9	90.5
[3]	97.2	92.1	88.1	85.2	92.2	91.4	88.7	90.7
[20]	97.9	93.6	89.0	85.8	92.9	91.2	90.5	91.6
[11]	98.1	93.7	89.3	86.9	93.4	94.0	92.5	92.6
[7]	98.3	94.5	92.2	88.9	94.4	95.0	93.7	93.9
[5]	98.2	94.9	92.2	89.5	94.2	95.0	94.1	94.0
[21]	98.3	95.9	93.5	90.7	95.0	96.6	95.7	95.1
FPSM+SCAM	98.3	95.8	95.7	94.3	95.7	96.5	95.4	96.0

Fig. 4. Examples of human pose estimation on MPII and LSP

5 Conclusion

In this paper, we propose to incorporate the Feature Pyramid Supervision Module and the Spatial and Channel Attention Module into an end-to-end framework. The Feature Pyramid Supervision Module (FPSM) extracts multi-scale features to improve the ability of human pose estimation against scale variations. In addition, the Spatial and Channel Attention Module (SCAM) is used to automatically learn and infer the semantic information which is beneficial for accurate estimation of body joints. The effectiveness of the Feature Pyramid Supervision Module and the Spatial and Channel Attention Module is evaluated on validation set of MPII Human Pose dataset. We evaluate our method on two basic benchmarks and the results show that our method is useful for improving the performance of human pose estimation.

Acknowledgement. This work is supported by National Natural Science Foundation of China (Grant No. 61871046).

References

1. Toshev, A., Szegedy, C.: Deeppose: human pose estimation via deep neural networks (2013)
2. Wei, S., Ramakrishna, V., Kanade, T., Sheikh, Y.: Convolutional pose machines (2016)
3. Bulat, A., Tzimiropoulos, G.: Human pose estimation via convolutional part heatmap regression. In: Leibe, B., Matas, J., Sebe, N., Welling, M. (eds.) ECCV 2016. LNCS, vol. 9911, pp. 717–732. Springer, Cham (2016). https://doi.org/10.1007/978-3-319-46478-7_44
4. Newell, A., Yang, K., Deng, J.: Stacked hourglass networks for human pose estimation. In: Leibe, B., Matas, J., Sebe, N., Welling, M. (eds.) ECCV 2016. LNCS, vol. 9912, pp. 483–499. Springer, Cham (2016). https://doi.org/10.1007/978-3-319-46484-8_29
5. Chou, C.J., Chien, J.T., Chen, H.T.: Self adversarial training for human pose estimation (2017)
6. Ke, L., Chang, M.C., Qi, H., Lyu, S.: Multi-scale structure-aware network for human pose estimation (2018)
7. Yang, W., Li, S., Ouyang, W., Li, H., Wang, X.: Learning feature pyramids for human pose estimation. In: 2017 Computer Vision and Pattern Recognition (2014)
8. Andriluka, M., Pishchulin, L., Gehler, P., Schiele, B.: 2d human pose estimation: new benchmark and state of the art analysis. In: Computer Vision and Pattern Recognition (2014)
9. Johnson, S., Everingham, M.: Clustered pose and nonlinear appearance models for human pose estimation. In: British Machine Vision Conference (2010)
10. Insafutdinov, E., Pishchulin, L., Andres, B., Andriluka, M., Schiele, B.: DeeperCut: a deeper, stronger, and faster multi-person pose estimation model. In: Leibe, B., Matas, J., Sebe, N., Welling, M. (eds.) ECCV 2016. LNCS, vol. 9910, pp. 34–50. Springer, Cham (2016). https://doi.org/10.1007/978-3-319-46466-4_3
11. Chu, X., Yang, W., Ouyang, W., Ma, C., Yuille, A.L., Wang, X.: Multi-context attention for human pose estimation. In: Computer Vision and Pattern Recognition (2017)
12. Cao, Z., Simon, T., Wei, S.E., Sheikh, Y.: Realtime multi-person 2D pose estimation using part affinity fields (2016)
13. Tsung, Y.L., Dollar, P., Girshick, R., He, K., Belongie, S.: Feature pyramid networks for object detection (2016)
14. Itti, L., Koch, C.: Computational modelling of visual attention. Nat. Rev. Neurosci. **2**(3), 194–203 (2001)
15. Zhang, H., Goodfellow, I., Metaxas, D., Odena, A.: Self-attention generative adversarial networks (2018)
16. Li, H., Xiong, P., An, J., Wang, L.: Pyramid attention network for semantic segmentation (2018)
17. Fu, J., Liu, J., Tian, H., Fang, Z., Lu, H.: Dual attention network for scene segmentation (2018)
18. He, K., Zhang, X., Ren, S., Sun, J.: Deep residual learning for image recognition. In: Proceedings of the IEEE Conference on Computer Vision and Pattern Recognition, pp. 770–778 (2016)

19. Tieleman, T., Hinton, G.: Lecture 6.5-rmsprop: divide the gradient by a running average of its recent magnitude. COURSERA: Neural Netw. Mach. Learn. **4**(2), 26–31 (2012)
20. Ke, S., Lan, C., Xing, J., Zeng, W., Dong, L., Wang, J.: Human pose estimation using global and local normalization. In: IEEE International Conference on Computer Vision (2017)
21. Tang, W., Yu, P., Wu, Y.: Deeply learned compositional models for human pose estimation. In: Proceedings of the European Conference on Computer Vision (ECCV), pp. 190–206 (2018)

Short-Term Traffic Flow Prediction Based on SVR and LSTM

Yi Wang[✉], Jiahao Xu, Xianwu Cao, Ruiguan He, and Jixiang Cao

School of Logistics, Wuhan University of Technology, Wuhan, Hubei, China
isisrun@126.com

Abstract. To alleviate traffic congestion and support the development of real-time traffic and public transport, this paper conducts research on adopting support vector regression (SVR) and long short term memory (LSTM) to predict traffic flow of the lane, and then compares the results with that using the quadratic exponential smoothing. The consequence shows that SVR and LSTM have better prediction accuracy, about 1%–3% in terms of MAPE, than quadratic exponential smoothing, and SVR is slightly better than LSTM. Furthermore, in order to improve the predictive accuracy of model, we compare the performance of grid search, whale optimization algorithm (WOA) and genetic algorithm (GA) respectively in the respect of optimizing models' parameters. The optimization effect of WOA-SVR and WOA-LSTM is better than the other two models respectively, about 0.9% and 2.52% better than GA-SVR and GA-LSTM while 0.29% and 2.32% better than GridSearch-SVR and GridSearch-LSTM considering MAPE.

Keywords: Short-term traffic flow prediction · Support vector regression · Long short term memory · Whale optimization algorithm · Genetic algorithm

1 Introduction

In recent years, with the rising of people's living standard and the increase of their traveling demand, the amount of owned motor vehicles throughout the country is increasing year by year. According to the *Statistics Bulletin on National Economy and Social Development of the People's Republic of China in 2019* [1] and the *Statistics Bulletin on National Economy and Social Development of Wuhan in 2019* [2], both the owned civilian cars nationwide and in Wuhan are growing fast, having reached 261.5 million and 3.14 million respectively by the end of 2019. The rapid growth of owned cars will inevitably impose a huge burden on the existing road, drawing forth problems such as traffic congestion, traffic accidents and energy consumption that not only reduces the maneuverability of city's operation, but also generates huge negative impacts on the development of the city and regional economy.

With continuous development of traffic information collecting technology, most of domestic cities have begun to construct Intelligent Transportation System (ITS). ITS is a comprehensive, efficient and real-time intelligent traffic management system that can effectively mitigate traffic congestion, improve the pass rate of road's network and

© Springer Nature Switzerland AG 2021
Q. Zu et al. (Eds.): HCC 2020, LNCS 12634, pp. 338–348, 2021.
https://doi.org/10.1007/978-3-030-70626-5_36

provide information service for people's traveling. Two core technologies of ITS are traffic control and traffic guidance. As one of core technologies of ITS's development, short-term traffic flow prediction can provide basic data for them. Therefore, how to effectively predict the short-term traffic flow has significant research values and it has received wide attention of scholars.

Short-term traffic flow generally refers to the time series data whose time interval is less than 15 min. Traditional linear traffic flow prediction models (e.g. Kalman filter model, time series model) are based on statistics theory. This kind of model has great interpretability, simple modeling process and low complexity [3–5]. However, it possesses low prediction accuracy and could not reflect the uncertainty and nonlinear features or satisfy the requirements of complex and dynamic traffic conditions.

Prediction models based on nonlinear system theory (e.g. wavelet analysis and chaos theory) possess relatively high accuracy [6–8]. It could satisfy nonlinear features of traffic system but with high computational complexity.

In recent years, with the development of machine learning and deep learning, scholars convert their research points to utilize intelligent algorithms to build prediction models. Some scholars utilized least square support vector machine to predict [9]. Others proposed different improved neural networks to predict [10–12]. Intelligent prediction models based on knowledge discovery have strong data fitting ability and could obtain satisfactory prediction effects when sample data is abundant. However, it has high computational complexity and is difficult to set appropriate parameters. Therefore, the application of these models still has large research space.

Composition prediction models are to combine two or more models together to predict traffic flow, which could make the best of every model's advantages and achieve better prediction effects [13–17]. Composition models could improve prediction performance through complementarity. However, how to combine different methods is still required to be studied because inappropriate combination may lower model's performance.

Support vector machine (SVM) is a supervised learning model. At the beginning, it is used to solve the classification problems. With the introduction of the insensitive loss function, SVM could solve nonlinear regression estimate problems, which is called SVR. Based on structural risk minimization (SRM), SVR has been proved to have better generalization ability than neural network which is based on empirical risk minimization (ERM). By far, SVR has been successfully applied in the fields of traffic flow prediction.

In the application of using neural network to predict time series data, recurrent neural network (RNN) could acquire data's short-time features. However, it was found that long-time historical information also generated huge influences on the predictions. Therefore, an improved RNN called long short term memory (LSTM) is created, which could maintain data's long-time features.

Whether it is SVR or LSTM, the selection of model's parameters plays a pivotal role on model's prediction ability, so it is a must to select proper parameters to gain great prediction effect. Based on this, the paper will study the application of SVR and LSTM in short-term traffic flow prediction with emphasis of the parameter selection problems. It makes comparisons among three selection methods and demonstrates the feasibility and robustness of optimized models through case study.

2 Representation of Data

2.1 Collection of Data

In this paper, the data is acquired from Youyi Avenue and Yuanlin Road intersection's N31831G geomagnetic collector. The data set includes traffic flow data about the west-to-east lane of the intersection from April 1 to April 30, 2018 and was collected every 15 min from 0:00 to 24:00, 2880 records in total.

2.2 Analysis and Preprocessing of Data

Quality Analysis of Data. Data quality analysis is the precondition of data preprocessing including analysis of missing values and abnormal values. Missing data mainly includes the missing of record and field information. In the collected data, 26 records are missing, mainly because of the failure of collection equipment, storage media or transmission media. As for abnormal values analysis, the paper uses the box plot theory for the analysis because it is based on quartile and interquartile range which possesses robustness and can generate objective results. From box plot, the mean of traffic flow is 397, the median is 428, the upper and lower bounds are 991 and 0. A total of 13 abnormal values is aggregated.

Preprocessing of Data. According to the analysis above, the paper regards numbers bigger than 991 as abnormal values and processes them as missing values. On the basis of the degree, methods to process missing values can be divided into three categories: deleting, interpolating and no processing. This paper uses interpolating to process because there are not so many of them. For single missing data, average interpolating is used to process it, the formula is as follows.

$$y(t) = round\left(\left[y(t-1) + y(t+1)\right]/2\right) \tag{1}$$

For continuous multiple missing data, weighted average interpolating is used.

$$y(t+1) = round\left(a * y(t-1) + (1-a) * y^{(k-1)}(t)\right) \tag{2}$$

$round()$ is the rounding fusnction. a is the weighting coefficient, representing the proportion of the historical trend data and the current data's contributions to data repairing, and the greater it is, the greater the current data will influence the repaired data. The paper makes $a = 0.5$. $y(t-1)$ represents the data at the last moment of missing moment. $y^{(k-1)}(t)$ represents the data at the same moment of the previous day. The line chart of repaired data in three days is shown in Fig. 1.

23 days' traffic flows ($23 \times 96 = 2208$) are considered as training set and 7 days' data ($7 \times 96 = 672$) are considered as test set.

The paper uses z-score to standardize the data because of two reasons. One is that different orders of magnitude of data will easily affect the robustness of the model because SVR is based on the assumption of normally distributed data. The other is that

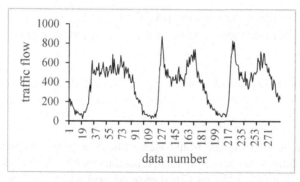

Fig. 1. Repaired data in three days

standardized data can improve the convergence speed of the algorithm. The formula is shown below.

$$z_i = \frac{x_i - \mu}{\delta} \tag{3}$$

x_i represents traffic flow at i time, μ and δ represents the mean and variance of data set respectively, z_i represents the standardized traffic flow.

2.3 Evaluation Indexes

Prediction performance evaluation indexes adopted in the paper are Mean Absolute Percentage Error (MAPE), Mean Absolute Error (MAE), Root Mean Square Error (RMSE) and Equal Coefficient (EC). The smaller MAPE, RMSE and MAE is, the higher the prediction precision is and the closer the EC gets to 1, the stronger the linear relationship is.

3 Prediction Using SVR or LSTM Based on Grid Search to Optimize Parameters

The experiment is based on the 64-bit Intel i5-7300HQ processor of Windows10 system with 8 GB RAM and realized in Python 3.7 environment. The code is written and executed in PyCharm.

3.1 GridSearch-SVR Model and GridSearch-LSTM Model

GridSearch-SVR Model. The paper mainly considers how traffic flows at last few moments influence those in the future. It uses past traffic flow time series $\{y_1, y_2,..., y_q\}$ to predict the traffic flow y_{q+s} at moment $q+s$ in the future. y_q is the traffic flow at moment q. q is the delay order, which presumes that traffic flow at first q time periods is associated with future predictions. s is the prediction step length. In order to make the input adaptable for the model, it is needed to construct the structure of data set.

Provided that it is given a data set $D = \{(X_i, y_i) | i = 1, 2, \ldots, l - q - s + 1\}$, where $X_i = (x_i, x_{i+1}, \ldots, x_{i+q-1})$ are the input variables and $y_i = x_{i+q+s-1}$ is the output variable, which is shown below.

$$X = \left[(x_1, x_2, x_3, \ldots, x_{l-q-s+1})^T, (x_2, x_3, x_4, \ldots, x_{l-q-s+2})^T, \ldots, (x_{q-1}, x_q, x_{q+1}, \ldots, x_{l-s})^T \right] \tag{4}$$

$$Y = \left[y_{q+s}, y_{q+s+1}, y_{q+s+2}, \ldots, y_l \right]^T \tag{5}$$

The paper uses SVR module from Sklearn library in python to predict the traffic flow. The first thing is to choose the related parameters of SVR including the kernel function, regularization coefficient C and error channel width ε etc. Several common kernel functions in python are linear, polynomial, sigmoid, and RBF kernel. The paper adopts RBF kernel function whose formula is as follows.

$$k(x_i, x_j) = \exp\left(-\frac{\|x_i - x_j\|^2}{2\sigma^2} \right) \tag{6}$$

Grid Search is adopted to determine the three parameters in RBF kernel function, where $C \in [0.1, 10]$, $\varepsilon \in [0.1, 1]$, $\gamma \in [0.1, 1]$. To prevent overfitting, the search step length of the parameters are set to 0.1. The search process is to minimize MAPE and the optimized parameters are $C = 0.9$, $\varepsilon = 0.1$, $\gamma = 0.3$. The operation time is 18.87 min. The convergence curve of MAPE is shown in Fig. 2.

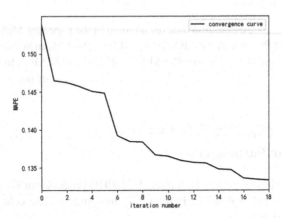

Fig. 2. Convergence curve of MAPE in GridSearch-SVR model

The paper takes delay order $q = 10$ and the prediction step length $s = 1$. Then, the paper uses the data in test set to predict with the previously trained SVR model. The left part of Fig. 3 is the fitting chart of predicted values and actual values in testing set while the right part is the fitting chart of them on a day and the scatterplot of their residuals. As is seen below, the curve to fit the test set basically conforms to the curve of actual values. When traffic flow undergoes large changes, it remains relatively consistent, proving the prediction is practical. Because of the interference of random factors, there is residual error between the predicted values and the actual values and it shows the characteristics of random fluctuations. Four indexes used to evaluate the performance of SVR are MAPE = 13.31%, MAE = 40.09, RMSE = 54.97 and EC = 93.63%. It is found that the prediction precision is above 86% and the linear relationship between two kinds of values is above 0.93, which demonstrates that SVR is feasible and effective in short-term traffic flow prediction.

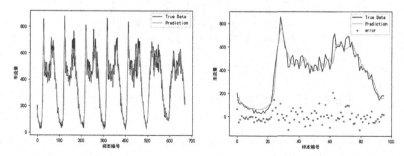

Fig. 3. Fitting chart and residual scatterplot in GridSearch-SVR model

GridSearch-LSTM Model. The paper builds LSTM model with the same data set, standardized method as SVR to predict short-term traffic flow. It utilizes LSTM module from Keras library using Tensorflow backend in python to build the model.

One kind of the parameters in LSTM is auto-regulatory in the process of training, such as weight and bias. The other is the parameters needing to manually set, including the number of input layer, hidden layer and output layer and neural node in each layer, activation function etc. Currently, there is no certain method to choose the artificial parameters and the most used method is empirical method. Similar to the method to select parameters in SVR, the paper uses grid search to determine dropout, the number of hidden layer and its neural node. The input length and prediction step length is set the same as SVR. The batch size, epoch, optimizer and loss function are determined with empiricism.

Considering the running time of grid search, the paper sets the range of hidden layer's number between [1, 5] with step length 1 and makes the number of neural node in every hidden layer equal, between [30, 50] with step length 5. Dropout is set between [0.1, 0.3] with step length 0.1 to prevent overfitting. The changes of MAPE is shown in Fig. 4. It is found that the curve fails to converge in this model and the search is terminated when the traversal is over.

The batch size is set to 128 and epochs to 20. Rmsprop optimizer and mse loss function is used in LSTM model. Taking the minimum MAPE as the optimization goal, the optimized parameters are: The number of hidden layers are 1 and its nodes are 40, dropout is 0.1. The operation time is 58.62 min. The four indexes are MAPE = 15.36%, MAE = 43.13, RMSE = 58.34 and EC = 93.31%. According to the indexes, the precision is about 85%, the linear relationship is above 0.93, which illustrated that LSTM has good memory when processing long time series and does not generate 'gradient

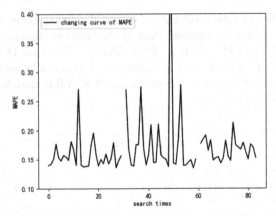

Fig. 4. Changes of MAPE based on GridSearch-LSTM model

explosion'. Furthermore, LSTM has great feasibility and portability in short-term traffic flow prediction. The Fig. 5 is shown below.

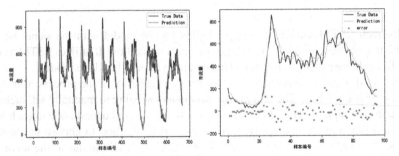

Fig. 5. Fitting chart and residual scatterplot in GridSearch-LSTM model

3.2 Comparison with Quadratic Exponential Smoothing

To find the difference between intelligent and traditional theory in terms of prediction performance, the paper compares prediction results between two kinds of intelligent methods and the traditional quadratic exponential smoothing.

The smoothing coefficient $\alpha = 0.5$ and $q = 10$, $s = 1$. The last 7 days of data set is regarded as the prediction object. The Fig. 6 is shown below.

As is seen from the figure, the prediction results roughly conform to the variation trend of traffic flow, but in the daily prediction chart, the predicted values pull ahead of the actual values. When traffic flow goes through large changes, the residual error is growing larger. Evaluation indexes of three different prediction methods are shown in Table 1.

As is seen from the table, compared with the quadratic exponential smoothing, SVR and LSTM are significantly better on four indexes. This phenomenon shows that SVR

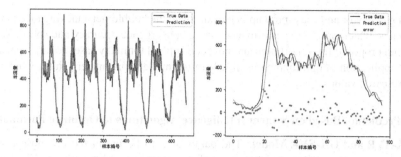

Fig. 6. Fitting chart and residual scatterplot in quadratic exponential smoothing model

Table 1. Evaluation indexes of three different prediction methods

Prediction methods	MAPE	MAE	RMSE	EC
GridSearch-SVR	13.31%	40.09	54.97	93.63%
GridSearch-LSTM	15.36%	43.13	58.34	93.31%
Quadratic exponential smoothing	16.48%	51.17	71.59	91.76%

and LSTM have superior performance over the quadratic exponential smoothing because intelligent prediction methods are more capable to capture the random disturbance factor and make accordingly adjustments compared with the traditional methods based on statistical theory. Furthermore, it still has great prediction effect when traffic flow changes drastically. Between the two intelligent methods, SVR is better than LSTM, about 2.05% in MAPE, and has faster calculation speed.

4 Prediction Using SVR or LSTM Based on Swarm Intelligence Algorithms to Optimize Parameters

To achieve more accurate prediction and accelerate the convergence, the paper utilizes two swarm intelligence algorithms, whale optimization algorithm and genetic algorithm to optimize the parameters of two intelligent models.

4.1 Swarm Intelligence Algorithms

Genetic Algorithm (GA) and Whales Optimization Algorithm (WOA) are two kinds of swarm intelligence algorithms. GA is an adaptive global optimization search algorithm that is formed by simulating organisms' genetic and evolutionary in the natural environment, which has been widely applied in machine learning, pattern recognition, neural network and other fields. WOA [18–20], presented by Australian academics Mirjalili and Lewis, is formed by simulating humpbacks' hunting behaviors which is to search, encircle and attack the prey to find the best object. Different from other swarm intelligence algorithms, it uses random or the best search iteration to simulate whales' hunting behavior and uses spiral update to simulate the bubble-net attacking mechanism of whales, which has good performance on the convergence accuracy and speed. WOA has been proposed for a short time and not been applied for many cases so far, especially in the application of optimizing machine learning model's parameters, thus rendering it high research values.

4.2 Prediction Based on Swarm Intelligence Algorithms to Optimize Parameters

WOA-SVR and GA-SVR Model. The parameters to be optimized compose a one-dimensional vector. WOA and GA is used to optimize regularization coefficient C, error channel width ε, kernel coefficient γ, delay order q and prediction step length s respectively. Upper and lower bound is the range of whale's position and also the range of search space. They are $C \in [10^{-5}, 10]$, $\varepsilon \in [10^{-5}, 1]$, $\gamma \in [10^{-5}, 1]$, $q \in [1, 50]$, $s \in [1, 10]$.

$$Maximum\ generation = \frac{\log\left(\frac{1}{\text{pop\_size}}\right)}{\log(\text{sub\_prob\_ratio})} \tag{7}$$

The optimized MAPE is 13.92%. Optimized parameters are as follows: $C = 2.94$, $\varepsilon = 0.11$, $\gamma = 0.04$, $q = 29$, $s = 1$. The operation time is 5.68 min.

WOA-LSTM and GA-LSTM Model. The paper uses WOA or GA combined with empirical method to determine LSTM model's artificial parameters. Considering the time cost and equipment factors, the paper uses WOA or GA to determine the number of hidden layer and its nodes, dropout, input length and prediction step length and uses empirical method to determine batch size, epochs, optimizer and loss function which are the same as the previous –LSTM models.

The five important parameters are arranged into a one-dimensional vector. The corresponding relation is hidden layers $\in [1, 5]$, each hidden layer's nodes $\in [20, 50]$, dropout $\in [0.01, 0.3]$, input length $\in [10, 50]$, prediction step length $\in [1, 10]$.

After optimization, the number of hidden layer and its nodes, input length and prediction step length are rounded down. In WOA-LSTM model, the paper sets the number of whales to 5 and the solving generation to 10. Optimized parameters are: The number of hidden layers are 1 and their nodes are 20, dropout is 0.05, input length is 26, prediction step length is 1. The operation time is 29.35 min. In GA-LSTM model, the number of chromosome of GA is set to 300 while the other parameters are the same as GA-SVR model. Optimized parameters are: The number of hidden layers are 2 and their nodes are 34, dropout is 0.09, input length is 26, prediction step length is 1. The operation time is 31.98 min.

Comparison Among Three Methods. The comparison among three methods to optimize the parameters of SVR model is represented in Table 2.

As is seen from the table, The convergence speed of WOA-SVR and GA-SVR is faster than GridSearch-SVR. Among three methods, WOA performs the best, four indexes of which are better than the other two methods and the convergence time is also the shortest. Although GA performs poorer than grid search in terms of four indexes, the convergence time is far shorter than grid search.

Table 2. Comparison among three parameter optimization methods of SVR model

Prediction methods	MAPE	MAE	RMSE	EC	Time (minutes)	Convergence
GridSearch-SVR	13.31%	40.09	54.97	93.63%	18.87	Slow
WOA-SVR	13.02%	40.03	54.53	93.69%	3.46	Fast
GA-SVR	13.92%	41.52	56.13	93.34%	5.68	Fast

The comparison among three methods in parameters optimization of LSTM model is represented in Table 3.

Table 3. Comparison among three parameter optimization methods of LSTM model

Prediction methods	MAPE	MAE	RMSE	EC	Time (minutes)	Convergence
GridSearch-LSTM	15.36%	43.13	58.34	93.31%	58.62	No convergence
WOA-LSTM	13.04%	40.83	55.61	93.52%	29.35	Fast
GA-LSTM	15.56%	45.30	59.76	93.07%	31.98	Fast

As is seen from the table, WOA performs the best when optimizing LSTM model's parameters, which possesses the greatest indexes and shortest search time. The optimization effect of GA is a little poorer than grid search and WOA. The search time of WOA and GA are faster than grid search and GridSearch-LSTM does not converge.

In conclusion, swarm intelligence algorithm could accelerate the convergence. WOA shows fabulous results when optimizing the parameters of SVR and LSTM, possessing best indexes and shortest operation time. It represents great optimization speed and simple coding and has a good applicability in engineering practice. However, the standard GA may not be quite suitable in this kind of optimization.

5   Conclusion

On the basis of data collected from the intersection in Wuhan, the paper conducts research on how to improve the accuracy of short-term traffic flow prediction. It adopts SVR and LSTM respectively to make predictions and then compares the results with that using quadratic exponential smoothing. Furthermore, as for model's parameters optimization, the paper compares two swarm intelligence algorithms - WOA and GA with grid search. The results demonstrate that two intelligent models have greater accuracy than traditional quadratic exponential smoothing in the field of short-term traffic flow prediction. WOA and GA is faster than grid search in parameters optimization. Besides, WOA is the best among three methods in all aspects.

References

1. National Bureau of Statistics [EB/OL]. http://www.stats.gov.cn/tjsj/zxfb/202002/t20200228_1728913.html. Accessed 17 Nov 2020
2. Wuhan Bureau of Statistics [EB/OL]. http://tjj.wuhan.gov.cn/tjfw/tjgb/202004/t20200429_1191417.shtml. Accessed 17 Nov 2020
3. Kumar, S.V., Vanajakshi, L.: Short-term traffic flow prediction using seasonal ARIMA model with limited input data. Eur. Transp. Res. Rev. 7(3), 1–9 (2015)
4. Cai, L.R., Zhang, Z.C., Yang, J.J., Yu, Y.D., Zhou, T., Qin, J.: A noise-immune Kalman filter for short-term traffic flow forecasting. Phys. A Stat. Mech. Appl. 536, 122601 (2019)
5. Zhou, T., Jiang, D.Z., Lin, Z.Z., Han, G.Q., Xu, X.M., Qin, J.: Hybrid dual Kalman filtering model for short-term traffic flow forecasting. Iet Intell. Transp. Syst. 13(6), 1023–1032 (2019)
6. Huang, Y.F.: Short-term traffic flow forecasting based on wavelet network model combined with PSO. In: International Conference on Intelligent Computation Technology and Automation on Proceedings, pp. 249–253 (2008)
7. Chen, Q.X., Song, Y., Zhao, J.F.: Short-term traffic flow prediction based on improved wavelet neural network. Neural Computing and Applications (2020)
8. Adewumi, A., Kagamba, J., Alochukwu, A.: Application of chaos theory in the prediction of motorised traffic flows on urban networks. Mathematical Problems in Engineering, vol. 2016 (2016)
9. Luo, C., et al.: Short-term traffic flow prediction based on least square support vector machine with hybrid optimization algorithm. Neural Process. Lett. 50(3), 2305–2322 (2019)
10. Yasdi, R.: Prediction of road traffic using a neural network approach. Neural Comput. Appl. 8, 135–142 (1999)
11. Vlahogianni, E.I., Karlaftis, M.G., Golias, J.C.: Optimized and meta-optimized neural networks for short-term traffic flow prediction: a genetic approach. Transp. Res. Part C Emerg. Technol. 13(3), 211–234 (2005)
12. Mou, L.T., Zhao, P.F., Xie, H.T., Chen, Y.Y.: T-LSTM: a long short-term memory neural network enhanced by temporal information for traffic flow prediction. IEEE Access 7, 98053–98060 (2019)
13. Hou, Q.Z., Leng, J.Q., Ma, G.S., Liu, W.Y., Cheng, Y.X.: An adaptive hybrid model for short-term urban traffic flow prediction. Phys. Stat. Mech. Appl. 527, 121065 (2019)
14. Zhang, H., Wang, X.M., Cao, J., Tang, M.N., Guo, Y.R.: A multivariate short-term traffic flow forecasting method based on wavelet analysis and seasonal time series. Appl. Intell. 48(10), 3827–3838 (2018)
15. Cheng, A.Y., Jiang, X., Li, Y.F., Zhang, C., Zhu, H.: Multiple sources and multiple measures based traffic flow prediction using the chaos theory and support vector regression method. Phys. Stat. Mech. Appl. 466, 422–434 (2017)
16. Luo, X.L., Li, D.Y., Yang, Y., Zhang, S.R.: Spatiotemporal traffic flow prediction with KNN and LSTM. Journal of Advanced Transportation (2019)
17. Zhou, J.M., Chang, H., Cheng, X., Zhao, X.M.: A multiscale and high-precision LSTM-GASVR short-term traffic flow prediction model. Complexity, vol. 2020 (2020)
18. Mirjalili, S., Lewis, A.: The whale optimization algorithm. Adv. Eng. Softw. 95(5), 51–67 (2016)
19. Nguyen, H., Bui, X.-N., Choi, Y., Lee, C.W., Armaghani, D.J.: A novel combination of whale optimization algorithm and support vector machine with different kernel functions for prediction of blasting-induced fly-rock in quarry mines. Nat. Resour. Res. 30(1), 191–207 (2020). https://doi.org/10.1007/s11053-020-09710-7
20. Gao, Y., Chen, K., Gao, H., Zheng, H.M., Wang, L., Xiao, P.: Energy consumption prediction for 3-RRR PPM through combining LSTM neural network with whale optimization algorithm. Mathematical Problems in Engineering, vol. 2020 (2020)

A Logging Overhead Optimization Method Based on Anomaly Detection Model

Yun Wang[⊠] and Qianhuizhi Zheng

School of Computer Science and Engineering, Southeast University, Key Lab of Computer Network and Integration, Ministry of Education, Nanjing 211189, China
ywang_cse@seu.edu.cn

Abstract. Logs play an important role in system anomaly detection. However, in today's large-scale software development and production, the cost of logging is non-negligible, and intensive logging in actual production processes will generate a large amount of redundant logs which are useless for anomaly detection. However, the current method of solving related problems is not ideal, and it is only applied when the overhead of the logging has affected the quality of service. Therefore, this paper proposes a method for optimizing logging records for this problem. Under a given budget (defined as the maximum volume of logs allowed to be output in a time interval), using an anomaly detection model based on deep learning and a two-phase filtering mechanism, the method determines whether to log according to the utility score of the log for anomaly detection to save useful logs and discard less useful logs during the system running process. The experimental results show that the proposed method alleviates the logging overhead problem without reducing the logging effectiveness.

Keywords: Logging · Anomaly detection · Overhead optimization

1 Introduction

With the rapid development of distributed computing and cloud computing, large-scale systems composed of hundreds of software components are running on thousands of computing nodes. Their runtime information is continuously collected and stored in log files. The log information is widely used in fault diagnosis, anomaly detection, performance monitoring, and data analysis. Although the role of logs is significantly indispensable, logging must bring additional overhead, such as disk I/O bandwidth, CPU, and memory consumption. As the system becomes larger and larger, the number of logs is very considerable (e.g. A large-scale service system can generate 50 GB (120–220 million) logs per hour [1]. The logging overhead problem cannot be ignored, so blindly recording too many logs is not desirable. In the data center, a large amount of log information is generated every day, some of which are irrelevant, and others are early warning of failures. Abnormal log information is found in time can reduce the occurrence of failures and ensure the smooth operation of business. Finding useful information for

© Springer Nature Switzerland AG 2021
Q. Zu et al. (Eds.): HCC 2020, LNCS 12634, pp. 349–359, 2021.
https://doi.org/10.1007/978-3-030-70626-5_37

anomaly detection in such a large amount of data scale is tantamount to finding a needle in a haystack. How to effectively remove useless redundant logs is urgently needed.

Recently, a few studies have focused on designing tools that systematically and automatically enhance and improve log contents [2–6]. In general, the state-of-art methods for logging optimization problems are all based on static analysis or machine learning to establish statistical models to improve the content of the log information. However, these methods only analyze and learn the source code to give suggestions for adding or modifying the logs, which will eventually bring about problems such as code recompilation and redeployment, which affects efficiency and does not take into account the overhead cost caused by logging. Existing technologies for reducing logging overhead, includes manually deleting part of the log print statement, changing the log level, and using sampling to output logs [7–9] are all aimed to reduce the number of output logs to alleviate overhead, which do not take into account the effectiveness of logs. For example, random sampling output is likely to miss important log contents, resulting in the possible loss of key information.

Therefore, the challenge of the problem is that how to balance the volume of logging information and logging enough information to analyze system abnormality. The main contribution of this paper is summarized as following:

1) A noval mechanism based on machine learning and data mining is proposed for logging overhead optimization to reduce the output of irrelevant logs.
2) Two-phase filtering is presented to evaluate logging information in order to effectively alleviate the resource overhead in the scenario of anomaly detection based on log analysis.
3) Experimental results show the efficiency of the aforementioned methods which are ultimately achieve the purpose of reducing the overhead of logging while preserving logging effectiveness.

2   Overview of Approach

In the scenario of anomaly detection based on log analysis, this paper proposes a new approach to solve the problem about logging overhead with the large logs in a distributed data center. This mechanism uses the knowledge about machine learning and data mining to analyze and train models offline on the previously saved logs. The trained model is used for online optimization and adjustment about logging during the system operation. Due to the budget for resource overhead by logging, the optimization mechanism will adaptively determine whether to log according to the budget constraint, and in the meanwhile, it should also preserve logging effectiveness (i.e., the key log cannot be lost). The overview of the proposed approach is shown in Fig. 1.

Specifically, firstly, the log-based anomaly detection model is trained offline by analyzing the historical log data, which mainly includes log analysis, feature extraction, and model training functions. Then the trained model will be used as input for local filtering. The log stream pass through the model and get a utility score. Then two-phase filter is used to determine whether to log online during the system operation. In the local filter phase, a large number of irrelevant logs are discarded efficiently and the logs that

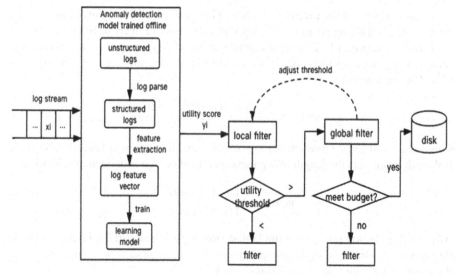

Fig. 1. Overview of the proposed approach.

are strongly related to anomaly detection are retained. A utility threshold is set to filter logs, and if the utility score of logs greater than this threshold, the logs can pass the local filtering phase to the global filtering phase. Finally, a budget is predefined for resource overhead costs in the global filtering phase. The cached logs are sorted according to utility. The global filtering phase can periodically trigger disk write request, and output some high-utility logs with a total cost less than the budget to the disk. In addition, the global filtering phase can dynamically adjust the utility threshold, and iteratively update it according to the amount of cached logs and the corresponding threshold in each time interval during operation.

3 Methodology

3.1 Log Parsing

Log parsing is usually the first and foremost step of various log-based analysis tasks, because the raw log data is unstructured and free-text, which often cannot be used directly for statistical analysis. It's necessary to process raw logs. Usually a log message is composed of a constant part and a variable part. The constant part often reveals the event template of the log, and the variable part carries the information of the dynamic runtime. Each log is composed of a fixed template and corresponding parameters (i.e., variable parts). The purpose of log parsing is to parse each free-text log into a structured event template.

Specifically, the process of traversing the parse tree according to the special-designed rules of internal node generally includes three steps:

Firstly, starting from the root node, the first layer of nodes represents log groups whose log messages are of different log message lengths. The length of log messages refers to the number of tokens in a log message.

Then grouping by the tokens in the beginning position of the log message, because tokens in the beginning positions of a log message are more likely to be constants.

Finally reaching a leaf node, it contains a list of log groups. To select the most suitable log group, calculate the similarity between the log message and the log event of each group simSeq:

$$simSeq = \frac{\sum_{i=1}^{n} equ(seq1(i), seq2(i))}{n} \tag{1}$$

where seq 1 and seq 2 represent the log message and the log event respectively; seq(i) is the i-th token; n is the length of log message; function equ is defined as following:

$$equ(t_1, t_2) \begin{cases} 1 & if \ t_1 = t_2 \\ 0 & otherwise \end{cases} \tag{2}$$

After finding the log group with the largest simSeq, it is compared with the similarity threshold st. If simSeq > st, the group is returned as the most suitable group. Otherwise, the group is not found and a new group is created.

After the log parsing is completed, the event templates are numbered, and add the corresponding log template number to each raw log to generate the structured log data.

3.2 Anomaly Detection Model

Feature Extraction. After the log is parsed into the corresponding event template, it needs to be further encoded into a vector with numerical features to be suitable for the input of the machine learning model. To this end, it is necessary to use different grouping techniques to slice the raw logs into a set of log sequences. This paper is based on the HDFS dataset and selects the session windows technique for feature extraction. The session windows technique is based on different identifiers to group logs. In some log data, identifiers can be utilized to mark different execution paths. Taking the HDFS system log data as an example, each block has a unique block_id. The HDFS logs use block_id to record various operations of a certain block. Logs are grouped into different sessions by block_id, and each session group is the life cycle of one block.

For traditional machine learning models (such as, SVM, logistic regression, etc.), the log sequence in each group needs to be further processed to extract the event count vector. They need to count the number of appearances for each distinct event within each session. Eventually, all vectors are constructed to be an event count matrix, which can be fed into traditional machine learning models. For the deep learning model, the weights and sub-sequences of each session window can be directly trained without statistically extracting event count vectors.

Anomaly Detection Model Based on Deep Learning. The anomaly detection model based on deep learning is constructed to perform anomaly detection in an online, streaming way. The key intuition behind it is: view the log sequence as a natural language that follows certain patterns and grammar rules. In fact, the system log is produced by a program that follows a rigorous set of logic and control flow generated by the program. LSTM (Long Short-Term Memory) networks have the ability to remember long-term dependencies over sequences [10]. To this end, LSTM is used to model the log sequence to obtain a deep neural network. The network can automatically learn the log patterns from the normal system execution path and mark the deviation from the normal execution as anomalies. In this paper, we use GRU (Gated Recurrent Unit) neural network for online anomaly detection over system logs. GRU is a very effective variant of the LSTM network with simpler structure. The GRU neural network model has the advantages of few parameters and fast training, which can improve the running speed while achieving high detection accuracy.

The following describes in detail how to use the log sequence to detect the abnormal execution path. Since the total number of distinct log print statements in the source code is constant, so is the total number of distinct log events. Let $K = \{k_1, k_2, \ldots, k_n\}$ be the set of distinct log events. Let m_t denote the event type at position t in a log sequence. Clearly, m_t should obviously belong to one of the n event types from K, and is strongly dependent on the most recent events that occurred prior to m_t. The anomaly detection in a log sequence is modeled as a multi-class classification problem, where each distinct log event defines a class. The input is a history of recent log events, and the output is a probability distribution over the n log events from K, representing the probability that the next log event in the sequence is $k_i \in K$. Figure 4 describes the classification setup. Suppose t is the sequence id of the next log event to occur. The input for the model is a window w of the h most recent log events, which is $w = \{m_{t-h}, \ldots, m_{t-2}, m_{t-1}\}$, where $m_i \in K$. Note that the same log event may occur several times in w. The output is the conditional probability distribution, which is $Pr[m_t = k_i|w], k_i \in K(i = 1, \ldots, n)$. The detection phase uses this model to make a prediction and compare the predicted output against the observed log event value that actually occurs (Fig. 2).

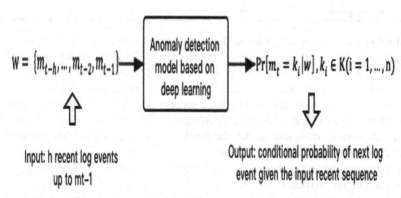

$$W = \{m_{t-h}, \ldots, m_{t-2}, m_{t-1}\} \rightarrow \boxed{\text{Anomaly detection model based on deep learning}} \rightarrow Pr[m_t = k_i|w], k_i \in K(i = 1, \ldots, n)$$

Input: h recent log events up to mt-1

Output: conditional probability of next log event given the input recent sequence

Fig. 2. Overview of the anomaly detection model based on deep learning.

In practice, there are many possibilities for the next log event to be normal, so all possible log events are sorted based on their probabilities $Pr[m_t = k_i|w]$, and treat an event as normal if it's among the top g candidates. A log event if flagged as from an abnormal execution otherwise.

Two-Phase Filter. Two-phase filtering is designed to determine whether to log during the system operation. After the log stream passes through the anomaly detection model, the corresponding utility score is obtained, and whether to be filtered is determined accordingly. The following is the overall description of the filtering algorithm. The input of Algorithm 1 are scores (a list of utility scores), threshold (utility threshold), and B (resource overhead budget).

Algorithm 1 Two-phase filter

Input: scores[], threshold, B
Output: The log id that should be output after filtering
1. scolen $\leftarrow$ length of scores
2. j $\leftarrow$ 0
3. local_index, global_index $\leftarrow$ empty list
4. V_{n-2}, V_{n-1} $\leftarrow$ log volume in the previous two iterations
5. G_{n-1}, G_{n-2} $\leftarrow$ utility threshold in the previous two iterations
6. **for each** i $\in$ [0, scolen] **do**
7.   **if** scores[i] > threshold **then**
8.   local_index[j] $\leftarrow$ i
9.   j $\leftarrow$ j+1
10.   **end if**
11. **end for**
12. V_n $\leftarrow$ length of local_index
13. V_{n-2} $\leftarrow$ V_{n-1}
14. V_{n-1} $\leftarrow$ V_n
15. global_index $\leftarrow$ After sorting the scores in descending order, take the index of the top B values
16. G_n $\leftarrow$ G_{n-1} − (V_{n-1}-B) * (G_{n-1} - G_{n-2}) / (V_{n-1} - V_{n-2})
17. G_{n-2} $\leftarrow$ G_{n-1}
18. G_{n-1} $\leftarrow$ G_n
19. **return** global_index

Local Filter. The major task of local filter component is to discard the most less useful logs and keep the useful ones for anomaly detection. Predefine a utility threshold, and the utility scores of logs are larger than this threshold will pass through the local filtering component to global filtering component. The utility score measures the usefulness of a log for anomaly detection, and the larger the score is if the log can indicate an anomaly. The main function of local filter phase is to calculate the utility score for each log. With the trained anomaly detection model, the newly arrived logs are fed into the model to calculate the corresponding utility score.

Specifically, using the anomaly detection model based on deep learning, when a new log event m_t arrives, $w = \{m_{t-h}, \ldots, m_{t-2}, m_{t-1}\}$ will be used as the input of this model to obtain the output of conditional probability distribution $Pr[m_t|w] = \{k_1 : p_1, k_2 : p_2, \ldots, k_n : p_n\}$. Then according to log event m_t is $k_i(k_i \in K)$, the corresponding probability p_i is obtained. Since the classification rule in the detection stage is: if the next log event is one of the top g candidates, it can be treated as normal. Therefore, the larger p_i is, the more likely to be a normal event. Therefore, define $1 - p_i$ as the utility score of the log, that is, the larger the utility score, the more abnormal is indicated. Compare the utility score of each log with the threshold, if score < threshold, then filter it in this phase, otherwise keep it to the global filtering phase.

Global Filter. There are two major tasks of global filtering component. One is to output the useful logs while complying with logging budget. The other is to adjust the utility threshold dynamically.

Predefine a resource budget for logging in the global filtering phase, and sort the caches logs according to the utility score, then flush the top ranked logs to disk so that the total log volume does not exceed the budget. The budget for logging is defined as "log bandwidth," which is the maximum volume of logs allowed to be output in a time interval (such as 1 KB/s or 1000 logs/s). In the practice production, I/O bandwidth is the most concerning overhead. In general, most logging overhead (such as disk storage, network I/O, and CPU usage) will be directly or indirectly affected by the I/O bandwidth. Therefore, it is reasonable to choose the resource budget such as "log bandwidth".

The utility threshold can effectively control the volume of logs that pass the local filtering phase. It is important to set a proper threshold. If the threshold is set too low, then massive logs can pass the local filtering phase, resulting in the larger overhead. In addition, due to budget constraints, most of logs will still be discarded at the global filtering phase. On the contrary, if the threshold is set too high, only a small amount of logs could pass the local filtering phase, thus important log information may be missed, resulting in unacceptable logging effectiveness. Therefore, the optimal situation is to choose a suitable threshold so that the volume of logs that can pass the local filtering phase is exactly the budget volume. This paper models the process of threshold adjustment, and designs an iterative algorithm to dynamically adjust the threshold. Each iteration duration is defined as adjust interval. Obviously, when the volume of logs in the previous adjust interval is higher than budget, then threshold should be increased. Conversely, threshold should be decreased when the volume of logs in the previous adjust interval is lower than budget. Such a process can be modeled as follows: given utility threshold g, the volume of all logs passing through the local filtering phase is:

$$b(g, t) = \{all\ the\ logs\ with\ utility\ scores\ larger\ than\ g\}$$

The equation is:

$$B = \int_{g^*}^{\infty} f(V_t, X)dX \tag{3}$$

where $f(V_t, x)$ represents probability distribution of utility score; t represents time; V_t represents some unknown parameter vector to model the probability distribution of utility score. The goal is to find a g^*, so that the volume is just the budget B.

Let G_n, V_n denote the threshold and log volume in the n-th adjust interval, and B denote logging budget. The threshold adjusting mechanism is as follows:

$$G_n = G_{n-1} - (V_{n-1} - B)\frac{G_{n-1} - G_{n-2}}{V_{n-1} - V_{n-2}} \tag{4}$$

To avoid dividing 0 in formula $\frac{G_{n-1}-G_{n-2}}{V_{n-1}-V_{n-2}}$, add 1 if $V_{n-1} - V_{n-2}$ is close to 0. When $G_{n-1} - G_{n-2}$ is equal to 0, threshold will trap to a certain number and never changes regardless to any dynamics. To avoid this, a very small offset 0.01 is added under such situation.

4 Evaluation

4.1 Setup

Environment. The experimental environment in this paper is the host with macOS Catalina 10.15.3 operating system, 3.6 GHz dual-core Intel Core i5 processor, and 4 GB memory. The version of Python is 3.7, the version of pytorch is 1.4.0, and the version of tensorboard is 2.0.2.

Dataset. Publicly available production logs are scarce data because companies rarely publish them due to confidential issues. The HDFS log dataset was successfully obtained by exploring an abundance of literature and contacting the corresponding authors. It is generated by setting up a Hadoop cluster on 203 EC2 nodes and running sample Hadoop map-reduce jobs for 38.7 h, and labeled by Hadoop domain experts. This log dataset is widely used in log analysis research and is representative. The specific information of the dataset is as follows (Table 1):

Table 1. Dataset Information.

System	Time span	Data size	Log messages	Anomalies
HDFS	38.7 h	1.55G	11175629	16838

4.2 Performance of Log Parsing

For the performance evaluation of the log parsing method, it mainly evaluates the accuracy and efficiency. As the log volume increases, the accuracy decreases slightly (because the type of log templates becomes more complex and diverse as the log volume increases), but it's relatively stable, achieving an accuracy over 90%. The result indicates that the log parsing method can achieve a stable accuracy and robustness. Obviously, the parsing time increases linearly with the raising of log size. When the data size is 100 KB (about 2000 logs), the parsing time is 0.25 s.

4.3 Performance of Anomaly Detection Model

The anomaly detection model based on deep learning only needs a small fraction of normal log events to train its model. In the case of HDFS log, only less than 1% of normal sessions are used for training. Specific training and test data settings are shown in the following Table 2:

The parameters for training and test are as follows: g = 5, h = 10, epoch = 100, num_layers = 2, hidden_size = 32. Use standard metrics such as Precision, Recall and F-measure to evaluate the model. The model has achieved a better overall performance, with an F-measure of 93.776%.

Table 2. Data settings.

System	Dataset division	Normal session	Abnormal session
HDFS	Training data	4855	0
	Test data	553366	16838

4.4 Logging Effectiveness

For the evaluation of logging effectiveness, this paper counts how many abnormal logs are actually covered in the resultant logs. According to the actual production environment, each time interval(30 s) there are about 4000 logs arriving. Compare the optimization method with sampling-based method, which is simply to randomly output part of the logs. Figure 3 shows how the logging effectiveness increases as the budget size/sampling rate increases.

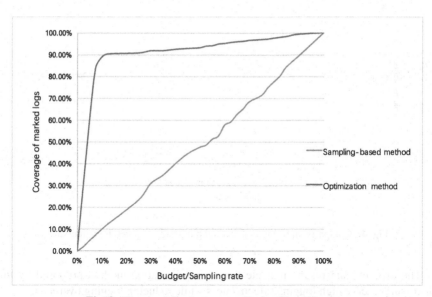

Fig. 3. Logging effectiveness vs. Budget/Sampling rate.

Obviously, the optimization method achieves high effectiveness, i.e. the coverage of marked logs increases quickly to almost 90% when the budget size starts to increase. However, the effectiveness of sampling-based method is approximately proportional to the sampling rate, which is much lower than what optimization method achieves. The results indicate that optimization method has ability to preserve high logging effectiveness.

4.5 Logging Throughput

For the evaluation of logging throughput, this paper counts how many logs are output in each time interval(30 s). The logging throughput directly affects the system overhead. The logging throughput is associated with the predefined budget, i.e., the volume of logs allowed to be output per time interval. The lower the budget, the lower the logging throughput. Otherwise, the higher the logging throughput.

Figure 4 shows the number of logs output per time interval using the optimization method and the traditional logging system, respectively. The budget is set to 400 logs/interval (equivalent to the sampling rate of 10%). Figure 4 shows that the logging throughput is significantly reduced using optimization method. The average number of output logs per time interval is about 400, while it is 4000 for the traditional logging system. The reduction on logging throughput is about 90%.

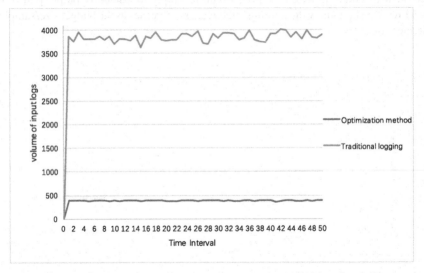

Fig. 4. Comparision of logging throughput. (budget = 400 logs/interval).

The experimental results indicate that the optimization method proposed by this paper can preserve high logging effectiveness while reducing logging overhead.

5 Conclusion

This paper proposes an optimization method for the logging overhead problem. Under a given budget (the maximum number of logs allowed to be output in a time interval), an anomaly detection model based on deep learning and a two-phase filtering mechanism are designed to determine whether logs are output according to the logs' utility scores for anomaly detection during the operation of the system. The experimental results show that the log parsing method and the deep learning-based anomaly detection method implemented in this paper have good performance. The cost-aware logging optimization method in this paper can alleviate the resource overhead for logging while preserving logging effectiveness.

Acknowledgement. This research is partially supported by National High-Tech Program with the Grant No. 315055101.

References

1. Mi, H., Wang, H., Zhou, Y., et al.: Toward fine-grained, unsupervised, scalable performance diagnosis for production cloud computing systems. IEEE Trans. Parallel Distrib. Syst. **24**(6), 1245–1255 (2013)
2. Li, H., Shang, W., Zou, Y., Hassan, A.E.: Towards just-in-time suggestions for log changes. Empirical Softw. Eng. **22**(4), 1831–1865 (2016). https://doi.org/10.1007/s10664-016-9467-z
3. Li, H., Shang, W., Hassan, A.E.: Which log level should developers choose for a new logging statement? Empirical Softw. Eng. **22**(4), 1684–1716 (2017)
4. Zhu, J., He, P., Qiang, F., et al.: Learning to log: helping developers make informed logging decisions. In: IEEE/ACM IEEE International Conference on Software Engineering (2015)
5. Xu, Z., Rodrigues, K., Yu, L., et al.: Log20: fully automated optimal placement of log printing statements under specified overhead threshold. In: The 26th Symposium (2017)
6. Li, H., Chen, T.H., Shang, W., et al.: Studying software logging using topic models. Empirical Softw. Eng. **7**, 1–40 (2018)
7. Sigelman, H.B., Barroso, L., Burrows, M., et al.: Dapper, a large-scale distributed systems tracing infrastructure (2010)
8. Hauswirth, M., Chilimbi, T.M.: Low-overhead memory leak detection using adaptive statistical profiling. ACM Sigplan Not. **39**(11), 156–164 (2004)
9. Arnold, M., Ryder, B.: A framework for reducing the cost of instrumented code, vol. 36 (2003)
10. Wei, X., Ling, H., Fox, A., et al.: Detecting large-scale system problems by mining console logs. In: ACM Sigops Symposium on Operating Systems Principles (2009)

Multi-objective Collaborative Optimization of Multi-level Inventory: A Model Driven by After-Sales Service

Mingxuan Ma, Yiping Lang, Xia Liu, Wendao Mao[(✉)], Lilin Fan, and Chunhong Liu

School of Computer and Information Engineering, Henan Normal University, Xinxiang 453007, Henan, China
maowt@htu.edu.cn

Abstract. To improve the quality of after-sale service that is a new aspect for all manufacturing enterprises, the allocation of inventory reserves as well as reasonable dispatch between inventories have become the key to meet customer demand and reduce inventory cost. In this paper, a multi-stage safety inventory optimization model is constructed for after-sales service demand. The order quantity of inventory in the model is set to consider the changes of customer demand and fault loss under the influence of different quarters and regions. The cost and transportation time are also optimized by using multi-objective particle swarm optimization algorithm at the same time. Simulational results show that the proposed model can not only respond to the demand changes in different regions and different quarters timely, but also reduce the cost and time loss to meet customer demand. Compared with the methods that merely considers time and cost respectively, the proposed model is more suitable to solve the multi-stage inventory optimization problem across regions.

Keywords: Multi-level inventory optimization · Cross-regional relocation · Particle swarm optimization · Multi-objective optimization

1 Introduction

With the quick development of manufacturing industry, the market share of manufacturing enterprises has risen to the bottleneck. Improving the level of after-sales service and consolidating customer loyalty have become a new point to lift profit. However, the stock inventory is often difficult to match the customer changeable demand timely. The difference of season and region will also lead to different parts failure rates from region to region [1]. As the demand can fluctuate, increase and decline, it is difficult to coordinate the inventory management scheme and unreasonable storage among various inventories. A reasonable allocation of storage location can not only shorten the delivery time, but also improve the competitiveness of the enterprise [2]. At present, the key issue of parts chain management is multi-level inventory management. Compared with single inventory optimization, it is prone to local inventory optimization. Multi-level inventory management aims to realize global resource allocation, improve the inventory allocation

© Springer Nature Switzerland AG 2021
Q. Zu et al. (Eds.): HCC 2020, LNCS 12634, pp. 360–371, 2021.
https://doi.org/10.1007/978-3-030-70626-5_38

efficiency of enterprises, reduce costs and expenses, and achieve the goal of improving business efficiency of enterprises.

In this paper, we focus on the multi-level safety inventory optimization problem. The aim is to set a reasonable parts storage quantity in all levels of inventory to realize a comprehensive and multi-level inventory optimization. Enterprises can then meet the complex and changeable customer needs more calmly. Meanwhile, in order to improve the service quality, the inventory cost and transportation time can be optimized to help enterprises obtain higher service value.

In recent years, there have been a lot of inventory optimization works proposed. By optimizing the dynamic inventory of a serial supply chain with stochastic andlead-time sensitive demand [3], customers can be provided the best services response time and meet customer needs; By constructing a multi-level inventory model and using multi-objective evolution technology, WEI Z et al. [4] can achieve performance and sensitivity improvements compared to single-objective optimization. However, this model requires data encoding, which is not easy to be combined with practical problems. Liu et al. [5] investigated the dynamic integration and optimization of inventory classification and inventory control decisions to maximize the net present value (NPV) of profit over a planning horizon. Meanwhile, it can manage the inventory both dynamically in the face of nonstationary demand and holistically. Chinello et al. [6] analyzed the driving factors of inventory optimization by analyzing a real case of a toy manufacturer, but due to the lack of more general data, it is difficult to verify the generality on the other companies; Zhao Baiqiang [7] put forward the inventory optimization model based on time competition and reduced the time cost from the perspective of multi-objective. Steenbergen et al. [8] presented a novel demand forecasting method denoted by DemandForest, which combines K-means, Random Forest, and Quantile Regression Forest. This method integrates the historical sales data of previously introduced products and product characteristics of existing and new products to make prelaunch forecasts and inventory optimization. But the disadvantage of this method is that it is not suitable for companies lacking historical data. H. Rau [9] studies an Integrated inventory model for deteriorating items, validating the decision result from global view is optimal, but the model lacks the scheduling of peer members. Durán et al. [10] proposed a optimization model based on activity-based life cycle costing. This model minimizes long-term management costs from a global perspective, but the disadvantage is that it only considers cost optimization, which may reduce delivery time during the optimization process. While these works achieved multi-level inventory optimization to some extent, there are some own limitations. The inventory services based on after-sales parts cannot only consider reducing inventory costs, but also need to consider the timeliness of transportation in the allocation of inventory. Therefore, a multi-level inventory model needs to achieve multi-objective optimization results.

Based on the above discussion, this paper proposes a multi-level inventory collaborative optimization model driven by after-sales service in manufacturing enterprises.This model utilizes the influence of warehouse transfer decision and the difference of after-sales service demand to realize the adjustment of the maximum order quantity between different inventories more dynamicly. As a result, this model can reduce the total inventory cost and transportation time, and ensure all levels of inventory running smoothly and efficiently. Moreover, the proposed model simulates the process of multilevel inventory and leverage the multi-objective particle swarm optimization algorithm to achieve the multi-objective optimization of cost and time. The experimental results show that the model can effectively balance the effects of inventory cost and time.Compared with the optimization model which only considers cost and time, this model can adjust the optimization target according to the actual demand and more suitable for creating higher service value for enterprises.

2 Model Analysis

In the multi-level inventory model, we define the functions of inventory at all levels in the parts management process as follows:

- The central storage is the core storage with highest management authority in the supply chain, manages the delivery and storage of parts, examines and approves parts allocation requests from lower-level storages, and has the function of improving after-sales parts service.
- The regional storage is the subordinate storage of the central storage, which accepts the management of the central storage uniformly. The regional storage maintains a certain number of parts to ensure the requirement of the node storages. There are three regional storages in this model.
- The node storage is located at the lowest level of the inventory model, which is the production point of the parts demand, the parts demand of each node storages is affected by seasons and regions. The node storage itself does not hold inventory, Its demand is first satisfied by the nearest regional storage, otherwise by subsequent regional storage. Timely supplement of spare parts will improve the service quality of after-sales spare parts.

2.1 Business Logic and Symbol Definition

Business logic and symbol definition:

(1) The stocking cycle of warehouse parts in the model is t, and the total inventory cycle is T, In order to meet the demand for parts in the next month, we set t as the advance purchase cycle of parts, which means that parts are ordered one inventory cycle in advance;
(2) The Replenishment strategy for regional storage is (s,S);
(3) The demand for parts is automatically generated by Poisson distribution and different parameters are set, so that the accessory demand data has seasonal and regional fluctuation trend.
(4) The number of faulty parts is affected by seasons and regions, and the settings will be as realistic as possible.

2.2 Multi-level Inventory Model Operation Process

The multilevel inventory model flow is as follows: (1) Calculate the reorder point; (2) Central storage accepts parts; (3) Regional storage accepts parts; (4) Central storage place an order; (5) Delivery to node storage; (6) Emergency replenishment. The flow chart is shown in Fig. 1 below:

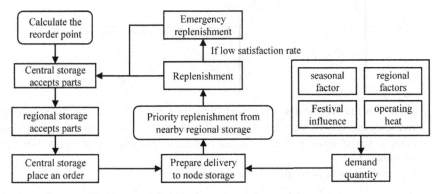

Fig. 1. Multi-level inventory model operation processes

The Multi-level inventory model operation processes are as follows:

(1) Calculate the reorder point

Calculate the reorder points of the central storage and each regional storage based on the data of each cycle as shown below:

$$s_i = K * \sqrt{t} * \partial \tag{1}$$

Where K is the safety factor set to improve the satisfaction rate, its value is 1.65. The calculation formula of δ is as follows:

$$\delta = \sqrt{\frac{\sum (d_j^t - \overline{d_j})^2}{n}} \tag{2}$$

where d_j^t represents the actual demand of the node storage j in the period t. $\overline{d_j}$ represents the average demand of the node storage j in the period t.

(2) Central storage accepts parts

At the beginning of each cycle, the parts ordered in the last cycle arrive and are colected into the central storage.

$$I_0^t = I_0^{t-1} + Y_t \tag{3}$$

Where I_0^t is the inventory level of central storage in period t, the central storage named by number 0; Y_t is the order quantity of central storage in period t.

(3) Regional storage accepts parts

If the inventory in each regional warehouse is lower than the order point s_i, replenish the inventory to S_i^t, follow the following rules when replenishing: (a) If the central storage is sufficient, all regional storage will be met; (b) If the inventory is insufficient, replenishment will be carried out according to the proportion of each regional warehouse demand in the total demand.

$$
\begin{cases}
I_i^t > s_i, \ no \ replenishment \\
I_i^t < s_i, Z_i^t = S_i^t - I_i^t,
\begin{cases}
\sum_{i=1}^{3} Z_i^t < I_0^t, \ regular \ replenishment \\
\sum_{i=1}^{3} Z_i^t > I_0^t, Z_i = \dfrac{Z_i^t}{\sum_{i=1}^{3} Z_i^t} * I_0^t
\end{cases}
\end{cases}
\tag{4}
$$

where s_i is the order point of regional storage i, $i \in \{1, 2, 3\}$, S_i^t is the maximum inventory quantity of regional storage i, I_i^t is the real inventory quantity of regional storage i in the period t, Z_i^t is the actual replenishment quantity of regional storage i in period t.

The holding cost of inventory is shown below:

$$
C_h^t = \sum_{i=0}^{3} h_c * I_i^t
\tag{5}
$$

where C_h^t is the holding cost of the central storage and the regional storage in period t, h_c is the holding cost of the individual parts in period t, It consists of fixed inventory costs, inventory management costs, depreciation costs, etc.

The cost of transportation between the central storage and the regional storage is shown below:

$$
C_{di}^t = \sum_{i=1}^{3} d_c * Z_i^t * time_i^0
\tag{6}
$$

where C_{di}^t is the transportation cost for regional storage i in period t, d_c is the transportation cost of spare parts per unit of time, $time_i^0$ is the transportation time from the central storage to the regional one.

(4) Central storage place an order

When the inventory of central storage is below the order point S_0^t, it will make purchase orders, the purchase quantity is $S_0^t - I_0^t$, and the arrival time is at the beginning of the next cycle.

$$
\begin{cases}
I_0^t > S_0^t, \ no \ replenishment \\
I_0^t < S_0^t, Y_{t+1} = S_0^t - I_0^t
\end{cases}
\tag{7}
$$

where I_0^t is the inventory of the central storage in period t.

The order cost of the central storage is shown below:

$$
C_{b0}^t = Y_t * b_c
\tag{8}
$$

where b_c is the ordering cost of unit parts.

(5) Delivery to node storage

The regional storage starts to supply parts to the node storage. The replenishment priority is set according to the size of the distribution time between each node storage and the regional storage. Each node storage has priority to get parts replenishment from the nearby regional storage.

(6) Emergency replenishment

Make statistics on the satisfaction rate of distribution in each node storage. If the rate is lower than the minimum satisfaction rate, emergency replenishment will be carried out, which will lead to higher procurement and transportation costs. The formula for calculating the satisfaction rate and the quantity of emergency replenishment are as follows.

$$CS_j^t = \frac{sup_j^t}{d_j^t} \tag{9}$$

$$\begin{cases} CS_j^t > 0.95 \\ CS_j^t < 0.95, \; ES_j^t = d_j^t * 0.95 - sup_j^t \end{cases} \tag{10}$$

Where CS_j^t is the satisfaction rate of node storage j in period t, sup_j^t is the actual arrival quantity of node storage j in period t, d_j^t is the actual demand of node storage j in period t, ES_j^t is the quantity of emergency replenishment of node storage j in period t.

The transportation cost of each node storage is shown below:

$$C_{dj}^t = \sum_{j=1}^{6} d_c * sup_j^t * time_j^i + d_{ce} * ES_j^t * time_e \tag{11}$$

Where $time_j^i$ is the transportation distance between the regional storage i and the node storage j, d_{ce} is the unit transport cost incurred by emergency replenishment of parts, $time_e$ is the transportation time for emergency replenishment.

$$C_{be}^t = ES_j^t * (b_c + b_e) \tag{12}$$

Where b_e is the order cost of emergency replenishment.

The total transportation time of the distribution process from regional storage to node storage is shown as follows:

$$Time^t = ES_j^t * time_e + \sum_{j=1}^{6} sup_j^t * time_j^i \tag{13}$$

2.3 Establish Multi-objective Optimization Model

For manufacturing enterprises that want to improve service quality, they not only need to meet the requirements of supply, but also need to optimize the response time of cross-regional inventory replenishment. This model is composed of two optimization objectives: the minimum total cost and the minimum distribution time of the whole system.

According to the formula listed above, the total cost of the multi-level inventory system in the whole cycle can be calculated as:

$$Cost = \sum_{t=1}^{T} (C_h^t + C_{di}^t + C_{dj}^t + C_{b0}^t + C_{be}^t) \tag{14}$$

The total transportation time of the multi-level inventory system in the whole cycle can be calculated as:

$$Time = \sum_{t=1}^{T} Tim^t \tag{15}$$

The optimization problem can be expressed as:

$$Min(Cost, Time) \tag{16}$$

$$CS_j^t > 0.95 \tag{17}$$

$$d_j^t = demand_j^t + dam_j^t \tag{18}$$

$$S_0^t \geq s_0^t \tag{19}$$

$$S_i^t \geq s_i^t \tag{20}$$

$$S_0^t, s_0^t, Z_i^t, I_0^t, I_i^t, \sup_j^t, d_j^t \in Z \tag{21}$$

Equation (18) shows the monthly actual demand of the node storage is the sum of the required parts and fault parts. Equation (19) and (20) show the maximum order quantity is always greater than or equal to the reorder point. Equation (21) shows that the maximum order quantity, reorder point and other data are constant.

3   Simulation Example

Combined with the above formulas, this section generates a set of simulation data with meeting the conditions listed in Eq. (17)–(21). Then the optimal solution can be determined by using particle swarm optimization to minimize the total cost and transportation time simultaneously.

3.1 Basic Data

The monthly number of requirements and failures are random Numbers generated by the Poisson distribution. Since the loss of parts is often related to the climate and operating heat, we adjusted the parameters of different months to make the data fit the actual situation. In summer, the high temperature will cause the failure rate of spare parts to be significantly higher than that of other seasons, so the storage needs to increase the corresponding spare parts reserve. The operating heat will be affected by the holiday. For instance, before the Chinese Spring Festival holiday, there is often a surge in damage to parts caused by rush work. When the holiday enters, the operating heat will decrease quickly, meanwhile, there will be regional differences in parts failure between different nodes. We set the first three node storages as low-risk areas, while the other three are high. The settings of the fault parameters are shown in Table 1, and the parameters for demand quantity of each quarter are shown in Table 2.

Table 1. Random number parameters for monthly failures

Month	1	2	3	4	5	6	7	8	9	10	11	12
Parameters for monthly failures	3	1	2	1	1	3	3	3	1	2	1	1

Table 2. Random number parameter of each quarter demand

Quarter	Quarter 1	Quarter 2	Quarter 3	Quarter 4
Parameter of each quarter demand	6	10	5	4

The settings of transportation time between each regional storage and node storage are shown in Table 3. The priorities of inventory distribution are shown in Table 4. Here we define the time from the central storage to the regional storage as 9, the emergency replenishment time as 15. The settings of calculation factor are shown in Table 5.

Table 3. Settings of transportation time between each regional storage and node storage

	Node 1	Node 2	Node 3	Node 4	Node 5	Node 6
Regional 1	7	6	9	12	13	7
Regional 2	8	13	5	6	10	15
Regional 3	11	9	12	8	7	9

Table 4. Priorities of inventory distribution

Priority	Node 1	Node 2	Node 3	Node 4	Node 5	Node 6
Regional 1	1	1	2	3	3	1
Regional 2	2	3	1	1	2	3
Regional 3	3	2	3	2	1	2

Table 5. Settings of calculation factor

Factor	Setting
h_c	150
b_c	2000
b_e	1000
d_c	60
d_{ce}	100

3.2 Optimization Results

The simulation data are input into the model, and the optimal ordering strategy is the calculated by the multi-objective particle swarm optimization algorithm. The results are shown in Fig. 2.

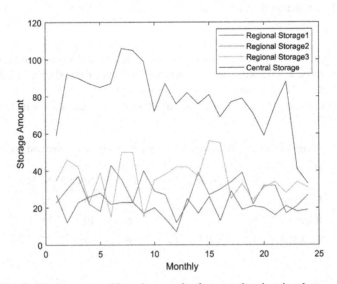

Fig. 2. Maximum monthly order quantity for central and regional storage

It can be seen from Fig. 2 that the inventory quantity of the regional node has significant seasonal variation. In the first quarter of the first year, the high wastage in January caused each warehouse to increase its own inventory. In the following two quarters, the central node and each regional node adjusted the storage quantity according to their respective demands. In the third quarter, despite a reduction in demand, the wear and tear on spares caused by bad weather led warehouses to increase stock for future use. In the following year, the model adjusts its inventory by minimizing the burden of inventory to meet demand during periods of high start heating and bad weather. Due to the difference in demand and failure rate among storehouses, they often improve the inventory level of nearby regional warehouses to ensure the timeliness of distribution. As seen from Fig. 2, regional node 3 has a higher inventory level and can provide better services for this area.

At the same time, in order to verify the effectiveness of the model, the multi-objective optimization scheme is compared with the model which only considered the cost or the time, the results were shown in Table 6 below:

Table 6. Results of three schemes

	Cost	Change ratio	Time	Change ratio
Multi-objective	3994105		1121	
Cost-based	3710405	−7.10%	1417	+26.40%
Time-based	4322495	+8.18%	1012	−9.72%

From the above comparison, it can be found that the single-objective optimization method performs well in optimizing costs, but the cost of other aspects rises sharply. For example, although the cost-based optimization method reduces the total cost by 7.10%, it increases the time deficit by 26.40%. Compared with our model, the aforementioned approach performs worse in terms of time loss. Although time-based optimization method does not cause too much cost loss, compared with multi-objective optimization, its results lack selectivity and it is difficult for enterprises to adjust flexibly according to specific requirements.

In Fig. 3, we counted and visualized the optimal values of each generation of the three schemes in the Multi-objective particle swarm optimization algorithm optimization process.

The optimization process is shown in Fig. 4. Multi-objective optimization can get a relatively good result sets, the results compared with the single objective optimization to more flexible, enterprise can according to their own preferences and service strategy, selecting suitable inventory program in the collection, saving inventory costs, improve the quality of service, to consolidate the core customer group, more long-term economic benefits for the enterprise.

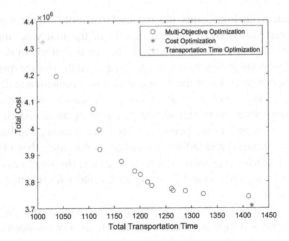

Fig. 3. The optimal nodes of the three schemes

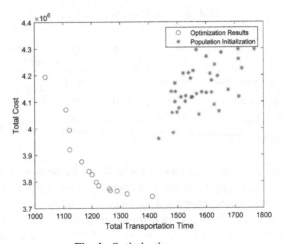

Fig. 4. Optimization process

4   Conclusions

Due to the complex market demand nowadays, multilevel inventory optimization not only considers the impact of their own storage and after-sales service costs, but also considers the efficiency of cross-regional scheduling. Considering inventory cost and scheduling time, this paper constructs a multi-stage inventory multi-objective optimization model considering the influence of quarterly and regional factors. It aims at improving customer satisfaction rate and realizing dynamic control of inventory by formulating inventory quantity in each cycle. This model can provide customers with faster and better service, with reducing the operating cost as much as possible. It has a good performance in inventory simulation under complex circumstances. In the future work, more detailed business logic will be considered. The allocation between regional storage will be taken into consideration to further share the inventory pressure of central storage and make the operation of the entire inventory network more nimble.

Acknowledgement. This work was supported by the National Key R&D Program of China (2018YFB1701400), the National Natural Science Foundation of China (No.U1704158).

References

1. Neale, J.J., Willems, S.P.: Managing Inventory in Supply Chains with Nonstationary Demand. Interfaces **39**(5), 388–399 (2009)
2. Boyaci, T., Ray, S.: The Impact of Capacity Costs on Product Differentiation in Delivery Time, Delivery Reliability, and Price. Prod. Oper. Manage. **15**(2), 179–197 (2006)
3. Dziri, E., Hammami, R., Jemai, Z.: Dynamic inventory optimization for a serial supply chain with stochastic and lead-time sensitive demand. IFAC PapersOnLine **52**(13), 1034–1039 (2019)
4. Wei, Z.: Multi objective optimization model for collaborative multi-echelon inventory control in supply chain. Acta Automatica Sinica **33**(2), 181–187 (2007)
5. Yang, L., Li, H., Campbell, J.F., Sweeney, D.C.: Integrated multi-period dynamic inventory classification and control. Int. J. Prod. Econ. **189**, 86–96 (2017)
6. Chinello, E., Herbert-Hansen, Z.N.L., Khalid, W.: Assessment of the impact of inventory optimization drivers in a multi-echelon supply chain: Case of a toy manufacturer. Comput. Ind. Eng. **141**, 106232 (2020)
7. Baiqiang, Z.: Research on Multi-stage Inventory Optimization of Supply Chain Based on Time Competition. Shenyang University (2013)
8. van Steenbergen, R., Mes, M.: Forecasting demand profiles of new products. Decis. Support Syst. **139**, 113401 (2020)
9. Rau, H., Wu, M.: Integrated inventory model for deteriorating items under a multi-echelon supply chain environment. Int. J. Prod. Econ. **86**(2), 155–168 (2003)
10. Durán, O., Carrasco, A., Afonso, P.S., Durán, P.A.: Evolutionary optimization of spare parts inventory policies: a life cycle costing perspective. IFAC PapersOnLine **52**(13), 2243–2248 (2019)

The Research About Spatial Distribution of Urban Functions Based on POI Data

Jingwen Li[1,2], Yuan Ma[1(✉)], Jianwu Jiang[1,2], Wenda Chen[1], Na Yu[1], and Shuo Pan[1]

[1] Guilin University of Technology, Guilin 541004, China
1286893393@qq.com
[2] Guangxi Key Laboratory of Spatial Information and Geomatics,
Guilin 541004, Guangxi, China

Abstract. The distribution of urban functional space is an important factor to measure the development of a city. Its reasonable layout plays an important role in the development of urban economy and the optimization of urban spatial pattern. This paper is based on the third ring road of Nanning city Based on POI data, the functional land is divided into five categories: life service, business and finance, public service, leisure and entertainment, and residence. By using kernel density analysis, frequency density analysis and standard deviation ellipse analysis, the urban function and distribution characteristics of the Third Ring Road area in Nanning city are identified and analyzed. The results show that: 1. The overall spatial layout is affected by traffic and urban master plan Under the influence of other factors, the spatial pattern of "circle layer" appears. Taking one ring as the core, it has obvious central agglomeration. 2. The trend of urban development shows an obvious pattern of prosperity in the West and decline in the East.

Keywords: Urban function region · Function distribution · Standard deviational ellipse · Kernel density

1 Introduction

Urban spatial structure is the embodiment of various urban components and functional organizations in the geographical space, and a reasonable urban spatial structure can improve the city's resource carrying capacity and ecological efficiency, which is conducive to promoting the sustainable development of urban ecology and economy. The traditional identification of urban structure is mainly based on expert judgment [1], survey statistics [2] and other experience-based methods, subjective. Some scholars also assist in urban functional zoning through remote sensing technology, but the cost of data acquisition and processing is higher and the time-ability is poor [3]. With the advent of the information age, data for urban planning related research institutes continue to emerge, including traditional data, open data and other types of big data resources. Under the background of abundant data, functional area analysis based on urban life data has become more rapid and effective.

Considering the limitations of traditional data and routine census, POI provides an accurate and effective alternative to spatial identification and quantitative study of cities

Q. Zu et al. (Eds.): HCC 2020, LNCS 12634, pp. 372–381, 2021.
https://doi.org/10.1007/978-3-030-70626-5_39

in different functional types while exploring the overall spatial structure of cities [4]. POI data is the dot data of real geographical entities, which has spatial and attribute information, high precision, wide coverage, fast update and large amount of data, and is widely used in urban research [4].

For this purpose, based on Amap POI data, with the third ring in Nanning City as the research scope, using kernel density analysis, frequency density analysis, standard deviation elliptical analysis and other methods, in the overall and different functional types, step-by-step identification of the urban structure of the study area, to explore the spatial differences of different functions and statistical clustering characteristics. Through the study of Nanning's urban spatial structure, the paper summarizes the effectiveness and shortcomings of Nanning's urban development, provides reference for urban development and spatial structure optimization, and provides reference for urban planning.

2 Research Area and Data Selection

2.1 Research Area

At present, a large number of urban spatial distribution research mainly focused on Beijing, Shanghai, Guangzhou and other first-tier cities, mega-cities [5], and for the general provincial capital cities pay less attention to the choice of Nanning City as the object of study is more representative. Nanning is located in the southwest of Guangxi Zhuang Autonomous Region, is the capital of Guangxi Zhuang Autonomous Region, but also the State Council approved the determination of the core city of China Beibu Gulf Economic Zone. This paper selected the Nanning city center of the third ring areas. The third ring areas in Nanning City are the fast loop, the ring road around the city, the outer ring high-speed, as shown in Fig. 1. The blue line in the figure is the Weijiang River, and the right side of the second ring is the outer east ring area.

Because of the particularity of Nanning urban terrain, the complex urban function, and it is in the transition period of accelerated urbanization development. It's also shows a lot of different characteristics from other cities. Therefore, the paper selects this area as the research area.

2.2 Data Sources and Processing

The data from the study is derived from the Amap POI data captured online in October 2019. Because POI data is mainly used for navigation maps, including the spatial location and attribute information of most entity objects in the city, it is an abstract expression of entity objects on the map, so it can be approximated that POI data contains all the research objects in urban space [6]. After de-weighting and correcting the obtained data, a total of 155936 POI data were obtained. Finally, according to the different functions of the city combined with the GAUD map POI classification system, the POI data are divided into the following five categories: life services, business finance, public services, Leisure and residential (Table 1).

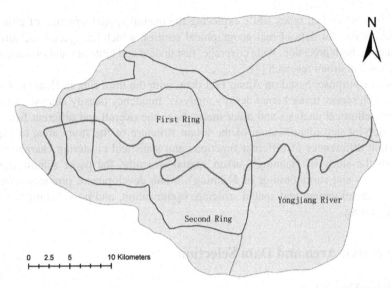

Fig. 1. A diagram of the three rings of Nanning. (Color figure online)

Table 1. Data types of POI in the third ring of nanning city.

POI classification	Content	Number	Proportion
Life services	Catering, Shopping, living facilities	93823	60.17%
Business finance	Banks, ATMs, insurance companies, securities companies, finance companies,	20827	13.36%
Public services	Health care services, government agencies and social groups, science, education and culture services	29172	18.71%
Leisure	Gymnasium, Locked Place, Holiday Place, Theatre, Park Square, Scenic Spot	3237	2.07%
Residential	Residential quarters, villas, dormitories and other residential-related buildings	8877	5.69%

3   Research Methods

3.1   Kernel Density Estimation

In recent years, kernel density analysis has been widely used in urban hot spot explo-
ration [7]. The basic principle of kernel density estimation is to calculate the density
contribution value of the sample point to the center point of the grid cell within a specified radius by the kernel function, using the secondary core function of Silverman and
so on:

$$d(x_i, y_i) = \frac{1}{nh} \sum_{i=1}^{n} k \frac{\sqrt{(x_i - x)^2 + (y_i - y)^2}}{h} \tag{1}$$

In the formula, $d(x_i, y_i)$ is the kernel density value at the spatial position (x_i, y_i); h is the bandwidth; n is the number of samples; $k()$ is the spatial weight function; $\sqrt{(x_i - x)^2 + (y_i - y)^2}$ is the Euclidean distance between the current feature point and the sample point (x_i, y_i) within the bandwidth.

3.2 Frequency Density Analysis

Based on the proportion of POI data in each cell in the study area, the frequency density is calculated and the frequency density is calculated for each cell. Establish type scales to identify the properties of different functional areas [8]. The formula is as follows:

$$F_i = \frac{n_i}{N_i}(i = 1, 2, 3, 4, 5) \tag{2}$$

$$C_i = \frac{F_i}{\sum\limits_{i=1}^{5} F_i} \times 100\%(i = 1, 2, 3, 4, 5) \tag{3}$$

In the class, i represents the type of POI, n_i represents the number of POI of the i type in the cell, N_i represents the total number of POI of the i type, F_i represents the frequency density of the first type POI in the total number of POI of that type, and C_i represents the frequency density of the type i POI as a proportion of the frequency density of all types of POI frequency density in the cell.

3.3 Standard Deviation Ellipse

Standard deviation ellipses were first proposed by Welty, a sociology professor at the University of Southern California. Le Fifer was proposed in 1926. By creating standard deviation ellipse, the spatial distribution, central trend and direction trend of each functional area in Nanning City are calculated to reveal the spatial development direction and distribution of the main spatial range. The standard deviation ellipse center coordinate formula is as follows:

$$SDE_x = \sqrt{\frac{\sum\limits_{i=1}^{n} \left(x_i - \overline{X}\right)^2}{n}} \quad SDE_x = \sqrt{\frac{\sum\limits_{i=1}^{n} \left(x_i - \overline{X}\right)^2}{n}} \tag{4}$$

The x_i and y_i cloth features are the coordinates of feature i, which represents the average center of the feature, and n is the total number of features.

4 The Spatial Distribution Characteristics of Nanning City

4.1 Overall Spatial Density Distribution

The standard deviation ellipse analysis of the first-level standard deviation parameters is used for five types of POI data in the third ring of Nanning City, and the standard deviation ellipse results are shown in Fig. 2. As can be found from the figure, the standard deviation ellipse long axis direction of these five types of data is basically the same, the overall distribution direction tends to southeast-northwest direction, the center distribution of each ellipse also tends to con center distribution, roughly located in Chaoyang Square area. Although there are differences between different individuals in terms of elliptical area, but the difference is not big, the short half axis of the ellipse is a long category of business finance and life services, the comprehensive performance of these two types of data is more discrete.

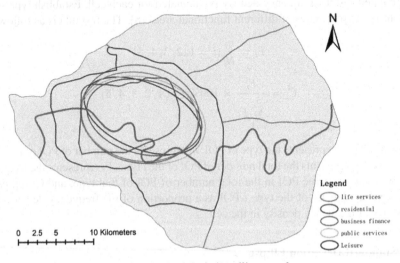

N

Legend
○ life services
○ residential
○ business finance
○ public services
○ Leisure

0   2.5   5   10 Kilometers

Fig. 2. Standard deviation ellipse result

4.2 Identification of Urban Spatial Structures Based on Kernel Density Analysis

Because in kernel density analysis, different search radius results in different results of kernel density analysis. The higher the search radius parameter value, the smoother and more generalized the resulting density raster. The smaller the value, the more detailed the resulting raster displays. The different bandwidths of 200 m, 400 m, 600 m, 800 m and 1000 m were set to compare the kernel density of POI in Nanning's third ring. The purpose of this study is to study and analyze the distribution of different functional areas of cities, and to select 600 m as the unified bandwidth of this study according to the existing study of urban spatial structure and the results of the overall and local effects of different search distances in the third ring areas of Nanning.

As can be known from Table 1, the number of POI of different types varies greatly. Among them, life services accounted for the largest category, accounting for 60.17% of the total, public services and business finance, respectively, accounted for 18.71% and 13.36%, residential, leisure and entertainment categories accounted for the lowest proportion, 5.69% and 2.07%, respectively. The distribution pattern and distribution density can be analyzed by kernel density analysis, and different types of poi data have different spatial distribution characteristics, the following are the kernel density map of living services, the kernel density map of public services, the kernel density map of commercial finance, the density map of residential kernel density and the kernel density map of leisure and entertainment (Figs. 3, 4, 5, 6 and 7).

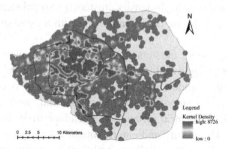

Fig. 3. The kernel density map of living services

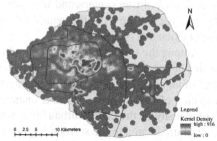

Fig. 4. The kernel density map of public services

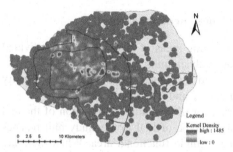

Fig. 5. The kernel density map of commercial finance

Fig. 6. The density map of residential kernel density

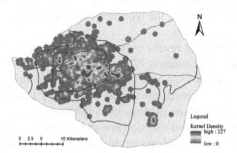

Fig. 7. The kernel density map of leisure and entertainment.

According to statistical analysis, 54.23% of poi points are distributed in first ring, 32.68% of poi points are distributed in second ring, and 13.09% of poi points are distributed in third ring. According to the results of the analysis of POI kernel density of each type, the main urban area of Nanning City as a whole formed a circle pattern centered on Chaoyang Road, and the distribution density showed a gradual decline from the inner city to the peripheral area. In general, various functional areas have formed a "large gathering, small dispersion" of urban spatial forms, distribution density from a central urban area along the main traffic gradually to the periphery of the trend of decline; This shows that Nanning's urban function is more concentrated, the same area in the city function has a high degree of re-assembly. Sub-density clusters are distributed around high-density cluster centers, and are connected with high-density clusters to expand step by step, decreasing from the central urban area to the outer ring layer, with a trend of finger extension outside the ring.

Among the five types of poi data, public service clusters are the most concentrated, with a larger concentration range than other categories, highly concentrated in one ring and small-scale clusters around Nanning Evergrande City in the second ring. The distribution of life services, business finance and residential categories is similar, with a high density in the central urban area, with the central urban area spreading from the central urban area to the surrounding area. The leisure and entertainment category has the smallest and most dispersed distribution due to the smallest number of poi points of all, with significant small-scale aggregation in the first, second and third ring.

4.3 Spatial Structure Recognition of Urban Functional Areas

Using ArcGis' fishing net function, the third ring areas of Nanning City are divided into 5825 400 m × 400 m square grids as analysis units, the number of POI points representing different city functions in each grid is counted, and the frequency density of each cell is calculated according to formula 2.2 and its type is identified.

Use a scale value of 50% as the criterion for determining the function of the cell area. When the frequency density ratio of one type of POI in the study unit reaches 50% or more, the unit is determined to be a single functional area. When there are 2 POI types and the frequency density ratio exceeds 40%, the unit is defined as dual functional area.when three POI types appear in the study unit and the frequency density ratio exceeds 30%, the unit is defined as multi-functional area. When the POI frequency density of all types in the cell is 0, the type area is dataless. The partition results for the function area are shown in Fig. 8.

According to the statistical results of the three functional area units, the maximum number of units without data zones is 2764, followed by the number of dual functional areas, the number of units is 1326, and the number of multi-function mixed zone units is the lowest, 640. On the whole, most areas of Nanning have more main functions, and there are more areas with no obvious main functions. Data-free zones are mainly distributed on the edge of the third-ring ring line and the outer east ring area, reflecting the slow economic development of urban marginal areas, low per capita income and weak consumption capacity, which is difficult to attract all kinds of living service industries. Multi-functional areas are mainly distributed in a ring, along the ring line and along both sides of the river, and the distribution is more concentrated, of which a large number of public and life-oriented functional areas.

The distribution of various types of functional areas is shown in Fig. 9. Nanning city third-ring area in the spatial layout of a total of 5 types of single functional areas, 10 types of dual functional areas, as well as 9 types of multi-functional areas. The final

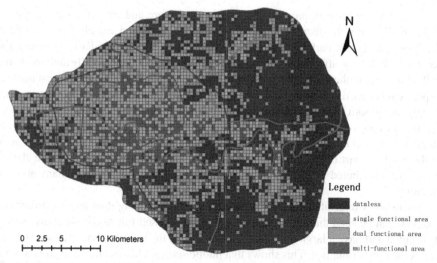

Fig. 8. Distribution map of single function area, double function area and multi-function area in Nanning

statistics have a total of 25 cases, because some types are small, so choose the number of POI more than 10 cells, and finally determine 19 types.

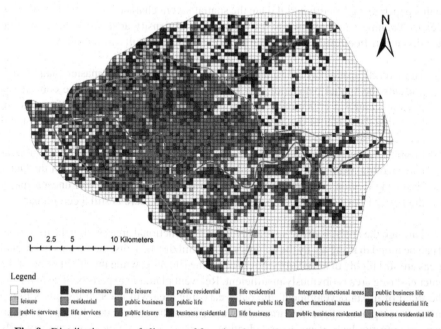

Fig. 9. Distribution map of all types of functional areas in the third ring road of Nanning.

In all the single-function areas, the number of leisure and entertainment categories is the smallest, but the distribution of this is more concentrated, mainly concentrated in Wu Xiangling Forest Park, Qingxiushan and Nanning Garden Expo Park. Secondly, the residential category, affected by the radiation of urban centers, is basically not distributed in the first ring, in the second ring, the third ring distribution is wide. The largest number of public services, distribution is also the most extensive, multi-concentrated distribution, mainly along both sides of the river and the ring line distribution. Business finance and life services are relatively small compared to public services, the number is relatively small, business classes are mostly distributed in the periphery of the central urban area, in the northern part of the third ring there is a small number of gathering. Life classes are evenly distributed within the third ring and are widely distributed in many areas of Nanning.

There are a total of 1966 hybrid function areas, including dual and multifunction zones. It can be seen that in the center of Nanning, mixed functional areas are widely distributed and have a large number. Of all the types, the public service class is the one that is the most contained. This shows that public service classes often exist mixed with other functional types in cities and play an auxiliary role. Life services were the most included, accounting for 31.3% of the total.This shows that the life service industry in Nanning is relatively developed to meet the daily needs of residents.

5 Conclusion

In this paper, using Gaudrem POI data, the spatial layout characteristics of the third-ring area of Nanning City are discussed by using kernel density analysis method, and the functional partition results of this area are obtained. The results are as follows:

1) The overall spatial layout is influenced by traffic and urban master plan, which presents the spatial pattern of "circle layering". With one ring as the core, it has obvious central a collectivity. Areas with diverse and highly mixed features are located in the center of the city, and the ring edge functional areas are less clustered and are mostly single-function or dual-function mixed zones.
2) From the present point of view, the development trend of the city shows a clear pattern of the decline of the west and east. At present, the development of the Outer East Ring Region is relatively slow, the overall development trend of uneven space, the level of development compared with other places there is still a certain gap.

Through the analysis of the spatial layout of the functional areas within the third ring, it can be found that, driven by the behavior of maximizing benefits, the various functional areas are still highly distributed in the center of the ring line, while the distribution of the outer east ring area is obviously insufficient. Relevant planning experts believe that from the trend of urban development, the outer east ring area will be Nanning city's future urban development of the key areas. However, from the current development point of view, due to the lack of necessary planning guidance, the region shows a disorderly development trend. For example, the development of the North Lake Industrial Zone along the Nanwu Highway, the town of Santang along the Nantang Highway, and the five-way University City along the Nanpu Highway have not coordinated the coordinated development of transportation system, infrastructure and industry, which has brought some hidden dangers to the expansion of urban space in the future. The government needs to use policy means to make reasonable planning and guidance on the spatial layout of the city and further strengthen the infrastructure construction in the area.

Overall, the use of POI data to analyze the spatial distribution of urban functional areas can truly reflect its overall development status, the future Nanning urban spatial layout and adjustment has a reference role. However, due to the limitation of POI data, it does not contain the geographical entity's range, mass and grade information, it is more difficult to carry out more in-depth analysis. Therefore, in the future research should be combined with other Open Data of the Internet, so as to explore the spatial distribution law of Nanning city and its internal operation mechanism in depth.

References

1. Dong, N., Yang, X., Cai, H., et al.: Research on grid size suitability of gridded population distribution in urban area: a case study in urban area of xuanzhou district, China. PLoS ONE **12**(1), e0170830 (2017)
2. Pan, H.Y., Gu, J.R., Zhou, J.M.: Research on the division for functional area of evaluation of intensive use based on delaunay triangulation and GIS – a case of functional area for residence in Qingyang. Adv. Mater. Res. **468–471**, 2749–2754 (2012)
3. Klug, K., Hogekamp, C.: André specht. per-pixel vs. object-based classification of urban land cover extraction using high spatial resolution imagery. Remote Sens. Environ. **115**(5), 1145–1161 (2011)
4. Tieying, Z., Hongwei, L.I., Donghao, X.U., et al.: POI data visualization based on DBSCAN algorithm. Sci. Surveying Mapp **41**(5), 157–162 (2016)
5. Xin, Y., WU, S. et al.: Analysis of the relevance of spatial distribution between urban parks and transportation infrastructure in Beijing based on POI data. J. Landscape Res. **10**(03), 30–37 (2018)
6. Ling, Z.: Research in POI classification standard. Bull. Surveying Mapp. (10), 82–84 (2012)
7. Peng, J., Zhao, S., Liu, Y., et al.: Identifying the urban-rural fringe using wavelet transform and kernel density estimation: a case study in Beijing City, China. Environ. Model. Softw. **83**, 286–302 (2016)
8. Chi, J., Jiao, L., Dong, T., et al.: Quantitative identification and visualization of urban functional area based on POI Data. J. Geomatics **41**(2), 68–73 (2016)

The Psychological Characteristics Changes Analysis of Su Shi Before and After the Wutai Poetry Case ——Based on the CC-LIWC

Rui Ma[1,2], Fugui Xing[1,2], Miaorong Fan[1,2], and Tingshao Zhu[1,2(✉)]

[1] Institute of Psychology, Chinese Academy of Sciences, Beijing 100101, China
tszhu@psych.ac.cn
[2] Department of Psychology, University of Chinese Academy of Sciences,
Beijing 100049, China

Abstract. Su Shi(hereinafter referred to as Su) is a well-known historical celebrity in China, the Wutai Poetry Case(WPC) was a famous literary inquisition in Chinese history, and was a turning point in Su's life. Pevious studies on the influence of WPC were based on Su's literatures, and there was a lack of empirical researches. In this paper, we conducted the Word Count Verbal Text Analysis on Su's literary works. [Methods] We firstly identified the writing year of Su's complete works, selected essays of five years before and after WPC, and used CC-LIWC(Classical Chinese version of Linguistic Inquiry and Word Count) to calculate the word frequency. Then, non-parametric tests were performed, and the CC-LIWC word categories with significant differences were obtained. Finally, the trend of word frequency changes of these categories were also reported. [Results] Before and after WPC, there are 9 categories with significant differences, which are the Third-person singular, Third-person plural, Tensem words, Focus present words, Negative Emotion words, Differentiation words, Perceptual Process words, See words and Time words. [Discussion] From the trend of the word frequency of the 9 categories, Su's psychological characteristics were consistent with the Buddhist pursuit of being free from foreign objects, paying attention to the present, practicing non-differentiation, and getting out of worries after WPC. This result indicated that Su has been more advocating Buddhism practice after experiencing the political blow, and prove that the method of quantitative analysis is feasible.

Keywords: Su Shi · The Wutai Poetry Case · LIWC · Digital psychology · Confucianism-Buddhism-Taoism

1 Introduction

Su Shi (hereinafter referred to as Su) is a well-known historical and cultural celebrity in China, a great litterateur and painter-calligrapher in Northern Song Dynasty. Among the writers of the Song Dynasty, only Su can be so widely loved and praised by later generation, this comes from his extremely high literary attainments and open-minded and noble personality charm. Su had been in diaspora throughout his life, and suffered several

© Springer Nature Switzerland AG 2021
Q. Zu et al. (Eds.): HCC 2020, LNCS 12634, pp. 382–388, 2021.
https://doi.org/10.1007/978-3-030-70626-5_40

setbacks in his career. The Wutai Poetry Case (WPC), a famous literary inquisition, was a turning point in Su's life, which is a consensus in academic.

The WPC took place in 1079 AD, the second year of the Yuanfeng reign of Emperor Shenzong of Song. Su moved to Huzhou, then he was impeached by the New Policy Group for "satirizing the new politics and criticizing the emperor" because of a few lines in the Memorial of thanks to the emperor. Su was arrested on July 28th and imprisoned on August 18th in the imperial jail, went through 130 days in jail, more than 20 charges, almost sentenced to death. After being rescued by many parties, the Emperor Shenzong finally did not execute Su. Su was released from prison on December 29th, and was demoted to the deputy official of the Huangzhou regiment as nominal title, virtual exiled, "may not sign official documents".

To identify how WPC affected Su, previous researches were mostly based on his literary works before and after WPC, but lack of quantitative empirical research. In this paper, we proposed a quantitative analysis method, to analyze the word count of Su's writings to quantify the psychological significance of WPC.

The ways people use words in their daily lives can provide rich information about their beliefs, fears, thinking patterns, social relationships, and personalities. The words have tremendous psychological value [1]. LIWC could calculate a percentage of words falling into 80 psychologically or linguistically meaningful categories. These categories cover several important psychological aspects of anindividual, including emotion, cognition, social contact and personal concerns [2].

In this paper, we selected the essays of Su's complete works with writing year as corpus, and analysised word frequency using the Classical Chinese Linguistic Inquiry and Word Count (known as CC-LIWC [3]) which is constructed by Computational Cyber-Psychology Lab, Institute of Psychology, Chinese Academy of Science based on Simplified Chinese LIWC (known as SC-LIWC [3]), for exploring the digital presentation of the changes in Su's psychological characteristics before and after the Wutai Poetry Case. The purpose is to contribute a quantifiable and verifiable new method to the psychological research of historical figures in order to corroborate the existing research results or put forward new arguments.

2 Methods

In this study, we selected essays from Su's complete works, took two sets of comparative texts before and after WPC to acquire word frequency using CC-LIWC, calculated and screened out the categories with significant differences by Difference Test. We analyzed the change trend of these categories, and interpreted the changes of several psychological indicators of Su before and after WPC.

2.1 Date Collection

Electronic resources for the Su's complete works are from wenxue360 website [4], based on the *collection of Complete Library of the Four Branches*. Because the poetry and lyrics are highly literary modifications, artistic expressions and non-verbal expressions that are not conducive to the LIWC analysis, so they were filtered out. Since the royal edicts were not the expression of Su's own will, and they were omitted as well. The genres retained as corpus are fugue, narrative, record, biography, discussion, strategy, memorial, state, epitaph, praise, miscellaneous writings, etc.

2.2 Date Preprocessing

In order to make a comparison of the data before and after WPC, the texts firstly need to be confirmed the year of writing. Only a few texts have a clear date of writing, so we referred to the records of *Annals of the Three Su father and sons by the Song People* [5] on the major events of Su's past years, the texts in Su's complete works were marked with the time of writing, and a total of 232 texts were obtained as corpus.

2.3 Data Analysis Before and After the Wutai Poetry Case

The WPC occurred in the second half of the second year of Yuanfeng reign of Emperor Shenzong (1079 AD), Su was released at the end of the year, and arrived in Huangzhou that the relegated land at the beginning of the following year (1080 AD). Su recommissioned to be the prefect of Dengzhou and officially returned to the court in the Eighth year of Yuanfeng (1085 AD). So from 1080 AD to 1084 AD, Su was relegated and stayed away from the political center for five years, we selected texts written in these five years as a set of target data.

In order to maintain parity in duration, we used texts from the five years before WPC (from 1075 to 1079 AD) as the control group to analyze the differences in linguistic and psychological characteristics before and after WPC.

From 1075 to 1079 AD, Su served as magistrate in Mizhou, Xuzhou, and Huzhou successively. Wherever he went, he did his best to serve the people. This is the period of his outstanding political achievements, which is in contrast with the relegation period after WPC (Table 1).

Table 1. Data groups information

Data groups	Period	Time range	Number of texts
Group a	Five years before WPC (Magistrate period)	From 1075 to 1079 AD	22
Group b	Five years after WPC (Relegation period)	From 1080 to 1084 AD	19

Using CC-LIWC, we calculated the word frequency of group a and group b to obtain two sets of data. Since both sets of data were non-normally distributed, the Mann-Whitney non-parametric test was performed on 81 categories of CC-LIWC to calculate the significance of the difference between the two sets of data, the p-value is less than 0.05 as significant difference.

3 Results

Using the above method, it is concluded that there are 9 categories with significant differences in the two sets of data as Table 2 shows,

Table 2. 9 categories with significant differences.

Category name	Simplified name	M		SD		p
		a	b	a	b	
Third-person singular words	Shehe	0.020	0.013	0.011	0.011	0.012*
Third-person plural words	They	0.02	0.012	0.011	0.01	0.010*
Tensem words	Tensem	0.025	0.037	0.009	0.012	0.000*
Focuspresent words	Focuspresent	0.007	0.014	0.005	0.008	0.003*
Negative Emotion words	Negative	0.074	0.059	0.023	0.022	0.039*
Differentiation words	Differ	0.056	0.039	0.017	0.018	0.004*
Perceptual process words	Percept	0.057	0.073	0.017	0.025	0.012*
See words	See	0.018	0.025	0.009	0.012	0.026*
Time words	Time	0.051	0.072	0.015	0.029	0.004*

Note. M is the mean, SD is the standard deviation, p is the p-value(p-value keep three decimal places); *$p < 0.05$; a refers to group a; b refers to group b.

In order to analyze the specific changes of each category with significant differences before and after WPC, we merged the two sets of texts to two texts, calculated word frequency using CC-LIWC again, and drew the comparison charts of word frequency before and after the WPC for the 9 categories with significant differences.

4 Discussion

From a psychological perspective, function words (pronouns, prepositions, articles, auxiliary verbs, adverbs, conjunctions, negative, tensem,.etc.) reflect how people are communicating, whereas content words(referred to as generally nouns, regular verbs, and many adjectives and adverbs) convey what they are saying (cf. Chung & Pennebaker, 2007) [6]. Function words are much more closely linked to measures of people's social and psychological worlds [7]. It can be seen from Figs. 1 that the function words with significant differences include third-person singular words, third-person plural words, tensem words(refers to as Tense marker words) and its sub-category focuspresent words.

The word frequency of third-person singular and plural words after WPC are significantly lower than before that. As function words, such as personal pronouns, also reflect attentional allocation. People who are experiencing physical or emotional pain tend to have their attention drawn to themselves and subsequently use more first-person singular pronouns (e.g., Rude, Gortner, & Pennebaker, 2004) [8]. Another study shows that participants used more first-person singular and fewer third person pronouns

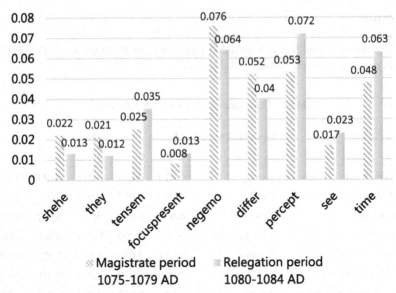

Fig. 1. Comparison chart of word frequency for 9 categories with significant differences Note. The data label on the histogram is word frequency for each category.

(e.g., "he," "she") while describing an event when they were being teased rather than they were teasing someone else [9]. The word frequency data shows that in the five years after WPC, Su used third-person pronouns less frequently. Accordingly, Su in relegation period has paid less attention to others in these five years.

Whereas personal pronouns provide information about the subject of attention, analyses of the tense of common verbs can tell us about the temporal focus of attention [10]. The frequency of *tensem words, focuspresent words* and *time words* increased. This reflects that Su not only tends to reduce expressions about others, but also expresses more frequently in terms of time, especially words in the present tense.

It is generally believed that having no foreign objects in the mind and paying attention to the present is an important part of Buddhist self-cultivation. *Huangzhou Anguo Temple Records* notes that after living in Huangzhou, Su spent most of his time living in the simple state, or practicing meditation or practicing Taoist bigu techniques, and spent more time studying Buddhism and Scripture. With no foreign objects in his mind, paying attention to himself and experiencing the present, Su showed more aloofness during this period, focusing on time flies and natural changes. The life state recorded in the literature of Su is reflected by the difference in word frequency now.

Research suggests that LIWC accurately identifies emotion in language use. For example, positive emotion words (e.g., love, nice, sweet) are used in writing about a positive event, and more negative emotion words (e.g., hurt, ugly, nasty) are used in writing about a negative event (Kahn,Tobin, Massey, and Anderson, 2007) [11]. LIWC ratings of positive and negative emotion words correspond with human ratings of the writing excerpts (Alpers et al., 2005) [12]. Therefore, the significantly decrease in the frequency of negative emotion words can be considered that Su's negative emotion did decrease after he moved away from the court. This result is in contrast to the texts criticizing the new policy and expressing anger prior to WPC.

The frequency of differentiation words also showed a decreasing trend. Differentiation is conceptually similar to the 2007 exclusive category [13]. According to the research of Yla R. Tausczik and James W. Pennebaker, exclusive words (e.g., but, without, exclude) are helpful in making distinctions. Indeed, people use exclusion words when they are attempting to make a distinction between what is in a category and what is not in a category [14]. The downward trend in the word frequency of negative emotion words and differentiation words is consistent with the Buddhist wisdom of shedding the relative phase, practicing non-differentiation, and being free from worries.

In a letter to his friend, Su described his living conditions, he said that the main theme of relegation period was to travel in the mountains and rivers, visit monkhood and self-sufficiency, the textual works of this period are exquisite in scenery writing and far-reaching in lyricism, which can be confirmed by the increasing trend of the frequency of Perceptual process words and its sub-category of See words. Su seemed very comfortable with the idle life of Buddhism and was happy to be unrecognized.

5 Limitation

This study has screened the articles in Su Shi's complete works, discarded articles which writing year cannot be confirmed, the text used for analysis is uneven in genre, and certain categories are naturally higher in certain genres, which may lead to deviations in word frequency results.

In addition, CC-LIWC is constructed on the basis of the 2015 edition of Simplified Chinese LIWC (known as SC-LIWC), in classical Chinese context, polysemy or diversity of a word is very common, which affects the classification of words in the dictionary [3]. Therefore, the accuracy of CC-LIWC also needs to be continuously improved in future research.

6 Conclusion

From the trend of the word frequency of the 9 categories with significant differences, Su's psychological characteristics were consistent with the Buddhist pursuit of being free from foreign objects, paying attention to the present, practicing non-differentiation, and getting out of worries after the WPC.

This result confirms the view put forward by previous studies that Su has been more advocating Buddhism practice after experiencing the blow of the WPC, and prove that it is feasible to quantitatively analyze psychological characteristics using CC-LIWC, objectified the interpretation of the psychological state of historical figures. With the continuous improvement of technology, it may be of considerable significance to apply it to more extensive psychological research.

References

1. Pennebaker, J.W., Boyd, R.L., Jordan, K., Blackburn, K.: The development and psychometric properties of LIWC2015, Austin, TX: University of Texas at Austin, p. 1, (2015) https://doi.org/10.15781/T29G6Z
2. Zhao, N., Jiao, D., Bai, S., et al.: Evaluating the validity of simplified Chinese version of LIWC in detecting psychological expressions in short texts on social network services, p. 2. PLoS ONE 11(6), 1–15 (2016)
3. Miaorong, F.: Critical Technology in Ancient Chinese Psychological Semantic Analysis Based on LIWC. Master Dissertation. University of Chinese Academy of Sciences, Beijing (2020)
4. https://www.wenxue360.com/archives/6711.html
5. Annals of the Three Su father and sons by the Song People. Edited by Wang Shuizhao, Published by Zhonghua Book Company (2015)
6. Chung, C.K., Pennebaker, J.W.: The psychological function of function words. In: Fiedler, K. (ed.) Social Communication: Frontiers of Social Psychology, pp. 343–359. Psychology Press, New York (2007)
7. Tausczik, Y.R., Pennebaker, J.W.: The psychological meaning of words: LIWC and computerized text analysis methods. J. Lang. Soc. Psychol. p. 29. 29, 24 (2010). originally published online 8 December 2009 https://doi.org/10.1177/0261927X09351676
8. Rude, S., Gortner, E.M., Pennebaker, J.: Language use of depressed and depression-vulnerable college students. Cogn. Emot. 18, 1121–1133 (2004)
9. Tausczik, Y.R., Pennebaker, J.W.: The psychological meaning of words: LIWC and computerized text analysis methods. J. Lang. Soc. Psychol. pp. 30–31. 29, 24 (2010). originally published online 8 December 2009.DOI: https://doi.org/10.1177/0261927X09351676
10. Tausczik, Y.R., Pennebaker, J.W.: The psychological meaning of words: LIWC and computerized text analysis methods. J. Lang. Soc. Psychol. pp. 31. 29, 24 (2010) originally published online 8 December 2009. https://doi.org/10.1177/0261927X09351676
11. Kahn, J.H., Tobin, R.M., Massey, A.E., Anderson, J.A.: Measuring emotional expression with the linguistic inquiry and word count. Am. J. Psychol. 120, 263–286 (2007)
12. Alpers, G.W., Winzelberg, A.J., Classen, C., Roberts, H., Dev, P., Koopman, C., et al.: Evaluation of computerized text analysis in an Internet breast cancer support group. Comput. Hum. Behav. 21, 361–376 (2005)
13. Pennebaker, J.W., Boyd, R.L., Jordan, K., Blackburn, K.: The development and psychometric properties of LIWC2015, Austin, TX: University of Texas at Austin, p.15 (2015). https://doi.org/10.15781/T29G6Z.Page16
14. Tausczik, Y.R., Pennebaker, JW.: The psychological meaning of words: LIWC and computerized text analysis methods. J. Lang. Soc. Psychol. p. 35. 29, 24 (2010) originally published online 8 December 2009. https://doi.org/10.1177/0261927X09351676

A Conference Publishing System Based on Academic Social Network

Jiongsheng Guo, Jianguo Li[✉], Yong Tang, and Weisheng Li

South China Normal University, Guangzhou 510631, Guangdong, China
gumpye@163.com

Abstract. Scholars and researchers participate in various academic conferences to share the latest research results and academic ideas. As more and more conferences are held, conference organizers need to promote their conference by setting up conference website. Therefore, conference organizers need a convenient and trusted tool to help them build a conference website. In addition, with the rapid growth of the Internet and the explosion of conference information, recommend conferences to scholars that interest them becomes important. For this consideration, we designed and implemented an academic conference publishing system, which can publish conference information quickly, implements multi-level management and recommends conferences that they are interested in for system users. In addition, by associating with SCHOLAT, our system presents conference members' personal academic information and provide an online communication platform. We simulate the actual use environment to deploy and verify the system, and it is proved that the system has excellent performance in security, availability, authority and other aspects.

Keywords: Conference publishing system · Hybrid recommendation · SCHOLAT

1 Introduction

With the rapid development of social science and technology, the way people obtain and exchange information has undergone tremendous changes, more and more scholars communicate by attending academic conference. The increasing number of conferences has led to an urgent need to build conference websites. More and more academic conference organizations have set up their own academic conference platforms, on which they can publish relevant information, promote the downloads and visits of conference papers, and enhance the influence of conferences.

However, for some organizers, they have no time to build a website, they need a conference publishing system to help them create a website easily, and there are many such systems: Easy Chair, Wiki Cfp, etc.

Although the above systems have been largely used to organize academic conferences, there are still some unresolved issues. Firstly, scholars do not have a platform for online communication after the conference. And traditional conference publishing system cannot provide the detailed information of conference organizers. Last but not least,

Q. Zu et al. (Eds.): HCC 2020, LNCS 12634, pp. 389–400, 2021.
https://doi.org/10.1007/978-3-030-70626-5_41

conference management platform generally adopts the mode of unique administrator, which cannot support multi-level management.

According to the above problem, we put forward a conference publishing system, which help conference organizer to create conference website, introduce their participants, and provide a communication platform to attendee.

In the following, we first review the related work in Sect. 2. In Sect. 3, We introduce the detailed design of the important function of our system. In Sect. 4, we describe some implemental detail. Finally, we conclude our work in Sect. 5.

2   Related Work

With the development of science and technology, more and more scholars design publish system to help conference organizers publish the conference information.

Yang et al. [1] explored the design and implementation of the conference publishing system in the early stage from three aspects: organization, management and problems that should be paid attention to.

In 2013, Xie [2] et al. conducted a research on the operation mode of a conference publish system. They found that the system implements the basic functions, but needed to be improved in following respects: resource acquisition and dissemination initiatives, influence surface and using range.

In addition to the basic functions, due to the increasing number of conference, the need to recommend conferences of interest to scholar become important. In the meantime, academic social media [3] develop rapidly. SCHOLAT is an authoritative academic social network. Since its launch in 2009, there have been more than 130,000 active scholars. We can refer to SCHOLAT to implement the recommendation. There is an example: Tang et al. [4] proposed a recommend model based on SCHOLAT, which provides two services: XPSearch and XSReco. XPSearch is a service that provides a vertical search for research papers with disambiguation of author names. XSReco use a topic-based approach to provide a list of "recommended scholars" to help users find potential collaborators who share the same research interests and may be interested in building partnerships. These two services improve the efficiency of scholars in retrieving papers and discovering research opportunities.

To sum up, the existing conference publishing system can help user publish conference basic information, but it has some deficiencies in promoting the communication between scholars. Therefore, we design a system that support multi-level management and strengthen the interaction between members of the conference.

3   System Design

3.1   Multilevel Management

Generally speaking, academic conference's website is generally created by a certain person in the name of the organization, but need more than one administrator to manage the website, so the conference needs to ensure the efficient management by set up multilevel levels to manage the conference.

The Spring Security framework is an enterprise-class Security framework based on the Spring ecosystem that provides declarative Security access control [5]. By using Spring Security framework, the process of multiple levels is as follows:

1. The permissions of a user's role have been designed by the system, and the conference administrator decides the permissions of other role by configuring the role of users.
2. When the system starts, the Spring Security Framework determines whether the current user has the authorization of the link according to user roles. If the user has the authorization, the request will be sent to the backend server; otherwise, the system prompt without the authorization will be returned.

3.2 Conference Recommendation Module

Nowadays, people's Daily life has been inseparable with data, how to help users obtain useful information from big data has become a hot field of scientific research [6], among which personalized recommendation has been widely concerned.

Collaborative filtering technology is a kind of algorithm based on association rules, widely used in common recommendation system and research, its core idea is similar users have similar interests. Recommendation method based on collaborative filtering technology research has a lot of, for example: Polatidis et al. [7] propose a multi-level recommendation method with its main purpose being to assist users in decision making by providing recommendations of better quality. The proposed method can be applied in different online domains that use collaborative recommender systems, thus improving the overall user experience.

For our system, one of the important tasks is to recommend conference to users what they are interested in. Conference recommendation module includes offline recommendation process, online recommendation process, and hybrid recommendation module.

Offline Recommendation Process. This process recommends the conference by compare user profiles with conference categories. The main steps of this module are as follows:

1. Extract the profile keywords of the users, and TF-IDF [9] algorithm is mainly adopted here. If certain words or phrases appear more frequently (TF) in the user profile, but less frequently in the others profile, these words are considered to be more representative of the user's personal characteristics.
2. Transform user keywords and conference category into word vectors by using the Bert model. Then calculate the similarity between the above keyword set, the formula is shown in Eq. 1.

$$sim(i,j) = \frac{\sum_{i=1}^{n}(x_i \times y_j)}{\sqrt{\sum_{i=1}^{n}(x_i^2)} \times \sqrt{\sum_{i=1}^{n}(y_i^2)}} \tag{1}$$

$sim(\theta)$ represents the similarity between user keywords and conference category, x_i represents the extracted list of user keywords, y_i represents the list of keywords in the conference category.

Online Recommendation Process. When user logs in, the system obtains user's historical conference access record, combines with the collaborative filtering algorithm [11–14] to make a recommendation list of the conference. The main steps are as follows:

1. Get user's log file according to the user Id, and then clean and format the data into a triple structure of user Id, conference Id and access time.
2. Deal with the triple structure, organize it into user conference relation table for discovering users with similar interests.
3. The cosine similarity formula is used to calculate the similarity between users and convert it into the relational degree matrix, so as to obtain users with similar interests.
4. Sort the users closest to the user by distance, remove the conferences that the user has been visited, and then sort them to get the recommendation list.

Hybrid Recommendation. In the hybrid recommendation module, the recommendation list obtained from the online recommendation module and the conference list obtained from the offline recommendation module are combined with the time attenuation function to generate the final recommendation list.

3.3 Associated with SCHOLAT

Link to User's SCHOLAT Homepage. The traditional conference system often lacks detailed academic introduction of a conference organizers, which makes it impossible for participants to acquaintance organizers' research field and achievements.

The SCHOLAT homepage provide detail academic information. With the help of SCHOLAT, we can display organizes' detailed information. The association model with SCHOLAT is shown in the Fig. 1.

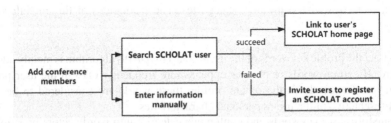

Fig. 1. The association model with SCHOLAT.

Create Communication Platform. During the conference, scholars did not have enough time to communicate, which greatly limits the development speed of academic technology. Therefore, how to facilitate communication among scholars is also an urgent problem to be solved.

The "TEAM" service provided by SCHOLAT, creates an online communication platform for scientific research teams. Users with certain qualifications can create teams and manage members, resources, research dynamics, etc. Each team has an independent domain name and page, which can meet the needs of the team to show the research process. In our system, by associating to SCHOLAT, we help conference organizers to create an online communication platform.

4 Implementation

In this chapter, combined with the system design and specific technical knowledge of the previous chapter, we will introduce the system development process and describe the implementation process of each module (Fig. 2).

Fig. 2. The homepage of the conference publishing system.

4.1 Multilevel Management

After the creator creates a meeting, a new administrator can be generated to assist in managing the meeting information. We define the creator as the first level administrator and the administrators generated by them as the second level administrator. The second level administrator is given the right to manage the conference. The design of permissions mainly includes the following steps:

1. Define roles and permissions.
2. Give the second level administrator a specific role.

The module is implemented by spring security. When a user logs in, the system obtains its permission and determines whether the current user has the right to access the URL to be visited (Fig. 3).

Fig. 3. An example to manage a conference.

4.2 Conference Recommendation Module

The recommendation module is divided into online module, offline module and hybrid recommendation module. The offline module mainly adopts TF-IDF to extract the keywords and BERT to get keywords' similarity for recommendation. The online module mainly adopts the user-based collaborative filtering algorithm to implement recommendations based on the user's historical access records. Hybrid recommendation module combines online and offline modules.

Offline Recommendation Process. Offline module use TF - IDF algorithm to extract the scholar profile keywords list: *SummaryKeywordList*. In order to improve the proportion of keywords in the scholar's profile and the accuracy of TF-IDF algorithm, we have acquired and processed some research directions of some scholars, and obtained academic direction keywords dictionary: *ResearchAreaDict*, part of the list of keywords in Table 1.

Table 1. Part of the list of keywords.

Keyword	Occurrences
Data mining	304
Cloud computing	221
Computer vision	74
The social network	34
The neural network	18
Recommendation system	17

The specific steps of keyword extraction are as follows:

1. To preprocess scholars' personal profile data, we remove the HTML tag codes and some useless stop words firstly.
2. We use "Jieba", a Chinese text segmentation, to handle the word segmentation procession, generating lexical groups with the same number of scholars.

3. Create the profile object, obtain the keyword dictionary and initially it into the "map", merge the generated vocabulary groups and re-store them into the "map".
4. Traversed all lexical groups, recorded the total number of vocabularies according to the "map", recorded the occurrence times of keywords in a brief introduction and in all profiles.
5. Use TF-IDF formula to calculate the "Top k keywords" of each scholar.
6. At last, according to BERT algorithm, we calculate the relationship between the keywords of the academic direction (*ResearchAreaDict*) and the keywords of the conference (*ConferenceKeywordList*). Then the list of conferences is sorted from large to small according to the degree of relationship and stored in the database.

Online Recommendation Process. The online recommend module use collaborative filtering algorithm to get another conference list: *OnlineHistoryList*, the main algorithm steps are as follows:

1. Format the JSON data file from the user log file and unrepeated the data to obtain a triplet structure of *userID*, *conferenceID* and *times*, normalize the number of times of the triplet structure, then we get a new triple: *userID,conferenceID* and *averTimes*.
2. This step aims to find users similar to the target user and calculate the similarity between them by using the cosine similarity formula. Assume that there are four users U_A, U_B, U_C, and U_D for the five conference C_A, C_B, C_C, C_D, C_E, C_F, and the relationship between users and the conference is shown in Table 2.

Table 2. Relationship between users and the conference.

USER ID	Conference and the degree of relationship
U_A	$(C_A,0.8)$ $(C_B,0.7)$ $(C_D,0.9)$
U_B	$(C_A,0.7)$ $(C_C,0.3)$
U_C	$(C_B,0.4)$ $(C_E,0.9)$
U_D	$(C_C,0.2)$ $(C_D,0.6)$ $(C_E,0.5)$

3. After getting the relationship table of user-conference, we set up an inverted table for users. For those who have visited the same conference, we calculate the average value of their "average Times" to the same conference, and fill the result in to Matrix. Then divide the average value by the square root of the product of the number of historical access meeting lists for two users. The user-conference relationship table is shown in Table 3.
4. After calculate the relationship between the user, obtain the "Top k" user from table according to the distance, represent the result with the set $S(u, c)$. Then obtain the conferences visited by "Top K" users, remove the conference which have been visited by user. Recycle formula 2 to calculate similarity between conferences with users, and sort by similarity to get a list of recommendations.

Table 3. Relationship between users and the conference.

	U_A	U_B	U_C	U_D
U_A	0	$\dfrac{0.75}{\sqrt{3\times2}}$	$\dfrac{0.55}{\sqrt{3\times2}}$	$\dfrac{0.55}{\sqrt{3\times3}}$
U_B	$\dfrac{0.75}{\sqrt{3\times2}}$	0	0	$\dfrac{0.25}{\sqrt{3\times2}}$
U_C	$\dfrac{0.55}{\sqrt{3\times2}}$	0	0	$\dfrac{0.7}{\sqrt{3\times2}}$
U_D	$\dfrac{0.55}{\sqrt{3\times2}}$	$\dfrac{0.25}{\sqrt{3\times2}}$	$\dfrac{0.7}{\sqrt{3\times2}}$	0

w_{uv} represents the similarity between users U and user V, the relationship between r_{vc} represent the similarity between user V and conference C.

$$p(u, c) = \sum_{v \in S_u,k} w_{uv} \times r_{vc} \tag{2}$$

Hybrid Recommendation. This module is mainly to integrate the recommendation results of offline recommendation module and online recommendation module, so as to further improve the accuracy of the recommendation results. Therefore, after getting the conference recommendation list based on user profile keywords and the conference list based on user collaborative filtering algorithm, we fuse it with the time attenuation function to get a hybrid recommendation result. The specific algorithm steps are as follows:

1. We get the *offlineRecommendList* and the *onlineRecommendList* by the process we mention above.
2. Make a linear combination of the two lists and fuse the exponential time attenuation function [16]. The final recommendation list calculation formula is shown in Eq. 3.

$$confList_i = (\alpha \times onlineList_i + \beta \times offlineList_i) \times e^{-\gamma(t-t_y)} \tag{3}$$

Where α is the weight of the online recommendation list, β is the weight of the offline recommendation list, $onlineList_i$ is the similarity of the online module of conference i, $offlineList_i$ is the similarity of the offline module of conference i, γ is the proportion of the time parameter, t is the current time, and t_y is the release time of the conference (Fig. 4).

4.3 Associated with SCHOLAT

Link to User's SCHOLAT Homepage. The conference details page contains all the member of the conference. User can visit conference member's SCHOLAT homepage by click the member in the conference detail page. The implementation details are as follows:

Fig. 4. The result of conference recommendation.

1. We designed an entity class for the conference organization members, which include the Java property: "scholat_name". When processing detail page request, system obtain all the members through Member Dao (designed tool to get the entity). The procession of getting members' information is shown in Fig. 5.

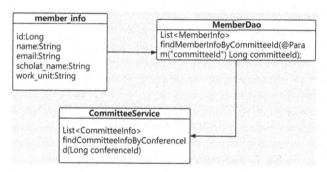

Fig. 5. Procession to get member's information.

2. In the conference detail page, members with the property of "scholat_name" are highlighted to guide users to see scholar details. Click the name to link to SCHOLAT homepage. Figure 6 show an example of the conference detail page.

Create a Communication Platform. The model of our system associated with SCHOLAT to create a team is shown in the Fig. 7.

When user use the team creation function, system provide the member list of a conference, the team roles define by user and team information. Then we Call the interface to create a SCHOLAT team. Figure 8 is an example to create a team.

Fig. 6. Conference details page.

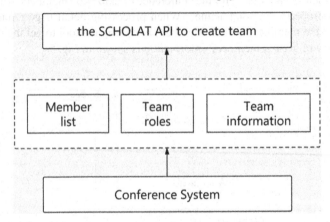

Fig. 7. The process of creating a SCHOLAT team.

Fig. 8. An example to create a team.

5   Conclusion

We design a practical, convenient conference publishing system based on SCHOLAT, which can help conference organizers publish conference website easily. Aiming at the problem that the academic conference is disorganized and difficult to find, this system integrates the conference information of each major conference website and defines the classification and label, providing the browsing mode of conference recommendation, conference classification browsing and academic conference searching. While satisfying the basic functions of the system, the communication between scholars is strengthened and the time for scholars to find conferences is greatly saved, which greatly improves the user's sense of experience.

Acknowledgments. Our work is supported by the National Natural Science Foundation of China (No. U1811263) and (No. 61772211).

References

1. Yang, G.: Organization and management of academic conferences. China Med. J. 106–108 (2007)
2. Xie, J., Li, L., Zhao, K.: Research on the operation mode of "Chinese academic conference online". Public Commun. Sci. Technol. (2), 292–294, 231 (2014)
3. Ovadia, S.: ResearchGate and Academia.edu: academic social networks. Behav. Soc. Sci. Librarian 33(3), 165–169 (2014)
4. Tang, F., Zhu, J., He, C., Fu, C., He, J., Tang, Y.: SCHOLAT: an innovative academic information service platform. In: Cheema, M.A., Zhang, W., Chang, L. (eds.) ADC 2016. LNCS, vol. 9877, pp. 453–456. Springer, Cham (2016). https://doi.org/10.1007/978-3-319-46922-5_38
5. Nguyen, Q., Baker, O.F.: Applying spring security framework and OAuth2 to protect microservice architecture API. JSW 14(6), 257–264 (2019)
6. Tuzhilin, A.: Towards the next generation of recommender systems. ICEBI 10 (2010)
7. Polatidis, N., Georgiadis, C.K.: A multi-level collaborative filtering method that improves recommendations. Expert Syst. Appl. 48(1), 100–110 (2016)
8. Hernando, A., Bobadilla, J., Ortega, F.: A non negative matrix factorization for collaborative filtering recommender systems based on a Bayesian probabilistic model. Knowl.-Based Syst. 97(1), 188–202 (2016)
9. Christian, H., Agus, M.P., Suhartono, D.: Single document automatic text summarization using term frequency-inverse document frequency (TF-IDF). ComTech Comput. Math. Eng. Appl. 7(4), 285–294 (2016)
10. Devlin, J., Chang, M.W., Lee, K., et al.: Bert: pre-training of deep bidirectional transformers for language understanding. arXiv preprint arXiv:1810.04805 (2018)
11. Mathew, P., Kuriakose, B., Hegde, V.: Book recommendation system through content based and collaborative filtering method. In: 2016 International Conference on Data Mining and Advanced Computing (SAPIENCE), pp. 47–52. IEEE (2016)
12. Gopalan, P., Hofman, J.M., Blei, D.M.: Scalable recommendation with hierarchical Poisson factorization. In: Proceedings of the Thirty-First Conference on Uncertainty in Artificial Intelligence, UAI 2015, Amsterdam, The Netherlands, 12–16 July 2015, pp. 326–335 (2016)

13. Chatzis, S.: Nonparametric Bayesian multitask collaborative filtering. In: 22nd ACM International Conference on Information and Knowledge Management, CIKM 2013, San Francisco, CA, USA, 27 October–1 November 2013, pp. 2149–2158 (2013)
14. He, X., Liao, L., Zhang, H., et al.: Neural collaborative filtering. In: Proceedings of the 26th International Conference on World Wide Web, WWW 2017, Perth, Australia, 3–7 April 2017, pp. 173–182 (2017)
15. Arthur, J., Azadegan, S.: Spring framework for rapid open source J2EE web application development: a case study. In: Proceedings of the 6th ACIS International Conference on Software Engineering, Artificial Intelligence, Networking and Parallel/Distributed Computing (SNPD 2005), Towson, Maryland, USA, 23–25 May 2005, pp. 90–95 (2015)
16. Wang, X., Zhao, Y.L., Nie, L., et al.: Semantic-based location recommendation with multimodal venue semantics. IEEE Trans. Multimedia 17(3), 409–419 (2015)

Software Component Library Management Mechanism for Equipment Parts Service Value-Net

Zhuo Tian[1], Changyou Zhang[1(✉)], Xiaofeng Cai[1,2], and Jiaojiao Xiao[1]

[1] Laboratory of Parallel Software and Computational Science, Institute of Software Chinese Academy of Sciences, Beijing, People's Republic of China
changyou@iscas.ac.cn

[2] School of Computer and Communication Engineering, University of Science and Technology Beijing, Beijing, People's Republic of China

Abstract. Software reuse realizes the sharing of software resources, and component-based reuse is the main form of software reuse. The classification, storage, retrieval, and release of a large number of component resources require efficient component library management methods. This paper proposes an open software component library management mechanism based on microservices. Taking manufacturing product service life cycle value chain collaboration as a case, it realizes a domain-oriented microservice granularity division mechanism and forms a domain-oriented software component representation, warehousing, discovery, selection, arrangement, filing, and other compatible open software component library management mechanisms; realization of the design of an open software component management system for manufacturing product operation and maintenance reengineering value chain collaboration, The simulation verifies the open ability and management efficiency of the management mechanism.

Keywords: Microservices · Software component library · Management mechanism · Value chain collaboration

1 Introduction

Software reuse is an effective way to improve software production efficiency and quality, and software component technology is the core technology that supports software reuse. As a reusable software asset, software components have relatively independent functions and reuse value [1].

This paper proposes an open software component library management mechanism based on microservices to realize domain-oriented microservice granularity division, development integration, and software component library management. An open software component library management mechanism based on microservices forms an open software component library management system oriented to manufacturing product operation, maintenance and reengineering value chain collaboration.

Q. Zu et al. (Eds.): HCC 2020, LNCS 12634, pp. 401–406, 2021.
https://doi.org/10.1007/978-3-030-70626-5_42

2 Background

The similarity between microservices and components is that they are the units that constitute the software system [2, 3], can realize specific functions, and both hide the internal implementation details. Both emphasize reuse, reduce the real-time cost of software systems, and follow the idea of high cohesion and low coupling in software design; implementation technology and design methods are not restricted, services or components can be implemented by object-oriented technology or structured programming.

Microservices and software components provide a loosely coupled and highly reusable software implementation method at different system implementation levels [4], which improves the efficiency of software design and development and shortens the software design and development cycle. The software component realizes the software system (application layer) by selecting and assembling the components (component layer), then the microservice is to add a service layer between the two layers, formed by the combination of fine-grained components (component layer) Unit services, and then assemble discrete unit services according to certain business rules to form a software system (application layer).

3 Component Storage Management for Manufacturing Service Value Chain Collaboration

The data structure stored by the component expresses its identification information and characteristic information to facilitate the retrieval and maintenance of the component. This article uses facet-based component classification description, describing and classifying the same component from multiple different facets. Each facet is orthogonal to each other, and each facet describes the information of the component from different levels

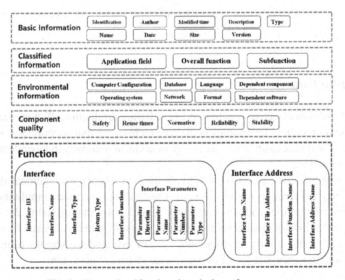

Fig. 1. Facet classification description of components

and different angles to improve Component retrieval efficiency. The component facet classification description uses five major facets: component basic information facet, component classification information facet, component environmental demand facet, component quality facet and component function facet as shown in Fig. 1.

4 Component Library Organization for Manufacturing Service Value Chain Collaboration

Component warehousing management is the core of component management. It completes the description of components. At the same time as the description, the components are preliminarily classified to provide support for future component retrieval and component assembly. The first step of component storage is to formulate component storage criteria. This requires component library managers to have very rich practical project experience [5, 6]. In addition, they must have the idea of software reuse.

4.1 Classification of Components

After the components are obtained and tested without error, the components can be classified and organized. Component classification is the basis of component retrieval. If the components in the component library are not carefully organized and classified, it will be difficult for software reusers to retrieve the required components [7–9].

4.2 Component Entry

Components entry into the component library is the first step of the component library management system [10, 11]. The metadata managed by the component library is the component. When the component is stored in the library, the facet description information of the component and the entity information of the component need to be stored for reuse of the component. The facet description information of the component is used for component retrieval. The component entity is the specific existence form of the component, which can be source code, related documents or other carriers, and is a direct manifestation of software reuse as shown in Fig. 2.

5 Component Library Maintenance for Manufacturing Service Value Chain Collaboration

The system administrator adopts the C/S structure, installs the component library management tool on the client, carries out the warehousing, storage, management and maintenance of the components, and connects with the local application Server [12–14]; the system user adopts the B/S structure to access through the browser The application system on the Web server performs component retrieval and component download as shown in Fig. 3.

Application	AMIs	Availability Zones	Status
BUSINESS	n/a (1)	(1)	UP (1) - f1841cebe46d.business:9081
COMPREVIEW	n/a (1)	(1)	UP (1) - 172.17.0.13:8097
PARTSEXAMINE	n/a (1)	(1)	UP (1) - 172.17.0.10:8094
PARTSINSTALL	n/a (1)	(1)	UP (1) - 172.17.0.7:8091
PARTSMAINTAIN	n/a (1)	(1)	UP (1) - 172.17.0.11:8095
PARTSRUN	n/a (1)	(1)	UP (1) - 172.17.0.8:8092
PARTSSCRAP	n/a (1)	(1)	UP (1) - 172.17.0.12:8096
PARTSSTANDBY	n/a (1)	(1)	UP (1) - 172.17.0.6:8090
PARTSUNLOAD	n/a (1)	(1)	UP (1) - 172.17.0.9:8093
RETROSPECTCONFIG	n/a (1)	(1)	UP (1) - 172.17.0.5:8089
SERVICE-GATEWAY	n/a (1)	(1)	UP (1) - 9af1dc9240c2:service-gateway:8082
USER-SERVICE	n/a (1)	(1)	UP (1) - 7f4fb2ffe8bc:user-service:9091

Fig. 2. List of registered components of the system

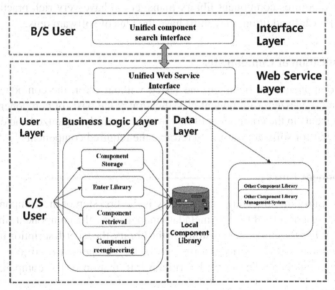

Fig. 3. Logical architecture diagram of component library management system based on Web Service

The component retrieval in the domain-oriented component library adopts a simple retrieval mode [15, 16], and submits relevant retrieval conditions through component facet information. The retrieval function obtains corresponding data through Web services and returns retrieval results.

The essence of component assembly is to establish associations between components, and coordinate their behaviors based on this association, and organize them into an organic whole. Existing component assembly usually uses a formal technology as the theoretical basis of semantic information description,describing the assembly structure of components from both static and dynamicaspects, supporting system evolution and modeling analysis.

The function of service composition mainly refers to the availability of service composition and the compatibility of interaction between network services. The performance of the service composition mainly refers to the optimization of the performance of the network service composition.

Martin Fowler proposed Micro Service in 2014. The micro service architecture is a lightweight Service-Oriented Architecture (SOA, Service-Oriented Architecture), which focuses on network-oriented services through componentized encapsulation and standards. The interface enables different network services to call each other.

6 Conclusion

This paper proposes an open software component library management mechanism based on microservices. Taking manufacturing product service life cycle value chain collaboration as a case, it realizes a domain-oriented microservice granularity division mechanism and forms a domain-oriented software component library management mechanism for component library storage, component library organization, and component library maintenance, and the design of an open software component management system that collaborates with the value chain of manufacturing product operation and maintenance reengineering. Simulation verifies the open management mechanism Competence and management efficiency.

Acknowledgement. The authors would like to express their thanks to the editors and experts who participated in the review of the paper for their valuable suggestions and comments. This research was supported by National Key R&D Program of China (2018YFB1701400), National Natural Science Foundation of China (61672508).

References

1. Mansouri, A.: Nooshin Soleymani Asl: Assessing mobile application components in Providing library services. Electron. Libr. **37**(1), 49–66 (2019)
2. Tan, D.-P., Chen, S.-T., Bao, G.-J., Zhang, L.-B.: An embedded lightweight GUI component library and ergonomics optimization method for industry process monitoring. Front. Inf. Technol. Electron. Eng. **19**(5), 604–625 (2018). https://doi.org/10.1631/FITEE.1601660
3. Christoforou, A., Odysseos, L., Andreou, A.S.: Migration of software components to microservices: matching and synthesis. In: ENASE 2019, pp. 134–146 (2019)
4. Boudeffa, A., Abherve, A., Bagnato, A., Di Ruscio, D., Mateus, M., Almeida, B.: Integrating and deploying heterogeneous components by means of a microservices architecture in the CROSSMINER project. STAF (Co-Located Events), pp. 67–72 (2019)
5. Gangemi, A., Chaudhri, V.K.: Representing the component library into ontology design patterns. In: WOP 2009 (2009)
6. Díaz-Casillas, L., Iglesias, C.A., Nieto, M.: A component library to improve the reusability in the development of converged services. In: iiWAS 2010, pp. 661–664 (2010)
7. Fumarola, M., Seck, M.D., Verbraeck, A.: A DEVS component library for simulation-based design of automated container terminals. SimuTools 2010, p. 71 (2010)
8. Korzonek, S.: Ewa dudek-dyduch:component library of problem models for ALMM solver. J. Inf. Telecommun. **1**(3), 224–240 (2017)

9. Dit, B., Moritz, E., Vásquez, M.L., Poshyvanyk, D.: Supporting and accelerating reproducible research in software maintenance using TraceLab component library. In: ICSM 2013, pp. 330–339 (2013)

10. Fernández-López, D., Cabido, R., Sierra-Alonso, A., Montemayor, A.S., Pantrigo, J.J.: A knowledge-based component library for high-level computer vision tasks. Knowl. Based Syst. **70**, 407–419 (2014)

11. Jiang, H., Fu, B., Zhang, M., Zhu, Y., Cao, W.: The Study on the Organization Approach of Agricultural Model Components Library Based on Topic Map. In: Li, D., Liu, Y., Chen, Y. (eds.) CCTA 2010. IAICT, vol. 346, pp. 186–197. Springer, Heidelberg (2011). https://doi.org/10.1007/978-3-642-18354-6_25

12. He, S., Chang, H., Wang, Q.: Component library-based ERP software development methodology. In: IESA 2009, pp. 34–38 (2009)

13. Dai, W.W., et al: Semantic integration of plug-and-play software components for industrial edges based on microservices. IEEE Access **7**, 125882–125892 (2019)

14. Gerking, C., Schubert, D.: Component-based refinement and verification of information-flow security policies for cyber-physical microservice architectures. In: ICSA 2019, pp. 61–70 (2019)

15. Jahromi, N.T., Glitho, R.H., Larabi, A., Brunner, R.: An NFV and Microservice Based Architecture for On-the-fly Component Provisioning in Content Delivery Networks. CoRR abs/1710.04991 (2017)

16. Paschke, A.: Provalets: component-based mobile agents as microservices for rule-based data access, processing and analytics. Bus. Inf. Syst. Eng. **58**(5), 329–340 (2016)

No-Reference Quality Assessment for UAV Patrol Images of Transmission Line

Xujuan Fan[1], Xiancong Zhang[1](✉), Jinqiang He[2], Yongli Liao[2], and Dengjie Zhu[2]

[1] Guangzhou Power Supply Bureau of Guangdong Power Grid Limited Liabilities Company, Guangzhou 510663, Guangdong, China
zhangxiancong33@163.com
[2] Electric Power Research Institute, China Southern Power Grid, Guangzhou 510663, Guangdong, China

Abstract. Uav patrol has gradually become the main operation mode of transmission line patrol task, however, due to the influence of the shooting mode and weather, the patrol images are inevitably distorted, resulting in quality degradation. In order to effectively evaluate the patrol image quality, this paper proposes a no-reference quality assessment method. Specifically, we first construct a dedicated patrol image quality assessment database, and then propose a no-referenced quality assessment model based on structure, texture and exposure. The experimental result shows that the method surpasses existing methods and is highly consistent with the quality labels.

Keywords: Transmission line · Uav patrol images · No-reference quality assessment

1 Introduction

In order to maintain the safe of transmission lines and to protect the stable supply of electricity, the transmission line patrol is very necessary. The traditional human-based patrol work is time-consuming and labor-intensive, and it is urgent to adopt new technical means to improve the quality and efficiency of line operation and maintenance. With the development of science and technology and the needs of smart grid construction, unmanned aerial vehicles (Uav) occupy an increasingly important position in the transmission line patrol work, which can not only overcome the influence of the geographical environment to a certain extent, but also expand the scope of the patrol, improve the efficiency and quality of patrol operations. After obtaining the patrol aerial images, often need to carry on a series of computer vision processing process such as target detection, segmentation, recognition and so on to detect the transmission lines in the images whether contain the defect as well as the defect localization. However, the images are affected by many factors such as bad weather (fog, rain, snow, etc.) and unbalanced lighting, and are also susceptible to distortion during images transmission and storage, which inevitably affects the accuracy of the later target detection and recognition methods. In order to improve the accuracy of object recognition and detection, qualitative and quantitative pre-evaluation of image quality is crucial.

© Springer Nature Switzerland AG 2021
Q. Zu et al. (Eds.): HCC 2020, LNCS 12634, pp. 407–418, 2021.
https://doi.org/10.1007/978-3-030-70626-5_43

Image quality evaluation (IQA) methods can be classified into three categories based on the degree of need for a reference image: full-reference (FR) IQA [1–4], reduced-reference (RR) IQA [5, 6], and no-reference (NR) IQA [7–10]. Although FR and RR are able to achieve superior performance in many cases, they are not applicable in real-world situations because they require that the reference image be fully or partially involved in the assessment process, as distortion-free reference images are not available in all scenarios. In contrast, NR methods are attracting more and more attention due to their ability to obtain better performance without reference images, which has become the mainstream direction of research this year.

In recent years, NR IQA is booming and many excellent algorithms have emerged for natural scene images. Dedicated image quality assessment algorithms for specific types of distortion assumes that the types of image distortion are known (common types of distortion include noise [11], JPEG compression [12], and blur [13]), and then the evaluation model is designed on this basis. [11] proposed a frequency mapping-based NR method for noise distortion, which measures the similarity between distorted and denoised images; [12] proposed a quality evaluation model for JPEG compression distortion based on quality correlation maps; Zhan et al. [13] found that maximum gradient is a valid index for characterizing image sharpness globally or locally, the maximum gradient and variability of gradients were used to predict the blurred image quality scores.

Although dedicated NR IQA methods can predict image quality more accurately, a priori knowledge of the type of distortion is often not available in real-world, making it more valuable to design general-purpose IQA methods that do not require prior knowledge of the type of image distortion. BRISQUE [10] assumes that distortion destroys image naturalness and computes natural scene statistical (NSS) features of the mean subtracted contrast normalized (MSCN) coefficients followed by a quality evaluation model using support vector regression (SVR). BLIINDS-II [14] extracts the NSS characteristics of the discrete cosine transform coefficient features and relied on a simple Bayesian inference model to predict the image quality score. NFERM [15] used free-energy-based brain theory and classical human visual system (HVS)-inspired features to extract three types of features and then use SVR to build IQA models. Based on the statistical theory of natural scenes, DIIVINE [16] realizes the evaluation of image verity and integrity across multiple distortion categories through two stages of distortion identification and quality evaluation against distortion. All of the above NR IQA methods need to use subjective human ratings in constructing assessment models, and there are other methods to evaluate image quality directly without human ratings, which we call the opinion-unaware method [17–19]. NIQE [17] first computes the NSS features of the distorted and natural image sets, and then computes the distance between their multivariate Gaussian models to achieve no-reference assessment of image quality. ILNIQE [18] is similar to NIQE in that it computes the distance of multivariate Gaussian model between distorted image patches and natural image patches to obtain the quality score of each image patch, and then average pooling to obtain the overall quality score of the image. QAC [19] evaluates the quality of each image block by using the centroids at each quality level as a code book, and then obtains the overall quality score.

In addition to the above-mentioned image quality assessment methods for natural scenes, many researchers have also focused on quality assessment problems for other image types in recent years, such as screen content images [20, 21], tone mapped images [22, 23], retargeted images [24], and UAV images [25]. Although the field of image quality assessment nowadays has covered a variety of image contents, to our knowledge, there is still little dedicated quality assessment algorithm for transmission line patrol images, and existing methods are not necessarily applicable to this problem, for which we believe there are two main reasons. Firstly, there is a lack of publicly available transmission line Uav patrol image quality assessment database, which is the basis of designing quality assessment method. Secondly, previous works did not consider the characteristics of the patrol image. Aiming at the problems summarized above, this paper starts to solve the problem of image quality assessment of transmission line Uav patrol images from two aspects of establishing a dedicated database and designing an objective algorithm. In terms of subjective image quality assessment, we build a large-scale transmission line Uav patrol image database as a benchmark for measure the performance of objective algorithms. In terms of objective quality assessment model design, we fully analyze the patrol image characteristics, extract the structure, texture and exposure features to build a no-reference patrol image quality assessment model. To our knowledge, this is the first dedicated quality assessment database and objective assessment algorithm for transmission line Uav patrol images. The experimental results show that the objective model has good performance and can accurately predict the quality of patrol image.

2 Method

2.1 Transmission Line Uav Patrol Image Database

The patrol images before, during and after shooting are susceptible to various factors, so the quality of images vary greatly. In order to establish a standard database, 128 high quality images were manually selected as reference images from thousands of patrol images. Since we mainly focus on the image quality of transmission line equipment, the criteria for selection is that transmission line equipment is clear and without distortion, while the environmental background does not play a major role. Considering that the common types of distortion are noise, blur, and compression, we apply four types of distortion to each reference image with five levels of Gaussian blur, Gaussian noise, JPEG2000 compression, and JPEG compression, resulting in 20 distorted images for each reference image, for a total of 2560 images. Since manual labeling of image quality is time consuming and expensive, in order to compensate for this deficiency and to satisfy the requirement that the labels need to be consistent with subjective human visual experience, we use a high-performance full reference method SSIM [1] similar to [26] to evaluate the distorted images and use results as the label for that image. The selected high-quality image is partially displayed in Fig. 1.

2.2 No-Reference Image Quality Assessment

Patrol images are inevitably affected by various distortions in the process of acquisition and transmission. The introduction of distortion changes the correlation between pixels,

Fig. 1. Sample images in database

and to a certain extent the original information of the image is lost, which will cause immeasurable impact on the recognition of image content. In order to enable detection and recognition algorithms to more accurately identify targets and reduce errors, it is necessary to predict the quality of the image. This paper intends to mine image structure, texture and exposure information, and establish a no-reference image quality assessment model. The algorithm framework is shown in Fig. 2. As a learning-based method, the model contains both training and testing steps. In the training step, we first extract features in terms of structure, texture and exposure. Then, the extracted features are pooled into a quality score reflecting the image quality using SVR. In the testing step, the features of the test images are captured and then an estimated quality score is obtained by a well-trained regression model. More details are given below.

Structure. Image structural information plays an important role in the recognition and understanding of objects. Image distortion changes the correlation between image pixels, causing changes in image content, disrupting image structure, and affecting content recognition and understand. Therefore, estimation of structural information changes can provide a good approximation of the perceived image distortion. We first capture the quality-sensitive features of an image by analyzing the image structure. For a given image I, its structural information can be reflected by the gradient amplitude G:

$$G = \sqrt{(I \otimes p_x)^2 + (I \otimes p_y)^2} \qquad (1)$$

Where $\otimes$ is the convolution operator, p_x and p_y represent the kernel functions of the horizontal and vertical filters, respectively, in this method, the Prewitt filter is applied.

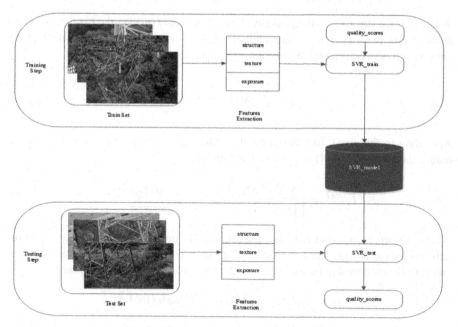

Fig. 2. The framework of proposed method

Patrol images are almost exclusively in a natural environment, and it has been shown that although natural images are diverse, they satisfy specific natural statistical laws. Experiments have shown that the gradient magnitude of natural images satisfies a two-parameter Weibull distribution [27], and distortion causes changes in the shape of this distribution. Therefore, we intend to quantify the image structure distortion using the parameters of the Weibull distribution. The probability function of the two-parameter Weibull distribution can be expressed as follows.

$$f(x) = \frac{\alpha}{\lambda}\left(\frac{x}{\lambda}\right)^{\alpha-1} \cdot e^{-\left(\frac{x}{\lambda}\right)^{\alpha}} \tag{2}$$

Where x represents the input image gradient amplitude; α is the shape parameter that controls the peak of the distribution; and λ determines the width of the distribution. These parameters reflect the correlation hidden in the Weber distribution, so we chose and to quantify the degree of distortion in the image.

Texture. The quality of structural information directly affects the effect of object detection, and texture information plays a crucial role for further object recognition. In this paper, we propose to evaluate the image quality from the perspective of texture distribution using the local binary model (LBP) to extract features [28]. For a pixel n_c, the relationship between it and its surrounding pixels n_i can be expressed as follows.

$$s(I(n_c), I(n_i)) = \begin{cases} 1, & if\ I(n_c) - I(n_i) \geq T \\ 0, & if\ I(n_c) - I(n_i) < T \end{cases} \tag{3}$$

Where T is the threshold value, usually 0, and $I(n_c)$ and $I(n_i)$ are the pixel gray values at the center pixel n_c and its surrounding pixels n_i, respectively. Then calculate the relationship between n_c and the surrounding Q pixels, and then binary code.

$$L_Q = \sum_{i=0}^{Q-1} s(I(n_c), I(n_i)) \cdot 2^i \tag{4}$$

According to the above formula, a pixel can have 2^Q numerical possibilities, and to reduce the number, Eq. (4) is improved as follows.

$$L_Q^{riu2} = \begin{cases} \sum_{i=0}^{Q-1} s(I(n_c), I(n_i)), & if \;\; \mu(L_Q) \leq 2 \\ Q+1, & otherwise \end{cases} \tag{5}$$

Where the superscript riu2 indicates the rotationally invariant uniform mode when the uniform measurement μ is less than 2. The number of changes used to calculate the comparative relationship between the middle pixel and the surrounding pixel.

$$\mu = \left\| s\big(I\big(n_{Q-1}\big), I(n_c)\big) - s(I(n_0), I(n_c)) \right\|$$
$$+ \sum_{i=0}^{Q-1} \left\| s(I(n_i), I(n_c)) - s(I(n_{i-1}), I(n_c)) \right\| \tag{6}$$

Based on the above processing, the comparison between a pixel and the Q pixels around it can be divided into Q+2 possibilities. We set Q = 8.

Figure 3(a) lists the images of the transmission line with glass insulator self-destruct defects, Fig. 3(b)–(f) shows the five levels of Gaussian blurred distortion images of Fig. 3(a), and Fig. 3(g) shows the LBP histograms of Fig. 3(a)–(f). It can be seen from Fig. 3(g) that distortion changes the distribution of the LBP histogram of the image, which has different distribution characteristics at different levels of blurring, which indicates that the LBP histogram distribution of the image is sensitive to distortion, and therefore LBP can be used as a feature to indicate image quality. Other types of distortion also have similar effects on LBP.

Not difficult to find, the previously acquired texture information is local feature. In order to fully characterize the image texture information, we further extract the global image texture features. For the image I with size $H \times W$, we first compress the gray value range to L level to obtain the processed matrix $M_{H \times W}$. For the pixel value i in $M_{H \times W}$, we calculate the number of occurrences of the pixel value j with distance d and direction θ. By traversing the whole matrix, the grayscale co-occurrence matrix $H_{d,\theta}$ can be obtained. Experiments show that extracting the contrast, energy and homogeneity from the grayscale co-occurrence matrix can reflect the image texture information well. Therefore, we first extract the contrast information G_c.

$$G_C = \sum_{i=1}^{L} \sum_{j=1}^{L} (i-j)^2 \cdot P_{d,\theta}(i,j) \tag{7}$$

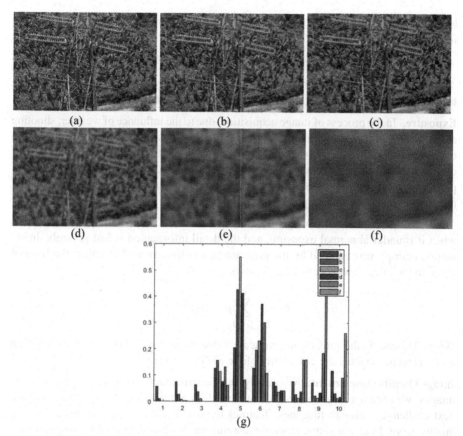

Fig. 3. Transmission line distortion images and corresponding LBP histogram distribution

Where $P_{d,\theta}(i,j)$ denotes the probability of an element $h_{d,\theta}(i,j)$ with position (i,j) in $H_{d,\theta}$, it can be defined as follows.

$$P_{d,\theta}(i,j) = h_{d,\theta}(i,j) / \sum_{i=1}^{L}\sum_{j=1}^{L} h_{d,\theta}(i,j) \tag{8}$$

Secondly, the energy information reflects the distribution of image texture information, describing whether the whole image has uniform distribution and its roughness, with a small energy value indicating that the image has uniform distribution and fine texture. Formally, it can be defined as follows.

$$G_E = \sum_{i=1}^{L}\sum_{j=1}^{L} P_{d,\theta}(i,j)^2 \tag{9}$$

Finally, homogeneity is a local image descriptor that reflects local gray-scale correlations. Specifically, homogeneity measures the local variation of an image. An image with small local variations will have a large homogeneity value. Homogeneity can be calculated as follows.

$$G_H = P_{d,\theta}(i,j)/(1 + |i - j - 2|) \tag{10}$$

After the above processing, this scheme proposes to apply Q+2 values to characterize local textures and apply $\{G_C, G_E, G_H\}$ to characterize global textures.

Exposure. In the process of image acquisition, due to the influence of weather, shooting angle, lighting factors, etc., there are exposure anomalies in the captured images to a certain extent. Whether overexposed or underexposed, it will have an impact on the comprehension and perceptual quality of the image content, thus affecting the subsequent processing results. We propose to further analyze the exposure intensity of the images, and then evaluate the image quality.

It was observed that the image tends to vary between 0 and 255 in grayscale values when it is under abnormal exposure, and the detail information is lost severely. Information entropy was adopted by the program as an effective tool to reflect the level of detail in the image and is defined as follows.

$$E = -\sum_{\mu=0}^{U} p_\mu \cdot \log p_\mu \tag{11}$$

Where U denotes the maximum grayscale value in the image (U = 255 for an 8-bit image) and p_μ denotes the probability of the μ-th pixel value.

Image Quality Assessment. After analyzing the structure, texture, and exposure of the images, we obtain a large number of features F that reflect image distortion, and the next challenge is how to fuse these features to provide feedback on the overall image quality score. In recent years, machine learning has been widely used in image quality assessment by virtue of its powerful ability to map high-dimensional space to low-dimensional space. We propose to use a support vector machine to establish the link $M = f(F)$ between the feature space f and the subjective rating value M. The support vector machine is used in this paper. The objective function is defined as follows.

$$\min_{w,\xi,\widehat{\xi}} \tfrac{1}{2}w^T w + C \sum_{i=1}^{z} \left(\xi + \widehat{\xi}\right) K(x_i, y_i) = e^{-\eta\|x_i - y_i\|^2}$$

$$s.t. \quad w^T \Phi(x_i) - y_i \le \varepsilon + \xi_i, \tag{12}$$

$$y_i - w^T \Phi(x_i) \le \varepsilon + \widehat{\xi}_i,$$

$$\xi, \widehat{\xi} \ge 0, i = 1, 2, ..., z$$

Where x_i denotes the feature vector of the i - th image and y_i denotes its corresponding subjective score value. $K(x_i, y_i) \equiv \phi(x_i)^T \phi(x_i)$ denotes the kernel function used, and this scheme proposes to use the RBF function as the kernel function.

$$K(x_i, y_i) = e^{-\eta\|x_i - y_i\|^2} \tag{13}$$

When the relationship f between the extracted features and the subjective score values is established, the quality score for a given image can be obtained by inputting the feature vector extracted for it into the model f.

3 Experiments

In order to verify the performance of the proposed algorithm, two generic no-reference image quality evaluation models NIQE [17] and ILNIQE [18] were selected for comparison on the database we created. These two algorithms do not require training and can directly compute the performance. In this paper, the algorithm is based on machine learning, so the database is randomly divided into training and test sets according to the ratio of 80%-20%, and the content of the two does not overlap. In order to ensure the reliability of the results, this paper performs 1000 times of training-testing process, and takes the median of the final results.

Before calculating the performance of the objective model, considering that the objective estimated score q_o and the quality label q_s may have different distribution characteristics, the five-parameter nonlinear equation is used to map the estimated score to the range of the quality label before the performance is calculated:

$$q_p = \beta_1 \left(\frac{1}{2} - \frac{1}{1 + \exp(\beta_2(q_o - \beta_3))} \right) + \beta_4 q_0 + \beta_5 \tag{14}$$

Where $exp(\cdot)$ represents the exponential function, $\beta_1, \beta_2, \beta_3, \beta_4$ and β_5 are all parameters of the regression equation. After nonlinear mapping, we use four common performance metrics to evaluate and compare the proposed approach with competing IQA metrics. Four performance evaluations were:

1. Pearson linear correlation coefficient (PLCC) measures the prediction consistency between q_p and q_s. It is defined as

$$PLCC = \frac{\sum_{i=1}^{N} \left(q_p^i - \overline{q_p} \right) \cdot \left(q_s^i - \overline{q_s} \right)}{\sqrt{\sum_{i=1}^{N} \left(q_p^i - \overline{q_p} \right)^2 \cdot \sum_{i=1}^{N} \left(q_s^i - \overline{q_s} \right)^2}} \tag{15}$$

Where, $\overline{q_p}$ and $\overline{q_s}$ represent the mean value of objective estimate and mean value of quality label of whole test set, q_s^i and q_p^i are the subjective score and the predicted score of the i-th image. N is the number of test images.

2. Spearman rank-order correlation coefficient (SROCC) can be calculated by

$$SRCC = 1 - \frac{6 \sum_{i=1}^{N} \Delta d_i^2}{N(N^2 - 1)} \tag{16}$$

Where Δd_i represents the difference in the order of score of the i-th image in q_p and q_s over the entire test set.

3. Kendall's rank-order correlation coefficient (KROCC) is another important rank correlation metric which can be given by

$$KRCC = \frac{2(N_c - N_d)}{N(N-1)} \tag{17}$$

Where N_c and N_d respectively represent the number of image pairs with consistent and inconsistent sorting in the data set.

4. Root mean-squared error (RMSE) is given by

$$RMSE = \sqrt{\frac{1}{N} \sum_{i=1}^{N} \left(q_p^i - \overline{q_s} \right)^2} \tag{18}$$

Researchers often use the absolute value of the above evaluation indexes to represent the performance of the objective algorithm. Therefore, the RMSE range is $[0, \infty]$, and the PLCC SRCC and KRCC range is $[0,1]$. For the objective algorithm with better performance, the RMSE value is closer to 0, and the PLCC SRCC and KRCC values are closer to 1. The comparison results are shown in Table1.

Table 1. Performance comparison of our method and existing general purpose NR metrics

Method	PLCC	SRCC	KRCC	RMSE
NIQE	0.6828	0.6181	0.4411	0.2263
ILNIQE	0.2128	0.3296	0.2206	0.3030
OURS	0.9794	0.9368	0.8133	0.0718

It can be seen from Table 1 that the performance of the NIQE and ILQE algorithms is very poor, indicating that they are not suitable for the quality assessment of transmission line patrol images. In response to this result, we think this may be because they are designed for natural scene images. Although the patrol images are also in natural scenes, they all contain transmission line facilities, so the natural characteristics are different from natural images, so the performance of NIQE and ILNIQE is poor. Whereas the algorithm we designed takes into account the characteristics of the patrol image, and the extracted image features can effectively represent the image quality.

4 Conclusion

In this paper, we analyze the Uav patrol image characteristics of transmission line, and propose a machine learning-based no-reference image quality assessment model, which can well predict the patrol image quality. Specifically, firstly, we constructed a patrol image quality assessment database as a benchmark for objective quality assessment.

Secondly, quality-sensitive structure, texture and exposure features are extracted to form feature vectors. And finally, SVR is applied to map the feature vectors into image quality scores. The experimental result shows that our proposed method has a high statistical consistency with the quality label and can be used for image quality assessment to help filter high quality images for the next target detection, recognition and other machine vision tasks.

References

1. Wang, Z., Bovik, A.C., Sheikh, H.R., et al.: Image quality assessment: from error visibility to structural similarity. IEEE Trans. Image Process. **13**(4), 600–612 (2004)
2. Zhang, L., Zhang, L., Mou, X., et al.: FSIM: a feature similarity index for image quality assessment. IEEE Trans. Image Process. **20**(8), 2378–2386 (2011)
3. Kim, J., Lee, S.: Deep learning of human visual sensitivity in image quality assessment framework, 1969–1977 (2017)
4. Liang, Y., Wang, J., Wan, X., Gong, Y., Zheng, N.: Image quality assessment using similar scene as reference. In: Leibe, B., Matas, J., Sebe, N., Welling, M. (eds.) ECCV 2016. LNCS, vol. 9909, pp. 3–18. Springer, Cham (2016). https://doi.org/10.1007/978-3-319-46454-1_1
5. Liu, Y., Zhai, G., Gu, K., et al.: Reduced-reference Image quality assessment in free-energy principle and sparse representation. IEEE Trans. Multimedia **20**(2), 379–391 (2018)
6. Golestaneh, S.A., Karam, L.J.: Reduced-reference quality assessment based on the entropy of DNT coefficients of locally weighted gradients, pp. 4117–4120 (2015)
7. Kang, L., Ye, P., Li, Y., et al.: Convolutional neural networks for no-reference image quality assessment. In: 2014 IEEE Conference on Computer Vision and Pattern Recognition. IEEE (2014)
8. Xu, J., Ye, P., Li, Q., et al.: Blind image quality assessment based on high order statistics aggregation. IEEE Trans. Image Process. **25**(9), 4444–4457 (2016)
9. Ye, P., Kumar, J., Doermann, D.: Beyond human opinion scores: blind image quality assessment based on synthetic scores. In: 2014 IEEE Conference on Computer Vision and Pattern Recognition (CVPR). IEEE (2014)
10. Mittal, A., Moorthy, A.K., Bovik, A.C.: No-reference image quality assessment in the spatial domain. IEEE Trans. Image Process. **21**(12), 4695–4708 (2012)
11. Yang, G., Liao, Y., Zhang, Q., et al.: No-reference quality assessment of noise-distorted images based on frequency mapping. IEEE Access **5**(99), 23146–23156 (2017)
12. Golestaneh, S.A., Chandler, D.M.: No-reference quality assessment of JPEG images via a quality relevance map. IEEE Signal Process. Lett. **21**(2), 155–158 (2014)
13. Zhan, Y., Zhang, R.: No-reference image sharpness assessment based on maximum gradient and variability of gradients. IEEE Trans. Multimedia **20**(7), 1796–1808 (2018)
14. Saad, M.A., Bovik, A.C., Charrier, C.: blind image quality assessment: a natural scene statistics approach in the DCT domain. IEEE Trans. Image Process. **21**(8), 3339–3352 (2012)
15. Gu, K., Zhai, G., Yang, X., et al.: Using free energy principle for blind image quality assessment. IEEE Trans. Multimedia **17**(1), 50–63 (2014)
16. Moorthy, A.K., Bovik, A.C.: Blind image quality assessment: from natural scene statistics to perceptual quality. IEEE Trans. Image Process. **20**(12), 3350–3364 (2011)
17. Mittal, A., Soundararajan, R., Bovik, A.C.: Making a 'Completely Blind' image quality analyzer. IEEE Signal Process. Lett. **20**(3), 209–212 (2013)
18. Zhang, L., Zhang, L., Bovik, A.C.: A feature-enriched completely blind image quality evaluator. IEEE Trans. Image Process. **24**(8), 2579–2591 (2015)

19. Xue, W., Zhang, L., Mou, X.: Learning without human scores for blind image quality assessment. In: 2013 IEEE Conference on Computer Vision and Pattern Recognition (CVPR). IEEE (2013)
20. Fang, Y., Yan, J., Li, L., et al.: No reference quality assessment for screen content images with both local and global feature representation. IEEE Trans. Image Process. **27**(4), 1600 (2018)
21. Gu, K., Qiao, J., Min, X., et al.: Evaluating quality of screen content images via structural variation analysis. IEEE Trans. Visual Comput. Graph. **24**(10), 2689–2701 (2018)
22. Yue, G., Hou, C., Zhou, T.: Blind quality assessment of tone-mapped images considering colorfulness, naturalness and structure. IEEE Trans. Ind. Electron. **66**(5), 3784–3793 (2019)
23. Yue, G., Hou, C., Gu, K., et al.: Biologically inspired blind quality assessment of tone-mapped images. IEEE Trans. Ind. Electron. **65**(3), 2525–2536 (2018)
24. Zhang, Y., Lin, W., Li, Q., et al.: Multiple-level feature based measure for retargeted image quality. IEEE Trans. Image Process. **27**(1), 451–463 (2018)
25. Guo, X., Li, X., Li, L., et al.: An efficient image quality assessment guidance method for unmanned aerial vehicle. In: 2019 Intelligent Robotics and Applications (2019)
26. Gu, K., Tao, D., Qiao, J.F., et al.: Learning a no-reference quality assessment model of enhanced images with big data. IEEE Trans. Neural Netw. Learn. Syst. **29**(4), 1301–1313 (2018)
27. Scholte, H., Ghebreab, S., Waldorp, L., et al.: Brain responses strongly correlate with Weibull image statistics when processing natural images. J. Vis. **9**, 29 (2009)
28. Ojala, T., Pietikälnen, M., Harwood, D.: A comparative study of texture measures with classification based on feature distributions. Pattern Recogn. **29**(1), 51–59 (1996)

An Inclusive Finance Consortium Blockchain Platform for Secure Data Storage and Value Analysis for Small and Medium-Sized Enterprises

Jiaxi Liu, Peihang Liu, Zhonghong Ou, Guangwei Zhang, and Meina Song[✉]

Beijing University of Posts and Telecommunications, Beijing, China
{ljx_228,mnsong}@bupt.edu.cn

Abstract. In the era of big data, people pay more and more attention to user privacy and data security. The market size of Small and Medium-sized Enterprises (SMEs) in China is sizable. Nevertheless, due to problems of data dispersion and lack of data features, it is very difficult to make use of the massive data of SMEs scattered in various institutions effectively, which leads to the inability to reflect the value of data. One of the problems is how to credit for SMEs better. Supported by federated learning for such scenarios, we present an inclusive Finance Consortium Blockchain platform in this paper. On the one hand, the platform combines decentralized identity and Blockchain as the underlying architecture, which guarantees authenticity of the data source and safe storage of user data on the chain. The smart contract mechanism provided by Blockchain can also assist secure storage and incentive mechanism of federated learning models effectively. On the other hand, we have innovatively introduced an asynchronous Federated Learning mode based on transfer learning, which encrypts the well-designed pre-training model and transfers it to each participant to guide training of the participant's local model. In the model reasoning stage, all the participants participate in a joint evaluation according to the local model and push the reasoning results on the Blockchain. The smart contract takes the weighted sum of the reasoning results provided by all the participants as the final result of shared model reasoning.

Keywords: Federated learning · Federated transfer learning · Blockchain · Decentralized identity

1 Introduction

China's market for Small and Medium-sized Enterprises (SMEs) is huge but their survival time is not optimistic. The huge SMEs market has created a high demand for financial services. Corporate entities, upstream and downstream companies, financial institutions (banks, small loan companies, etc.) and government departments (such as the Industrial and Commercial Bureau, tax bureau, etc.) together constituting a "financial services chain" in the supply chain. However, in this financial service chain, there is often a lack

© Springer Nature Switzerland AG 2021
Q. Zu et al. (Eds.): HCC 2020, LNCS 12634, pp. 419–429, 2021.
https://doi.org/10.1007/978-3-030-70626-5_44

of effective information coordination mechanism between each institution and there are many constraints in related credit interoperability and data sharing.

In this paper, we innovatively introduce Decentralized identity (DID) technology and Blockchain to jointly build the underlying architecture of the financial chain for SMEs. Which can guarantee the authenticity of the uploaded data from the source. Actually, the real identity information encrypted and stored at the user's center and only put the created digital identity (representing the user's unique identification) on chain which can provide individual information and form a complete on-chain routing for SMEs to promote enterprise credit interoperability and encrypted data sharing. On the other hand, the technology innovatively combines with federated learning which effectively utilizing the small data encrypted and stored by the participants in the chain to virtualize massive data for federated learning modeling. We also have improved the federated transfer learning algorithm, replacing synchronous updates with asynchronous updates in federated learning training, effectively utilizing the server computing resources and the data resources on the Blockchain, and providing a better solution to the large amount of data without the visibility of the real data. The data are analyzed for potential value, providing more accurate credit assessment of SMEs for financial institutions, and further solving problems such as difficulty in lending and long audit cycle for SMEs. The improved federated transfer learning algorithm is also resistant to the model inversion attack in federated learning training. Our system has effective defenses against malicious attacks during the interaction between federated learning training and credit evaluation models.

2 Background

Our work is closely related to Blockchain, Decentralized Identity and federated transfer learning and we give background knowledge in this section.

2.1 Blockchain

Blockchain is a technology that is jointly maintained by multiple parties, using cryptography to ensure transmission and enables consistent storage of data which is difficult to tamper with, also known as Distributed Ledger Technology (DLT). It is characterized by the fact that data is not stored and managed through a single central node, but is jointly maintained by multiple participants, ensuring that its storage process is monitored and authenticated by all participants and that all transactions are transparent and traceable. Blockchain's distributed structure provides strong security and privacy guarantees, and its own encryption algorithms and mathematical principles ensure that all transaction records cannot be tampered with or falsified. In this paper, we replace the third-party trusted authorities in federated learning with a Blockchain system to build distributed identities for the participants. Meanwhile, the integration of Blockchain and decentralized identity (DID) technology significantly reduce the cost of data governance and maintenance. We use Blockchain as the underlying technical architecture to build a set of consortium chain platform for data storage, management and value analysis for SMEs.

2.2 Decentralized Identity (DID)

In the Internet environment, how to construct a reasonable and credible digital identity is worth to explore. Internet technology companies usually make a set of business systems for handling authentication and access control, but the following pain points still exist.

- The problem of identity data of decentralization and duplicated authentication.
- Identity data privacy and security issues.
- Low fault tolerance and efficiency of centralized authentication.
- Proof of identity does not <u>cover</u> everyone.

A decentralized identity (DID) is a set of identities that are completely disintermediated and allow an individual or organization to have full ownership, management and control over their digital identity, which securely stores elements of their digital identity as well as true identity information and protects privacy [1].

We aims to provide a set of functions for the creation, verification and management of decentralized digital identity for SMEs in lending scenarios and achieve a more standardized management and protection of entity data. This scheme integrates the data resources of the same users in the chain and selectively disclose the information shared by users to ensure the authenticity and efficiency of information flow which can resist data security attacks such as de-anonymization faced by the system.

2.3 Federated Learning and Federated Transfer Learning

Transfer Learning [13]. Transfer learning is a learning technique that offers solutions for cross-domain knowledge transfer. In many applications, we only have small-scale labeled data or weak supervisory capabilities, which lead to reliable machine learning models that are not built [2,3]. In these cases, we can still build high-performance machine learning models using and debugging models for similar tasks or in similar domains. The essence of transfer learning goes to discover an invariance (or similarity) between a resource-rich source domain and a resource-scarce target domain, and to use that invariance to transfer knowledge between the two domains. Been prepared on the basis of the methods used to perform transfer learning, the literature [3] classifies transfer learning into 3 main categories: instance-based transfer, feature-based transfer, and model-based transfer.

Federated Learning. The idea of federated learning is to build a federated learning model based on distributed datasets, train one model for each organization that has a data source, and then let the organizations communicate with each other on their own models, and finally get a global model through model aggregation [14]. The process of exchanging model information between organizations will be carefully designed to ensure that no organization can guess the private data content of any other organization. Also, when constructing the global model, the data sources will be as if they have been integrated together. Federated learning falls into three categories: horizontal federated learning, vertical federated learning, and federated transfer learning [4].

Federated Transfer Learning. If the datasets owned by federated learning participants with only small amounts of data and weak supervision (fewer markers), federated transfer learning relies on transfer learning algorithms to protect data privacy by frameworks in federated learning. Model aggregation by currently existing transfer learning algorithms [3] enables them to help build effective and accurate machine learning models comply with data privacy and security regulations [5,12].

In our paper, we introduce a new federated transfer learning approach to transfer the risk control model of large enterprise loans as a pre-trained model to the local participants of the federated training of the risk control model of SMEs. The shift from synchronous to asynchronous updates for federated learning has significantly reduced the time and computational costs for various participants involved in the training.

3   Platform

3.1   Decentralized Identity (DID) and Blockchain

Distributed identities always use multi-tier architecture. In this paper, we use the most classic three-tier architecture to construct the infrastructure (Fig. 1).

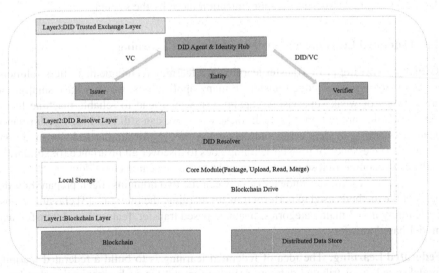

Fig. 1. This is a three-layer architecture for decentralized digital identity based on the Blockchain framework [15]. Users respectively need Blockchain to provide support for identity autonomy. In addition, a distributed data storage component is required to provide secure data management services when user data are stored in the cloud instead of local storage.

Blockchain Layer. This layer is utilized for storing DID documents. The most critical aspect within the DID documents is the correspondence between DID and public key. We selectively used distributed storage to store DID documents for data privacy.

DID Resolver Layer. This layer provides a unified DID resolution service to the upper layer, i.e., DID resolver. The layer node will package the DID-related commands from the upper level and create an on chain transaction. In order to make identity truly autonomous, a hash of that command in the transaction will be embed. The Blockchain supports key and identity discovery through identifier management system based on Public Key Infrastructures (PKI).

DID Trusted Exchange Layer. The DID Trusted Exchange Layer is the layer that helps participants establish secure identity and data exchange, including entity, issuer, verifier and user agent. The entities obtain decentralized identifiers (DIDs) and Verifiable Credentials (VC) by registering the identity on the Blockchain through a user agent and issuer, which can be provided to the verifier to complete the verification (Fig. 2).

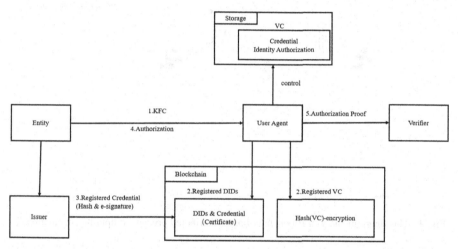

Fig. 2. This is the distributed digital identity authentication process [1, 6]. The "Entity" represents an entity object with an ID on Blockchain, and it can authorize the relevant agency to use its own relevant data. The "Issuer" is an organization or individual that issues and certifies data, and only data issued by an authority have authority. The "Verifier" is the organisation that uses the data and can verify that the data has been authenticated by Issuer or tampered with. The "User Agent" is a trusted authority that generates DIDs for users and provides KYC services, typically an authoritative authority through which entities interact with chain identities or data [6].

DID technology based on Blockchain can help to solve these problems. On the one hand, users can manage their own identity data and disclose it to relying parties after notification and authorization; on the other hand, enterprises can rely on verifiable user information to streamline their operations without bearing the associated costs and the risks of data leakage themselves.

3.2 Blockchain Combined with Federated Transfer Learning

Computing on Blockchain directly consumes more computing and storage resources, so that we divide the distributed privacy computing into off-chain computing and on-chain verification to achieve a balance between computational efficiency and verifiability (Fig. 3).

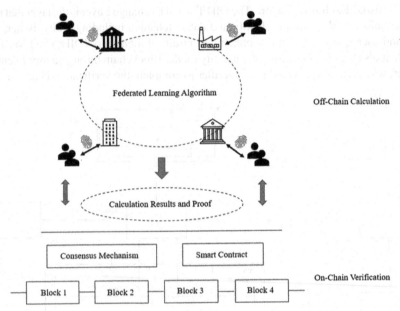

Fig. 3. This diagram shows a Blockchain-based distributed privacy computing schematic [7].

Off-Chain Calculation. We provide off-chain computing servers for federated learning training. Federated learning participants upload their encrypted data to the computing server for federated model training (Fig. 4).

- Prepare relevant domain dataset and public dataset. Relevant domain refers to the situation where there is little overlap of data features and samples, e.g., utilizing the domain expert model from the large enterprise credit data training to the SMES training model. It is a private dataset. The public dataset is an unlabeled dataset with n samples in this field [8].
- Providing domain expert model. The domain expert model is provided by the participant who initiates the task and trained from a private dataset of the relevant domain. This method, to a certain extent, can prevent malicious participants in the federated training from judging whether a particular sample is contained in the model's training set, since the malicious participants do not have prior knowledge of the model's training samples. The domain expert model is stored encrypted by a trusted computing server.

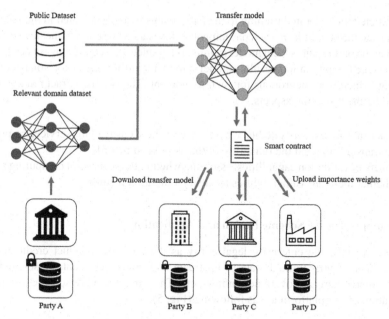

Fig. 4. The diagram shows how to obtain a more generalized transfer model, achieves model transfer and collect important weights by smart contract.

- Training transfer model. To obtain a more generalized transfer model and to counter model inversion attacks (malicious participants from back-extrapolating the original dataset with model parameter information) during the training, the off-chain comput-ing server generates labels for a public unlabeled dataset by using the domain expert model, and then the public dataset can be used to train a more generalized sharing model by applying differential privacy technology [9, 10].
- Federated transfer training process. The transfer model is allocated as a pre-training model to the participants in the federated training, and each participant uploads its own encrypted data to the computing server, which conducts federated training based on the pre-training model. In this process, each iteration does not need to wait for all participants to update the model gradient information after their local training, which greatly reduces the computational resource consumption and time consumption.

On-Chain Verification. We implement validation rules by writing smart contracts. The validation process can be made at any node on the Blockchain, the nodes we called workers [11]. Workflow is fixed as below.

The consensus algorithm competes for which the worker gets the right to check.

According to the smart contract, workers need to calculate the importance weight of the data features owned by each participant in the global model as the participant's contribution to the model to aggregate the weighted results and it can also distribute the benefits to the participants. The benefits included the benefit from the participant who initiated the federated training to the participants who participated in the federated training and the benefit from each participant to the users who supplied the licensed data.

After the participant's local model training is complete, the computing server uploads the encrypted computation results (model parameter information) for smart contract auditing, which is asynchronous for each participant. The smart contract remains to be audited by the following aspects.

—Service interface security audits: whether satisfy the security of the service, the privacy of data transmission and the reliability of the associated records.
—Security of validation capabilities: assessing whether the security of computing results and whether there are vulnerabilities to falsify validation results.

3.3 Comprehensive Scheme Design and Application

Suppose we take the enterprise legal entity as the core and conduct credit evaluating model for SMEs through data of upstream and downstream enterprises, industry and commerce bureau, government affairs, etc. The comprehensive scheme design and application of our platform are as following (Fig. 5).

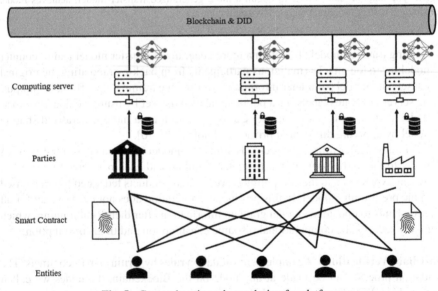

Fig. 5. Comprehensive scheme design for platform.

- First of all, the DID created on the Blockchain for enterprise legal entity is mapped to real verifiable resident documents (ID card number, passport number, pass number), so that when the enterprise legal entity go to government departments or enterprises (such as tax bureau, real estate bureau, industry and commerce bureau, small loan enterprises, upstream and downstream enterprises, etc.) to do business. They can be mapped to the identity on the chain through real documents to reach cross-agency account associations.

- Each organization encrypts the data of the enterprise legal entity, push its data extract on Blockchain and then their chain route can be formed.

- when the enterprise legal entity applies for bank loans, as long as he authorizes it, the banking institution can discover all the required data in the tax office, real estate office, industrial and commercial office, small loan enterprises, upstream and downstream enterprises, etc., which based on his digital identity through on-chain routing. These institutions, as participants participating in federated training, train the model through asynchronous federated training and conduct online reasoning under the premise of guaranteeing data and model privacy. Banks can obtain the credit score of micro-enterprises through legal compliance, which allowing for rapid validation or loan processing.

4   Conclusion

In this paper, we propose an inclusive Finance Consortium Blockchain platform by using a combination of federated transfer learning, decentralized identity and Blockchain technologies, which can help to solve the problem of credit difficulties for SMEs. Our platform gets a few advantages.

4.1   From Data Security Perspective

- The new federal transfer learning program can effectively respond to model inversion attacks and, avoid model gradient information leakage during joint modeling by participants.

- The combination of Blockchain and DID technology to build the underlying architecture can guarantee the authenticity of the data on the chain from the source and ensure that the data on the chain is not tampered with; the privacy computing service of "online storage and offline computing" provided by Blockchain ensures the security of the information exchange between the computing server and Blockchain. Therefore, our platform can effectively resist poisoning attacks, including data poisoning and model poisoning.

4.2   From the Value of Data and Platform Practicality Perspective

For SMEs

- Enhance credit worthiness based on real business data, which can help them access to financial services easier.

- Our platform will reduce the maintenance cost of human and material resources in the way of storing and certificating on the Blockchain, which can enhance the overall efficiency of the industry.

For Financial Service Providers

- By using artificial intelligence to dig deep into data on the premise of protecting data privacy, they can improve both the comprehensive assessment and risk control capabilities for the service principal, which greatly picking up the speed of auditing and lending to SMEs.
- Based on artificial intelligence, our platform can help to foster the development of their own more competitive financial services products to enhance market influence and actual revenue.

5 Future Works

In the long term, Blockchain and federated learning are playing an increasingly important role in data security and data value extraction. On the one hand, Blockchain and DID technologies enable data credibility enhancement, and make it possible to realize the automation and standardization of business through smart contracts. On the other hand, the data accumulated on the Blockchain ledger will grow as the business continues to advance, and the increased amount of data means that the value of the data will also grow exponentially. Using federated learning capability under the premise of protecting data privacy, effective data information can be discovered after learning, pooling and mining big data. And business can help financial institutions in marketing and risk control by artificial intelligence models and accumulation data models to discover behavioral patterns in the data. However, as the volume of business increases, the Blockchain needs to consume a large amount of storage resources, as well as the communication overhead and network speed on the Blockchain can be a big problem. The efficiency and accuracy of model training in federated learning under data encryption also need to be taken into consideration. The enhancement of privacy protection is bound to affect the accuracy of model training or incur larger communication overhead. On the other hand, we need to work out a scheme for learning transferable knowledge. The scheme can capture the invariance between participants well. The knowledge in our federated transfer learning is provided by participants (e.g., financial institutions), and then train more generalized models through a public dataset, but how to select different public datasets according to different domains to find a balance between the autonomy and generalization performance of the federal transfer learning models is the next focus of our work.

Acknowledgment. This work is supported by the National Key Research and Development Program of China (2018YFB143000); Engineering Research Center of Information Networks, Ministry of Education.

References

1. Decentralized Identifiers (DIDs) v1.0. https://www.w3.org/TR/did-core/, Accessed 31 Oct 2020

2. Yang, Q., Zhang, Y., et al.: Transfer Learning. China Machine Press, Beijing (2020)
3. Pan, S.J., Yang, Q.: A survey on transfer learning. IEEE Trans. Knowl. Data Eng. **22**(10), 1345–1359 (2010)
4. Yang, Q., Liu, Y., et al.: Federated Learning. Publishing House of Electronics Industry, Beijing (2020)
5. Yang, Q., Liu, Y., Chen, T., Tong, Y.: federated machine learning: concept and applications. ACM Trans. Intell. Syst. Technol. **10**(2), 1–19 (2019). Article 12
6. WeIdentity Document. https://fintech.webank.com/developer/docs/weidentity/, Accessed 29 Oct 2020
7. China Block Chain Technology and Industrial Development Forum: Blockchain——Privacy-preserving computing service guideline (2019)
8. Hamm, J., Cao, Y., Belkin, M.: Learning privately from multiparty data. In: Proceedings of the 33rd International conference on Machine Learning, pp. 555–563 (2016)
9. Dwork, C.: Differential privacy: a survey of result. In: Processsdings of the 5th Annual conference on Theory and Applications of Models of Computation, pp. 1–19 (2008)
10. Dwork, C., Roth, A.: The algorithmic foundations od differential privacy. Found. Trends Theor. Comput. Sci. **9**(3–4), 211–407 (2014)
11. Kang, J., et al.: Incentive mechanism for reliable federated learning: a joint optimization approach to combining reputation and contract theory. IEEE Internet Things J. **6**, 10700–10714 (2019)
12. Chen, X., Ji, J., Luo, C., Liao, W., Li, P.: When machine learning meets blockchain: a decentralized, privacy-preserving and secure design. In: IEEE International Conference on Big Data (Big Data), Seattle, WA, USA, pp. 1178–1187 (2018)
13. Wang, Y., Gu, Q., Brown, D.: Differentially private hypothesis transfer learning. In: Berlingerio, M., Bonchi, F., Gärtner, T., Hurley, N., Ifrim, G. (eds.) ECML PKDD 2018. LNCS (LNAI), vol. 11052, pp. 811–826. Springer, Cham (2019). https://doi.org/10.1007/978-3-030-10928-8_48
14. FederatedAI Document (WeBank AI Department). https://github.com/FederatedAI/DOC-CHN, Accessed 29 Oct 2020
15. China Decentralized Indentity Alliance: DIDA whitepaper (2020)

An Under-Sampling Method of Unbalanced Data Classification Based on Support Vector

Jinqiang He[1(✉)], Yongli Liao[1], Dengjie Zhu[1], Xujuan Fan[2], and Xiancong Zhang[2]

[1] Electric Power Research Institute, China Southern Power Grid, Guangzhou 510663, Guangdong, China
hejinqiang_jq@163.com

[2] Guangzhou Power Supply Bureau of Guangdong Power Grid Limited Liabilities Company, Guangzhou 510000, Guangdong, China

Abstract. To address the problem of unbalanced class distribution of power grid transmission line fault data, the number of fault classes is relatively smaller compared to the number of normal classes, an algorithm based on support vector under-sampling is proposed for transmission line fault classification. The method obtains the support vector on the original data, calculates the distance from the majority of the classes to the k nearest neighbor support vector, then calculates the class bit statistics to measure the local density information according to the distance, and finally under-samples based on the size of the sample class bit statistics. The bird nest dataset and insulator dataset are selected for performance evaluation, and the results show that this method has a good classification effect on unbalanced data, provides a theoretical reference for unbalanced data classification research, and has a certain practical value in the problem of grid transmission line fault classification.

Keywords: Unbalanced data · Grid fault classification · Under-sampling · Support vector machine

1 Introduction

With the rapid development of Chinese economy, society's demand for electricity is growing. To ensure the normal operation of transmission lines, the correct classification of common faults is of great significance. Using technologies such as the Electricity Internet of Things (IoT) and the unmanned aerial information acquisition system to collect faults on common transmission lines, the data show that the occurrence of transmission line faults by chance leads to the limited amount of defect data and the uncertainty of the number of occurrences of various types of faults, resulting in an unbalanced amount of data. It is of great research value to study more effective transmission line fault classification techniques in view of the characteristics of transmission line faults and the defects of existing fault classification techniques.

With the continuous development of computer and diagnostic technology in recent years, grid fault diagnosis technology has also been rapidly improved. Fault classification

Q. Zu et al. (Eds.): HCC 2020, LNCS 12634, pp. 430–440, 2021.
https://doi.org/10.1007/978-3-030-70626-5_45

techniques such as neural networks and machine learning, which have been used by scholars at home and abroad, have achieved some results, but they are harsh on data requirements. This paper uses the bird's nest dataset and insulator dataset normal samples occupy the majority, and fault samples only account for a small part, which constitute the unbalanced dataset.

The unbalanced data classification problem is solved in two main ways: at the data level and at the algorithm level. The data level includes up-sampling and under-sampling [1, 2], which reduces the imbalanced ratio and improves the classification effect by changing the data distribution; the algorithm level improves the algorithm or proposes new algorithms to improve the classification accuracy by analyzing the defects of existing algorithms in handling unbalanced data, such as cost-sensitive learning and integration learning [3].

For the unbalanced sample data for fault classification, Zhang et al. [4] proposed a fault diagnosis algorithm for rotating machinery based on fast clustering and support vectors, which reduces the data by fast clustering and uses support vector machines for training after balancing, with better classification results. Zhang et al. [5] proposed an integrated up-sampling and feature learning method for fault diagnosis of unbalanced data of rotating machinery, using weighted up Sampling method to balance the data distribution and feature selection with enhanced automatic coding can classify fault samples more effectively.

The unbalanced learning method has been used with good results in the field of fault classification, but is less applied in the field of transmission line fault classification. At present, the study of unbalanced data is mostly based on the SMOTE method [6] as an improved model [7, 8], but the SMOTE method is prone to overlap the generated samples of a few classes, because the generated samples are randomly generated from each of the few classes, ignoring the distribution characteristics of its neighboring samples. The number of samples generated by the Adaptive Synthetic Sampling Approach [9] (ADASYN) is calculated based on the density distribution of each small class, which generates more samples at the clustering of samples of a few classes and better model classification [10]. Svmsmote sampling [11] is based on the SVM Hyperplanes generate new samples; BorderlineSMOTE sampling [12] generates samples near minority class boundary points. The above up-sampling method reduces the imbalance between the classes by increasing the number of minority class samples. Japkow [13] found that the under-sampling method easily removes important samples from the majority class, while some current heuristics are not yet good enough to ensure that only redundant and noisy samples are removed.

Therefore, when addressing the unbalanced classification problem at the data level, it is important to measure the degree of imbalance in the data distribution between minority and majority classes, and it is of great research value to investigate how to increase valid minority sample data and remove redundant majority sample data. In this paper, we use the distance between the sample and the support vector to measure the local density information of the data sample, consider the imbalance degree of the data set from the distribution, propose the support vector-based under-sampling algorithm, and verify the effectiveness of the algorithm through experiments.

2 Methods

2.1 Support Vector Machines

The support vector machine is a general machine learning method based on statistical theory that tends to achieve better results on binary classification problems. Most machine
learning methods aim to minimize the loss function to achieve an optimal model, while
the SVM goal is the maximum interval hyperplane, where the interval is the distance
between the separated hyperplanes, and the samples close to the hyperplane are support
vectors, as shown in Fig. 1. In addition SVM can achieve nonlinear classification by
introducing a kernel function to map the inner product of the high dimensional space to
the low dimensional space.

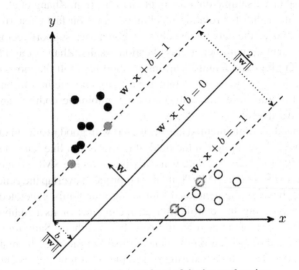

Fig. 1. Maximum hyperplane of the interval region.

Decision region spacing is positively correlated with model performance; a larger
spacing means that the model has better generalization capabilities, and a smaller
spacing can lead to overfitting problems. Before deducing how to determine the
maximized hyperplane, assume that given a training dataset on feature space $T =
\{(x_1, y_1), (x_2, y_2), \cdots, (x_N, y_N)\}$, where $x_i \in R^n$, is the feature vector of the ith sam-
ple, $y_i \in \{+1, -1\}, i = 1, 2, \cdots N$, y_i is the class label, with positive examples at $+1$
and negative examples at -1.

In a linearly separable sample space, the separation hyperplane is

$$w^T x + b = 0 \tag{1}$$

where: $w = (w_1; w_2; \cdots w_n)$ is the normal vector, which determines the direction
of the hyperplane; b is the displacement term, which determines the distance between
the hyperplane and the origin, and the set of positive and negative hyperplane linear

equations is expressed as follows

$$\begin{cases} b + w^T x_{pos} = 1 \\ b + w^T x_{neg} = -1 \end{cases} \tag{2}$$

where: x_{pos} denotes the hyperplane of the positive class, x_{neg} denotes the hyperplane of the negative class, and the difference between the positive and negative hyperplanes is the hyperplane distance to be optimized. Normalizing the hyperplane distance, $||w|| = \sqrt{\sum_{j=1}^{n} w_j^2}$, the hyperplane to be optimized is computed as follows

$$\frac{w^T (x_{pos} - x_{neg})}{||w||} = \frac{2}{||w||} \tag{3}$$

and because

$$\max \frac{2}{||w||} \Leftrightarrow \min \frac{1}{2}|w|^2 \tag{4}$$

Therefore, the constraint on the problem of linear classification is optimized as

$$\min_{w,b} \frac{1}{2}|w|^2 \tag{5}$$
$$s.t. \ y_i(wx_i + b) - 1 \geq 0$$

Introduce the Lagrange multiplier, solve for the constraint function, and according to the duality, finally equate to the optimization problem

$$\max w(\alpha) = \sum_{i=1}^{n} a_i - \frac{1}{2} \sum_{i=1}^{n} \sum_{j=1}^{n} a_i a_j y_i y_j (x_i, x_j) \tag{6}$$
$$s.t. \ \sum_{i=1}^{n} a_i y_i = 0, a_i \geq 0, i = 1, 2, \cdots n$$

The linear indistinguishability of the sample space implies that there are sample points that do not satisfy constraint (5). To solve this problem, a relaxation variable $\xi_i \geq 0$ is introduced for each sample and the constraint becomes

$$\min_{w,b,\xi} \frac{1}{2}||w||^2 + C \sum_{i=1}^{N} \xi_i \tag{7}$$
$$s.t. \ y_i(wx_i + b) \geq 1 - \xi_i$$

Equivalent parity problem by constructing Lagrange functions

$$\max w(\alpha) = \sum_{i=1}^{n} a_i - \frac{1}{2} \sum_{i=1}^{n} \sum_{j=1}^{n} a_i a_j y_i y_j (x_i, x_j) \tag{8}$$
$$s.t. \ \sum_{i=1}^{n} a_i y_i = 0, C \geq a_i \geq 0, i = 1, 2, \cdots n$$

For non-linear problems, the SVM needs to map the data into another space through the kernel function $(k(x_i, x_j) = \varphi(x_i)^T \varphi(x_j))$, as shown in Fig. 2, which transforms the

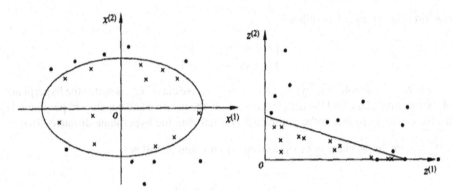

Fig. 2. Spatial transformation.

separated surface of the ellipse in the left figure into a straight line in the right figure through the space transformation.

Training a linear SVM model on a new feature space for classification operations, nonlinear classification optimization problem

$$\max w(\alpha) = \sum_{i=1}^{n} a_i - \frac{1}{2} \sum_{i=1}^{n} \sum_{j=1}^{n} a_i a_j y_i y_j k(x_i, x_j)$$

$$s.t. \sum_{i=1}^{n} a_i y_i = 0, C \geq a_i \geq 0, i = 1, 2, \cdots n$$

(9)

2.2 Under-Sampling Based on Support Vector

Set m_1, m_{-1} the number of positive and negative samples, respectively, $s_i = (s_{i1}, s_{i2}, \cdots, s_{in})$ to represent the support vectors of the unbalanced data set T. And any sample is represented by $x_i = (x_{i1}, x_{i2}, \cdots, x_{in})$, the Euclidean distance between the samples and the support vectors as in Eq. (10).

$$d(x_i, s_j) = \sqrt{\sum_{k=1}^{n} |x_{ik} - s_{jk}|^2}$$

(10)

Based on Eq. (11) for the *ith* sample of a given dataset, define a sample x_i position statistic based on a support vector.

$$\mu_k(x_i) = \frac{1}{k} \sum_{t=1}^{k} d_t(x_i, Q_k(s))$$

(11)

where $Q_k(s)$ is the k nearest neighbors set of x_i, denoted by (11), $\mu_k(x_i)$ is a local density measure that measures the average distance of the sample x_i from the support vector and expresses density information in the sample x_i neighborhood. The sample density-based statistics are shown in Eq. (12).

$$w_i = \frac{\mu_k(x_i)}{\sum\limits_{x_j \in c_l} \mu_k(x_j)} \tag{12}$$

c_1 and c_{-1}, which represent the minority and majority classes, respectively, w_i is proportions of the total density, and the intra-class density of each sample can be measured by density-based locality statistics, where the higher the density of the neighborhood, the smaller w_i and $\mu_k(x_i)$. In other words, w_i of a sample reflects its intra-class density.

In this paper, the majority of the samples are sorted by size w_i and divided into two parts according to the median: the smaller part, which indicates that the support vector around the sample is dense, under-sampling $(m_{-1} - m_1)m_1/m_{-1}$ samples in this part. And under-sampling too many samples in this part will reduce the diversity of majority classes and increase the risk of classification error. On the other hand, a larger w_i part indicates that the support vector around this sample is sparse, under-sampling $(m_{-1} - m_1)(1 - m_1/m_{-1})$ samples in this part. The appropriate down-sampling in this region can effectively reduce the number of majority classes and has a low impact on the classification accuracy.

The details of the support vector-based under-sampling of the dataset are shown in Table 1.

Table 1. Support vector-based under-sampling algorithm.

Input: data set T Output: down sampled data set T_{down}
1. Divide the set T by label into a small number of sets C_1 and a large number of sets C_{-1}, with the numbers denoted by m_1 and m_{-1}, initialized $T_{down} = \varnothing$ 2. The SVM algorithm is used to obtain the support vectors set S for the dataset T.   3. Calculate the distance $d(x_i, S)$ from all samples to the set C_{-1} and support vectors using equation (10).   4. Calculate position statistics for all samples in the set C_{-1} using equation (11)   5. Calculate the density-based statistics w_i for all samples in the set C_{-1} using equation (12).   6. Sort the majority sample by size w_i and divide the majority set C_{-1} into two parts according to median, under-sample $(m_{-1} - m_1)m_1/m_{-1}$ samples in the smaller part and $(m_{-1} - m_1)(1 - m_1/m_{-1})$ samples in the larger part, merging the under-sampled dataset together T_{down}

3 Experiments

3.1 Data Sets

Grid transmission lines exposed to the field for a long time, by the atmosphere, climate change and other factors, is fault-prone equipment in power production. The common types of faults include lightning strike, wind deflection, dirt flash, ice flash, ice cover, external forces, bird damage, etc. The occasional occurrence of faults in transmission lines leads to the limited amount of defect data and the uncertainty of the number of occurrences of various types of faults, resulting in an uneven amount of data for each type of fault. This paper illustrates the effectiveness of our algorithm by taking the bird nest dataset and insulator dataset as examples, as shown in Fig. 3. The positive class of the bird nest dataset is with nest data, and the negative class is without nest data; the positive class of the insulator dataset is insulator breakage data, and the negative class is normal insulator data, and the quantity distribution is shown in Table 2.

3.2 Evaluation Indicators

For the binary classification problem, the confusion matrix is usually used to evaluate the classification model, which divides the samples that are actually positive into two parts: the samples that are predicted to be positive and the samples that are predicted to be negative, and the negative samples into two parts that are predicted to be negative and the samples that are predicted to be positive, and their numbers are abbreviated as: TP,TN,FP,FN, respectively, as shown in Table 3.

The effectiveness of the classification is evaluated by the following indicators, respectively.

$Accuracy = \frac{TP+TN}{TP+FN+FP+TN}$: indicates the classifier's ability to classify the whole sample.

$Precision = \frac{TP}{TP+FP}$: indicates the accuracy of the classifier.

$Recall = \frac{TP}{TP+FN}$: indicates the completeness of the classifier and the accuracy of the sample for a small number of categories.

$F_1 = \frac{2TP}{Number\ of\ samples+TP-TN}$: considers both the accuracy and recall rate of the minority class.

In order to verify the validity of our algorithm, the SMOTE algorithm and the random under-sampling algorithm are selected for comparison on the bird's nest dataset and the insulator dataset. The classification evaluation metrics of the raw data and the data processed by the different algorithms are compared. Table 4 shows the results of comparison of evaluation metrics for the bird nest dataset on each algorithm and Table 5 shows the results of comparison of evaluation metrics for the insulator dataset on each algorithm.

From Tables 4 and 5, it can be seen that in the bird's nest dataset and insulator dataset, the various evaluation indexes of our algorithm for classification are higher than those of the original data and other algorithms, indicating that our algorithm performs better in the unbalanced data classification problem. The indicators of the SMOTE and random under-sampling algorithms are also improved in the bird's nest dataset compared with the original data, and the indicators of our algorithms are slightly better than those of the

Fig. 3. Example diagram of positive and negative classes in a dataset.

Table 2. Distribution of positive and negative classes in the data set.

Data set	Number of positive samples	Number of negative samples
Bird nest	300	600
Insulator	100	150

Table 3. Confusion matrix.

Category	Forecast Positive	Forecast Negative
Positive	TP	FN
Negative	FP	TN

Table 4. Comparison results of the bird nest dataset with different algorithms.

Algorithm	Accuracy	Precision	Recall	F_1
Raw data	0.62	0.76	0.62	0.57
SMOTE [18]	0.75	0.82	0.75	0.74
Under-sampling [19]	0.83	0.86	0.84	0.84
Our algorithm	**0.86**	**0.88**	**0.86**	**0.86**

Table 5. Comparison results of insulator datasets with different algorithms.

Algorithm	Accuracy	Precision	Recall	F_1
Raw data	0.66	0.70	0.66	0.64
SMOTE [18]	0.58	0.58	0.58	0.58
Under-sampling [19]	0.56	0.56	0.56	0.56
Our algorithm	**0.74**	**0.78**	**0.74**	**0.73**

random under-sampling algorithm, which proves the effectiveness of our algorithm in combining under-sampling and SVM. In the insulator dataset, due to the small difference between the positive and negative classes of the original data, the imbalanced ratio is not large, so the effect of SMOTE and random under-sampling algorithm is not good and the indexes are decreased, and the indexes of our algorithm are improved, but *Accuracy* is only 0.74. Therefore, the classification of power transmission line faults based our algorithm can be more accurately identified the types of faults in the power grid to enhance the stability and reliability of the safe operation of the power system.

4 Conclusion

In order to address the data imbalance phenomenon of power grid transmission line fault classification samples, this paper proposes an under-sampling method of unbalanced data classification based on support vector. The main contributions of this paper include (1) improving balanced ratio by under-sampling raw data based on support vectors; (2) improving the accuracy rate of minority classes and ensuring a high accuracy rate of majority classes; (3) using a real dataset for fault classification of power grid transmission lines with an unbalanced classification method, with a complete line of research and a more comprehensive consideration of normal data and characteristics of the fault data.

Experiments show that the proposed algorithm in this paper can effectively improve the performance of the under-sampling algorithm, but the mode is less effective in datasets with large imbalanced ratio and high positive and negative class similarity. This problem will continue to be investigated in the future and the algorithm in this paper will be validated on other fault classification datasets to further improve the accuracy of fault classification.

References

1. Lin, W.C., Ts, C.F., Hu, Y.H., et al.: Clustering-based under sampling class-imbalanced data. Inf. Sci. **409**, 17–26 (2017)
2. He, H., Bai, Y., Garcia, E.A., et al.: ADASYN: adaptive synthetic sampling approach for imbalanced learning. In: 2008 IEEE International Joint Conference on Neural Networks, pp. 1322–1328 (2008)
3. Galar, M., Fernandez, A., Barrenechea, E., et al.: A review on ensembles for the class imbalance problem: bagging-, boosting-, and hybrid-based approaches. IEEE Trans. Syst. Man Cybern. Part C Appl. Rev. **42**(4), 463–484 (2011)
4. Zhang, X.C., Jiang, D.X., Han, T., et al.: Rotating machinery fault diagnosis for imbalanced data based on fast clustering algorithm and support vector machine. J. Sens. **2017**, 8092691 (2017)
5. Zhang, Y.Y., Li, X.Y., Gao, L., et al.: Imbalanced data fault diagnosis of rotating machinery using synthetic oversampling and feature learning. J. Manuf. Syst, **48**, 34–50 (2018)
6. Chawla, N.V., Bowyer, K.W., Hall, L.O., et al.: SMOTE: Synthetic minority over-sampling technique. J. Artif. Intell. Res. **16**(1), 321–357 (2002)
7. Zhao, J., Lu, H., Jiang, J., et al.: An oversampling random forest algorithm for unbalanced data classification. Comput. Appl. Softw. **36**(4), 255–261, 316 (2019)
8. Peng, R., Yang, T., Kong, H., et al.: Research on class imbalance data classification algorithm based on CPD-SMOTE. Computer Applications and Software **35**(12), 259–262, 268 (2018)
9. He, H.B., Bai, Y., Garcia, E.A., et al.:ADASYN: adaptive synthetic sampling approach for imbalanced learning. In: 2008 IEEE International Joint Conference on Neural Networks (2008)
10. Puichung, L., Ming, H., Detian, H., et al.: An unbalanced classification algorithm based on the combination of ADASYN and Ada Boost SVM. J. Beijing Univ. Technol. **43**(3), 368–375 (2017)
11. Nguyen, H.M., Cooper, E.W., Kamei, K.: Borderline over-sampling for imbalanced data classification. Int. J. Knowl. Eng. Soft Data Paradigms **3**(1), 4–21 (2009)
12. Han, H., Wang, W.Y., Mao, B.H.: Borderline-SMOTE: a new over-sampling method in imbalanced data sets learning. In: 2005 International Conference on Advances in Intelligent Computing, pp. 878–887 (2005)

13. Japkowiczn, N., Stephen, S.: The class imbalance problem: a systematic study. Intell. Data Anal. J. **6**(5), 429–450 (2002)
14. Alcalá-Fdez, J., Fernández, A., Luengo, J., et al.: Keel data-mining software tool: data set repository, integration of algorithms and experimental analysis framework. J. Multiple-Valued Logic Soft Comput. **17**, 1–33 (2011)
15. Asuncion, A., Newman, D.: UCI machine learning repository. Univ. Calif. Irvine School Inf. Comput. Sci. **9**, 10–23 (2007)
16. Kullback, S., York, N.: Information theory and entropy. Model Based Inference Life Sci. A Primer Evid. **2008**, 51–82 (2008)
17. Tang, B., He, H.: GIR-based ensemble sampling approaches for imbalanced learning. Pattern Recogn. **71**, 306–319 (2017)
18. Chawla, N.V., Bowyer, K.W., Hall, L.O., Kegelmeyer, W.P.: Smote: synthetic minority over-sampling technique. J. Artif. Intell. Res. **16**, 321–357 (2002)
19. Mani, I., Zhang, I: KNN approach to unbalanced data distributions: a case study involving information extraction. In: Proceedings of Workshop on Learning from Imbalanced datasets, vol. 126 (2003)

Image Classification Algorithm for Transmission Line Defects Based on Dual-Channel Feature Fusion

Yongli Liao[1], Jinqiang He[1(✉)], Dengjie Zhu[1], Xujuan Fan[2], and Xiancong Zhang[2]

[1] Electric Power Research Institute, CSG, Guangzhou 510663, Guangdong, China
hejinqiang_jq@163.com
[2] Guangzhou Power Supply Bureau of Guangdong Power Grid Limited Liabilities Company, Guangzhou 510000, Guangdong, China

Abstract. The power system is of great significance to the normal production of society and the daily life of the people, so regular inspection of transmission lines is essential. However, transmission lines are usually exposed to the outdoors, and the surrounding terrain and environment are complex, which may lead to problems such as structural aging and mechanical strength reduction, which in turn may lead to large area power outages and cause huge economic losses. In this paper, a two-channel feature fusion classification method is proposed to address the transmission line image classification problem. Using a two-channel parallel network structure, a neural network model is constructed to fuse the overall and local feature information, and then determine whether there are defects in the transmission line images. The experimental results show that the classification accuracy of the two-channel parallel convolutional neural network based on ResNet32 is 82.24% and 77.87% for the bird's nest defect and insulator burst defect, respectively, on the actual transmission line image dataset, which exceeds the classification accuracy of other CNN models. This indicates that the classification accuracy can be effectively improved by fusing feature information.

Keywords: Transmission line defect classification · Deep learning · Feature integration · Convolutional neural network

1 Introduction

The power system as a basic resource related to people's livelihood, the normal production of society and the people's daily life is of great importance. But electrical lines mostly run in remote areas, the surrounding environment terrain is responsible, exposed to the outdoors will inevitably be affected by lightning, strong winds, bird damage and other external factors, once the problem is found but no immediate treatment, it is easy to cause widespread power outages, and then cause a lot of economic losses and other serious consequences. So based on the transmission line image defect identification is the essential link in the transmission and distribution lines [1]. At present the most common transmission line defect identification includes bird's nest defect identification and

© Springer Nature Switzerland AG 2021
Q. Zu et al. (Eds.): HCC 2020, LNCS 12634, pp. 441–451, 2021.
https://doi.org/10.1007/978-3-030-70626-5_46

insulator defect fault. In view of the transmission line defect background complexity, the defect is not obvious, the quantity is huge and other factors, in the absence of rich experience, only by the naked eye is not easy to distinguish the defect image efficiently [2–5], therefore the research automation transmission line defect recognition method when necessary and meaningful work.

At present, machine vision-based inspection is non-contact, low-cost, and fast-response, which offers the possibility of rapid online detection of transmission line image defects. In the literature [6], Wang et al. proposed the use of gamma filter and non-maximal suppression as a means to extract insulator features and input them into SVM for classification. In [7], Jabid et al. first rotated insulators in aerial images to positions parallel and perpendicular to the bottom edge, and then extracted insulator features using Hoff transform and local binary features as input to the SVM. Currently, deep learning has become an important method for solving image classification problems [8–10]. In particular, deep learning is based on Convolutional Neural Network (CNN). The method is able to approach the advanced semantic features of images more accurately and has made breakthroughs in image classification [11, 12]. In the literature [13], Zhao et al. used the Alexnet network to extract feature maps of the sixth and seventh fully connected layers to classify insulator images. In the literature [14], Pang et al. used the Faster R-CNN algorithm as the basis to extract the nest feature map using the ZF-NET network, corrected to output the matrix candidate region and use the detection window to complete the detection of the presence or absence of a nest. All of these studies, however, faced the problem of low classification accuracy.

This study focuses on the problem of classifying transmission line bird's nest defect images as well as insulator defect images, and based on the existing CNN model, we propose a two-channel parallel convolutional network model that fuses global and local information, which significantly exceeds the classification accuracy of existing methods on the image datasets of transmission line bird's nest defect faults as well as transmission line insulator defect faults.

2   Transmission Line Defect Classification Method

2.1   Construction of Transmission Line Data Set

Since there is usually no public dataset for electrical line faults, the original samples are high-definition aerial images provided by the Southern Power Grid, which vary greatly in scale, illumination, and angle. The construction of the dataset is divided into two main parts: the construction of the transmission line insulator image dataset and the construction of the transmission line bird nest image dataset. There are a total of 3340 images of transmission line images, including 2565 images of transmission line insulators and 675 images of transmission line birds' nests, as shown in Tables 1 and 2. All image sizes were processed in three sizes, 224*224, 256*256, and 600*400. In order to avoid the category imbalance problem that leads to the poor performance of the convolutional neural network classification, the ratio of positive and negative samples for the training set and the verification machine was set to 1:1. In the real case, the number of transmission line failures is relatively small, i.e., the number of positive samples is much larger than the number of negative samples, and to simulate the real situation, the ratio of positive and negative samples for the test machine was set to 14:1.

Table 1. Bird' Nest sample data distribution.

Sample classification	Train set	Validation set	Test set	Total
Bird's nest	380	190	95	665
No bird's nest	380	190	1330	1900
Total	760	380	1425	2565

Table 2. Insulator sample data distribution.

Sample classification	Train set	Validation set	Test set	Total
Self-detonating insulator	100	50	25	175
Normal insulator	100	50	350	500
Total	200	100	375	675

For the transmission line bird nest image, make the image containing the bird nest a negative sample and the opposite a positive sample. The two samples are shown in Fig. 1, where Fig. 1(a) is a negative sample and Fig. 1(b) is a positive sample. The image of missing insulators in the same transmission line is a negative sample, and the opposite is a positive sample. The two samples are shown in Fig. 2. Figure 2 (a) is a negative sample and Fig. 2 (b) is a positive sample.

2.2 Common Convolutional Network Models

In 1994, the earliest convolutional neural network LeNet [15] was born, which was the cornerstone of deep learning, but it was not suitable for complex image classification due to its low computational efficiency, shallow model depth, few parameters and single structure. In 2012, the Alexnet [16] network proposed by Alex et al. won the ILSVRC2012 champion network, which uses the Dropout operation to avoid the overfitting generated by training to some extent, and the computation amount is also greatly reduced, but its image description and extraction is very limited. In 2014, the Oxford University Visual Geometry Group proposed VGGNet [17] The ILSVRC2014 champion network is GooleNet [18], which introduces an Inception module that greatly improves the learning ability of the model. Efficiency of parameter utilization. From the development of the above CNN model, the trend is to adopt an increasingly deep network structure.

(a) negative sample

(b) positive sample

Fig. 1. Sample of bird's Nest transmission line.

In the models of convolutional neural networks, deepening the network depth is usually used to improve the expression of the model, but as the network depth deepens and the model becomes more complex, the optimization of Stochastic Gradient Descent (SGD) becomes more difficult, which will result in the model not reaching a better learning effect. For this reason, Keming Ho et al. proposed the ResNet [19] network. The core of ResNet uses a Residual structure, as shown in Fig. 3. This structure adds a constant mapping that converts the original learned function H(X) to F(X)+X. The model makes the network oh the model depth is unrestricted in a large range.

(a) negative sample

(b) positive sample

Fig. 2. Sample of transmission line insulators.

2.3 Dual-Channel Feature Fusion ResNet Model Construction

To address the problem of image classification of transmission line defects, this study builds a convolutional neural network model (DCFF-CNN) with two-channel feature fusion as shown in Fig. 4.

Overview of Network Model Structure. The DCFF-CNN model contains two parallel branches for extracting the global and local features of the transmission line image, and both branches use the ResNet network structure. The input of the first branch is the complete transmission line image (overall information), and its output is the overall features of the transmission line. The input of the second branch is the local transmission

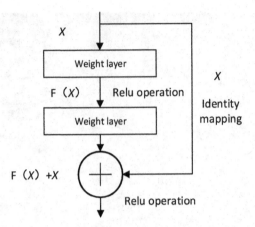

Fig. 3. Residual structure in ResNet.

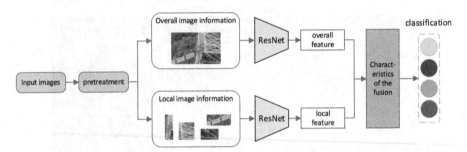

Fig. 4. CNN model of two-channel feature fusion.

line image, and its output is the local features of the transmission line image. Next, the overall and global features are fused at the feature fusion layer to obtain the final feature vector corresponding to the transmission line image. Finally, the fused feature vector is classified at the classification layer to determine whether the input image is a defective image.

Local Feature Extraction and Feature Fusion. The transmission line images are all normalized to 224*224 pixels, and then the images are pre-processed to obtain the specific locations of the transmission lines in the transmission line images and save the corresponding picture blocks. Local information is collected for the overall target and the local feature information is used as input for the second branch in the DCFF-CNN model, and the corresponding local features are obtained after the second branch processing.

Using the feature concatenation operation, the overall feature f_global and the local feature f_local of the transmission line image extracted from the DCFF-CNNN model are combined into a feature vector [f_global,f_local], where f_global and f_local are both 512-dimensional vectors.

Loss Function. Considering the problem of small sample size and possible unevenness in the number of transmis-sion line images, this study uses cross-entropy as a loss function. Suppose there are M classes of transmission line defect images to be classified. Then any transmission line sample image k, defining its true label t as a vector [a1,a2,...,aM], when sample k belongs to class i defects, ai = 1, otherwise ai = 0. The output predicted probability distribution of transmission line image sample k in the DCFF-CNN model is given as pred = [y1, y2,..., y M]. For transmission line sample k, the difference between its true class label and the output prediction distribution is represented by the cross-entropy loss_CE as follows:

$$loss\_CE = -\sum_{i=1}^{M} a_i \log(y_i) \tag{1}$$

For all N transmission line image samples, the total loss function L is as follows:

$$L = -\sum_{k=1}^{N}\sum_{i=1}^{M} a_{ki} \log(y_{ki}) \tag{2}$$

When the transmission line k belongs to category i defect, $a_{ki} = 1$, otherwise $a_{ki} = 0$. Where y_{ki} is the probability that the DCFF-CNN model predicts that the transmission line sample k belongs to category i defect.

3 Experiment

3.1 Evaluation Index

The indicators used in this study to judge the effectiveness of fault classification of transmission line defects include accuracy (ACC), precision(P), recall(R), false positive rate (FPR), false negative rate (FNR). Their formulas are shown in (3)–(7), and the parameters used in the formulas are shown in Table 3.

Table 3. Meanings of evaluation index parameters.

Parameter	Meaning
ALL	The total number of images in the test set
TP	The result of normal image classification is the number of normal images
FN	The result of normal image classification is the number of defective images
FP	The classification results of defective images are the number of normal images
TN	The classification results of defective images are the number of defective images

The formula of accuracy rate is expressed by ACC:

$$ACC = \frac{TP + TN}{ALL} \tag{3}$$

The formula of precision is expressed as P:

$$P = \frac{TP}{TP + FP} \tag{4}$$

The recall ratio formula is expressed as R:

$$R = \frac{TP}{TP + FN} \tag{5}$$

The formula of false positive rate (i.e., the omission rate of defects) is expressed by FPR:

$$FPR = \frac{FP}{TN + FP} \tag{6}$$

The formula of False negative rate (i.e., the error detection rate of defects) is expressed as FNR:

$$FNR = \frac{FN}{TP + FN} \tag{7}$$

3.2 Experimental Training Process

The transmission line image is divided into the transmission line insulator image and the transmission line bird's nest image. In the DCFF-CNN model, the input of the first ResNet model is the whole transmission line image, which is used to extract the global features of the transmission line image, and the input of the second ResNet model is the image block of the transmission line, i.e., the local information of the transmission line, which is used to extract the global features of the transmission line.

In the experiment both branches of the DCFF-CNN model were set to ResNet32, and the output of the full connection was set to a 2-dimensional vector to determine whether the transmission line image was defective. Random gradient descent was used to train the model. In order to more objectively reflect the effect of the comparison network model and the training mechanism on the classification accuracy, all the network models in the comparison experiments are classical network models, and the hyper-parameters are uniformly treated in all training experiments. The specific values of the hyperparameters are shown in Table 4:

Table 4. Super parameter Settings of DCFF-CNN model.

Parameter types	Parameter values
Learning rate	1e−3
Batch size	128
Dropout	0.5
Weight decay	5e−4
Momentum	0.9
Epoch	2000

3.3 Power Line Image Classification Effect

After the training number classifier, the classification results can be obtained by entering all the test set images into the DCFF-CNN model.

There are 1425 pictures in the test set of transmission line bird's nest images and 375 pictures in the test set of transmission line insulator images in this experiment. In order to be closer to the real sample, the ratio of positive and negative samples in the test set is set to 14:1.

The test results of the bird's nest images are shown in Table 5, and the test results of the insulator images are shown in Table 6.

Table 5. Image test results of transmission line Bird's Nest.

Algorithm	ACC	P	R	FPR	FNR
AlexNet	74.25%	98.68%	73.38%	13.68%	81.19%
ResNet32	79.80%	99.33%	78.87%	7.36%	76.15%
ResNet50	79.93%	99.53%	78.87%	5.26%	75.74%
DCFF-AlexNet	77.33%	98.93%	76.54%	11.58%	78.79%
DCFF-ResNet32	**82.24%**	**99.54%**	**81.35%**	**5.26%**	**73.37%**
DCFF-ResNet50	82.18%	99.44%	81.35%	6.32%	73.59%

Table 6. Test results of transmission line insulator image.

Algorithm	ACC	P	R	FPR	FNR
AlexNet	68.26%	91.69%	72.57%	92%	97.79%
ResNet32	74.41%	95.03%	76.57%	56%	81.17%
ResNet50	74.67%	95.69%	76.29%	48%	86.45%
DCFF-AlexNet	70.13%	94.07%	72.57%	64%	91.43%
DCFF-ResNet32	**77.87%**	**96.19%**	**79.43%**	**44%**	**83.72%**
DCFF-ResNet50	77.33%	95.85%	79.14%	48%	84.88%

For the experimental analysis of the nest images and insulator images, the following conclusions can be drawn.

(1) The DCFF-CNN model is simple in design, but can be applied to most of the convolutional neural network models, and the accuracy of the model has been improved significantly.
(2) For single-branch network, DCFF-CNN model with dual-channel feature fusion can extract richer feature information, and the test accuracy is about 3% higher than that of single-branch network.

(3) It is important to choose the depth of CNN model reasonably, theoretically the deeper the depth of the model, the better the expression ability of the model, comparing the experimental results of ResNet32 and ResNet50, we can find that the accuracy of ResNet50 is slightly higher than ResNet32. ResNet50's DCFF-CNN model. The situation is reversed. Considering the small number of training datasets and the fact that the structure of the DCFF-CNN model contains twice as many parameters as the single-branch CNN model, the DCFF-ResNet50 model is not sufficient for configuring the DCFF-ResNet50 model. Therefore, too few datasets are not sufficient for configuring the DCFF-ResNet50 model, and the capacity of the transmission line image dataset needs to be increased if the accuracy of the DCFF-ResNet50 model is to be further improved.

4 Conclusion

To address the problem of image classification of transmission line defects, this study designs a convolutional neural network model based on two-channel feature fusion. The model is able to better fuse the whole and local information of transmission line images. And it achieves good classification accuracy on a given transmission line defect image dataset of the Southern Power Grid. In addition, the experimental results also show that, in addition to the global information of transmission line images, the local information of images is also an important feature for identifying whether there are defects in transmission line images. The above findings provide a useful reference for subsequent work.

References

1. Li, N., Zheng, Q., Xie, G.W., et al.: Detection of defects in transmission line based on the unmanned aerial vehicle image recognition technology. Electron. Design Eng. 27(10), 102–106+112 (2019)
2. Qiu, L.H., Zhu, Z.T.: Research on fault detection of transmission line insulators based on deep learning. Appl. Res. Comput. 37(S1), 358–360+365 (2020)
3. Yang, X.H., Sheng, F., Xue, P., et al.: Defect detection for grid insulator using aerial image based on deep learning. Inf. Technol. 44(04), 37–40+45 (2020)
4. Wei, S.F., Huang, S., Cao, W.B., et al.: Identification and defect detection of transmission line insulator based on aerial images. Geotech. Invest. Surveying 48(04), 39–43+71 (2020)
5. Xu, J., Han, J., Tong, Z.G., et al.: A detection method of bird's nest on tower in uav image. Comput. Eng. Appl. 53(06), 231–235 (2017)
6. Wang, X., Zhang, Y.: Insulator identification from aerial images using Support Vector Machine with background suppression. In: International Conference on Unmanned Aircraft Systems, pp. 892–897. IEEE (2016)
7. Jabid, T., Uddin, M.Z.: Rotation invariant power line insulator detection using local directional pattern and support vector machine. In: International Conference on Innovations in Science, Engineering and Technology, pp. 1–4. IEEE (2017)
8. Luo, H.L., Chen, H.K.: Survey of object detection based on deep learning. Acta Electronica Sinica 48(06), 1230–1239 (2020)
9. Cao, Y., Huan, L., Wang, T.B.: A survey of research on target detection algorithms based on deep learning. Comput. Modernization (05), 63–69 (2020)

10. Lv, F., Lv, Q., Luo, R.Z.: Telecommunications science. Telecommun. Sci. **35**(11), 58–74 (2019)
11. Zhang, Y., Yang, H.T., Yuan, C.H.: A survey of remote sensing image classification methods. J. Ordnance Equip. Eng. **39**(08), 108–112 (2018)
12. Ge, H.: Research on Image Classification Method based on Deep Learning. School of Information and Communication Engineering (2020)
13. Zhao, Z., Xu, G., Qi, Y.: Representation of binary feature pooling for detection of insulator strings in infrared images. IEEE Trans. Dielectr. Electr. Insul. **23**(5), 2858–2866 (2016)
14. Pang, N.: Deep learning based detection and identification of power line tower nest. Autom. Instrum. (04), 195–198+204 (2020)
15. Lecun, Y., Bottou, L.: Gradient-based learning applied to document recognition. Proc. IEEE **86**(11), 2278–2324 (1998)
16. Krizhevsky, A., Sutskever, I., Hinton, G.: ImageNet classification with deep convolutional neural networks. In: Advances in Neural Information Processing Systems, vol. 25, no. 2 (2012)
17. Simonyan, K., Zisserman, A.: Very deep convolutional networks for large-scale image recognition. Comput. Sci. (2014)
18. Szegedy, C., Liu, W., Jia, Y., et al.: Going Deeper with Convolutions (2014)
19. He, K., Zhang, X., Ren, S., et al.: Deep residual learning for image recognition. In: IEEE Conference on Computer Vision & Pattern Recognition. IEEE Computer Society (2016)
20. Hu, Y., Luo, D.Y., Hua, K., et al.: A review and discussion of deep learning. CAAI Trans. Intell. Syst. **14**(01), 1–9 (2019)
21. Liu, X., et al.: Deep multiview union learning network for multisource image classification. IEEE Trans. Cybern., 1–13 (2020)
22. Cen, F., Zhao, X., Li, W., Wang, G.: Deep feature augmentation for occluded image classification. Pattern Recogn. **111**, 107737 (2021)

Research on Alarm Causality Filtering Based on Association Mining

Yuan Liu[1](✉), Yi Man[1](✉), and Jianuo Cui[2]

[1] Beijing University of Posts and Telecommunications, Beijing 100876, China
{liuyuanlyre,manyi}@bupt.edu.cn
[2] CRSC Research & Design Institute Group Co., Ltd., Building No. 18, Huayuan Yili,
Fengtai District, Beijing 100073, China
cuijianuo@crscd.com.cn

Abstract. Mining the association rules in the alarm data generated by network is an important method for operations to monitor and manage the network equipment. Analyzing the correlation of alarms through association rule mining algorithms can effectively simplify alarms and help locate network faults. Since the network alarm data has an obvious chronological relationship, it needs to be processed by the association rule mining algorithm based on time series. Through investigation, it is found that current association rule algorithms based on time series lack the determination of the realistic cause-and-effect relationship between successive alarms. Therefore, in order to improve the effectiveness of the association algorithm, this paper adopts an association mining algorithm based on the existing time series, which supports filtering the useless sequential associated items that have no causal relationship in the results. The experimental and analytical results show that the proposed method is effective.

Keywords: Association Rule mining · Time series · Causal relationship · Frequent sequence · Alarm data

1 Introduction

Association rule mining is a rule-based machine learning algorithm, which can discover the potential associations of data in large databases, and use some metrics to distinguish strong rules in the database [1]. Alarm data in telecom network is an important basis for operation engineers to monitor and manage network equipment. The relevant knowledge of alarms can be obtained through data mining methods such as association mining, which can compress massive alarm data and locate faults, to assist operations to manage and maintain network equipment better [2].

The representative algorithms are Apriori proposed by Agrawal et al. [3] and FP-Growth proposed by JiaWei et al. [4]. As alarm events contain time information and chronological relationship, it is necessary to consider the time sequence between the data to be mined when mining alarm data for association rules [5]. However, the current methods of association rule mining such as Apriori and FP-growth are limited by the time sequence.

© Springer Nature Switzerland AG 2021
Q. Zu et al. (Eds.): HCC 2020, LNCS 12634, pp. 452–458, 2021.
https://doi.org/10.1007/978-3-030-70626-5_47

This paper proposes an algorithm that supports filtering the useless frequent sequences that does not have causality relationship mined by association rule mining algorithm so that effectively compress massive alarm data further. In order to filter the rules, we combine all the alarm information in each alarm association rule in pairs, and count the occurrence number that one of the alarm information is the cause of the other alarm information and the result of the other alarm information in the actual alarm data. When the ratio of these two counts is within a certain range, it can be considered that although these alarms have a sequential relationship in time, there is no cause-and-effect relationship, so these purely statistical association rules will be filtered out on this basis.

The rest of this article is laid out as follows. The second part introduces the current association rule mining algorithms used in the processing of telecom network alarm data. The third part introduces in detail that the association rule mining algorithm that supports filtering frequent sequences which have no causal relationship in network alarm data. The fourth part verifies the feasibility of the proposed method through experiments. The fifth part summarizes the whole thesis.

2 Related Work

In order to solve the problem of mining the association rules in frequent sequences, researchers have proposed many related methods.

In [6] Jia Wang et al. present a weighted fuzzy association rules mining approach to discover correlated alarm sequences. This algorithm combines fuzzy sets, Apriori and alarm time series analysis to find out root causes of consequential alarms, it is potentially effective to remove redundant alarm sequences.

In [7] John K. Tarus et al. propose a hybrid recommendation approach combining context awareness, sequential pattern mining (SPM) and CF algorithms for recommending learning resources to learners. In this recommendation approach, the users' access content and sequence will be mined to obtain their access habits and the association between the recent accessed resources so as to get accurate recommendations.

In [8] Jiaqi G et al. suggest an efficient algorithm to mine sequential patterns from data with temporal uncertainty. By integrating it into the pattern-growth sequential pattern mining algorithm, probabilistic frequent sequential patterns will be discovered.

In [9] Zhai Liang et al. suggest a new algorithm "T-Apriori" based on time constraint on the basis of analyzing the related definitions and general steps of temporal association rule mining. The concepts of ecological event and sequence of ecological events are proposed in this algorithm. It can effectively extract temporal association rules based on the concept and sequence of events.

Through the above researches, this paper will firstly mine the correlation relationship contained in alarm data based on the association rule mining algorithm that can handle frequent sequence sets, and then verify the causal relationship between successive alarm events in mining results. Otherwise, this associated alarm pair will be eliminated. The effectiveness and reliability of association rule mining algorithms in processing time series data can therefore be improved.

3 Our Approach

Since PrefixSpan, the association rule mining algorithm which can scan the original sequence database only twice in the entire mining process and need not generate candidate sequences, is suitable for processing frequent sequences that have a causal relationship between various data. PrefixSpan can be used to efficiently mine the association rules of the alarms in network. Then, the mining results are verified and filtered by using causality filtering algorithm to fully consider whether there is a cause-and-effect relationship between the successive alarm events, so as to improve the usability of PrefixSpan for mining association rules in this scenario.

3.1 Algorithm Overview

PrefixSpan is a pattern mining algorithm using prefix projections. Since it considers the prefix of current subsequence and constructs the corresponding projection in this algorithm, it is suitable for processing time series data sets. PrefixSpan uses frequent subsequences in database as a prefix to partition the sequence database, and generates a projection database prefixed by these frequent subsequences.

Definition 3.1 Prefix and Postfix: Assuming that all items in the element are listed in a certain order. Given a sequence $\alpha = <\alpha_1, \alpha_2, \alpha_3, \ldots, \alpha_m>$, a sequence $\beta = <\beta_1, \beta_2, \beta_3, \ldots, \beta_n> (1 \leq m \leq n)$. α is called a prefix of β if and only if: (1) $\alpha_k = \beta_k (1 \leq k \leq m - 1)$; (2) $\alpha_m \subseteq \beta_m$; (3) all the items in subsequence $\beta'_m = \beta_m - \alpha_m$ are alphabetically after those in α_m.

Sequence $\beta' = <\beta'_m, \beta_{(m+1)}, \ldots, \beta_n>$ is called the postfix of β with respect to its prefix α, denoted as $\beta = \alpha \cdot \beta'$, where $\beta'_m = \beta_m - \alpha_m$.

For example, sequence $\alpha = <a(bc)>$ is a prefix of sequence $\beta = <a(bcd)e>$. Sequence $\beta' = <(\_d)e)>$ is the postfix of β with respect to its prefix α. The '$\_$' in sequence β' indicates that the item in the last item set of α and the item in the first item set of β' are in the same sequence element.

Definition 3.2 Projection Sequence: For the sequences α and β, let α be the prefix of β and γ be the subsequence of β. If γ is the postfix of β with respect to its sequence prefix α, it is said that γ is the projection sequence of β with respect to its sequence prefix α. If $\beta = \alpha \cdot \gamma$, and there is no super sequence γ' exceeding γ that satisfies $\beta = \alpha \cdot \gamma'$, then γ is the maximum projection sequence of β with respect to its sequence prefix α [10].

The algorithm firstly uses the frequent sequence pattern α obtained by scanning the sequence database as a prefix, then divides the sequence database by projection, scans the corresponding projection database $\mathcal{D}|_\alpha$, and obtains all its length-1 sequential patterns FL. Each length-1 sequential pattern γ is appended to the prefix α to form a new sequence pattern α'. Further, each newly obtained sequence pattern α' is used as a prefix to partition $\mathcal{D}|_\alpha$ to generate a new projection database $\mathcal{D}|_{\alpha'}$. Repeat the above operations until all frequent subsequence patterns in the sequence database are mined, then the algorithm ends.

3.2 Verification and Filtering of Causality Between Alarms

After obtaining the association rules in alarm data, it is necessary to verify and filter these statistical association rules. Aiming at the common problem that although alarm events in telecom network often appear together, there is no actual causal relationship between them, this paper proposes a causality filtering algorithm.

The main idea of the algorithm is that assuming that there are two events A and B in the alarm data set. If it is known that the occurrence of event B is caused by the occurrence of event A, then events A and B will satisfy the following conditions:

1) When performing the time slicing operation on massive alarm data, event A occurs before event B on most cases.

2) In all time slices, the probability of event A occurring before event B must be much greater than the probability of event B occurring before event A in the same time period.

Assuming that there are two statistically related events C and D in the association rules mined by PrefixSpan, and a filter factor $k (k \geq 1)$. If the count that event C occurs earlier than event D is recorded as m, and the count that event D occurs earlier than event C is recorded as n in a long time range, the above parameters have the following relationships:

1) If $1/k < n/m < k$, it can be determined that there is no causal relationship between event C and event D.

2) If $n/m \geq k$, there is an association relationship between these two associated events that the occurrence of event D will lead to the occurrence of event C.

3) If $n/m \leq 1/k$, which can also be denoted by $m/n \geq k$, there is an association relationship that the occurrence of event C will lead to the occurrence of event D.

Algorithm 1 Causality filtering algorithm:
Input: frequent pattern, and a filter factor k
Output: Frequent sequence patterns with real causal relationships among items

1.	**Causality filtering algorithm(frequent pattern, k);**
2.	for each $\alpha, \beta \in$ frequent_pattern:
3.	// α and β are two different events in this association rule, and α always appears
4.	// before β
5.	positive = count $(\alpha \rightarrow \beta)$;
6.	reverse = count $(\beta \rightarrow \alpha)$;
7.	if $1/k \leq$ positive/reverse $\leq k$:
8.	delete$(\beta \rightarrow \alpha)$;
9.	delete$(\alpha \rightarrow \beta)$;
10.	// At this time, α and β do not satisfy the condition of a causal relationship,
11.	// so the association rules between α and β will be determined as invalid rules
12.	// and removed
13.	**else if positive/reverse $> k$:**
14.	**delete$(\beta \rightarrow \alpha)$;**
15.	**else:**
16.	**delete$(\alpha \rightarrow \beta)$;**
17.	**End if**
18.	**End for**

Fig. 1. Causality filtering algorithm.

Therefore, through the above rules, the alarm pairs that do not satisfy the causal association relationship the purely statistical alarm information association rule can be eliminated so as to obtain the actual effective association rules (Fig. 1).

4 Performance Evaluation

In this experiment, 220,874 alarm data in a total of 5 regions within 48 h were used for verification. Firstly, the alarm data in different regions are processed by time slicing with the window size of 6 s and the step size of 3 s to obtain the sequence database required by the association rule mining algorithm. Then use PrefixSpan to analyze the statistical relevance of the sequence and get the unfiltered pure statistical frequent item sets. After that, verify the causal relationship by using causality filtering algorithm, eliminate the useless association rules, then obtain the final correlation analysis result of the alarm data. In this experiment, the value of the filter factor k is set to 2 and the minimum support threshold min_sup is set to 70.

Figure 2 shows the comparison of the number of alarm pairs in the association rule before and after using the causality filtering algorithm proposed in this paper. It intuitively shows that the causality filtering algorithm can effectively reduce the number of useless correlation pairs, so it completes the filtering of invalid rules and improves the validity and reliability of the association rules mining algorithm.

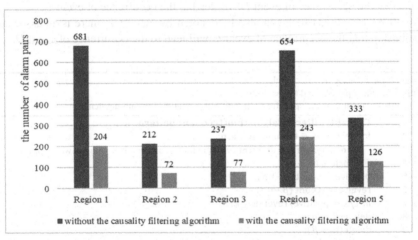

Fig. 2. Comparison of the number of alarm correlation pairs without and with the causality verification algorithm.

The speed of calculation of the causality filtering algorithm proposed in this paper is shown in the following Table 1. It can be seen intuitively that this algorithm can verify and filter the frequent sequence patterns in a relatively short time without affecting the efficiency of the overall association rule mining algorithm, so it can be applied to deal with the real telecom network alarm.

Table 1. Comparison of the execution time without and with adopting causality verification algorithm (second).

	Region 1	Region 2	Region 3	Region 4	Region 5
Execution time without causality verification	4.71	49.28	0.49	7.31	3.70
Execution time with causality verification	4.78	51.26	0.52	7.44	3.74
Percentage of time of causality verification	1.46%	3.86%	5.76%	1.75%	1.07%

The experimental verification above shows that in the application scenario of telecom network alarms analysis and detection, the use of association rule mining algorithms that support causality verification can effectively reduce invalid causality rules of alarm data to provide operations with more accurate alarm correlation information, so as to locate and trace the root fault rapidly. In addition, the algorithm proposed in this paper has high robustness and execution efficiency.

5   Conclusion

In this paper, we propose a method for mining the causal association in network alarm data based on PrefixSpan. By using PrefixSpan, alarm association rules that take the sequence order into account are obtained firstly, then all the rules will be processed by the causality filtering algorithm so that only if the association rule actually has the causality relationship can it be retained. Experimental results and analysis show that the method proposed in this paper is effective.

In the future, several directions are expected to explore, such as the optimal value of the filter factor k that maximizes the accuracy of filtering algorithm in different scenarios, how to further improve the filtering speed when dealing with larger amounts of alarms.

Acknowledgement. This work is supported by the National Natural Science Foundation of China under Grant No. 61771072 and the Industrial technology basic public service platform with project No. 2019–00899-3–1.

References

1. Zhang, G., Liu, C., Men, T.: Research on data mining technology based on association rules algorithm. In: IEEE 8th Joint International Information Technology and Artificial Intelligence Conference (ITAIC), pp. 526–530. IEEE (2019)
2. Du, J., Luo, H.: Network security situation analysis of weighted neural network with association rules mining. In: Proceedings of the 2016 5th International Conference on Advanced Materials and Computer Science (2016)
3. Yuan, X.: An improved Apriori algorithm for mining association rules. In: AIP Conference Proceedings. Vol. 1820. No. 1. AIP Publishing LLC (2017)

4. Chang, H-Y., et al.: A novel incremental data mining algorithm based on fp-growth for big data. In: International Conference on Networking and Network Applications (NaNA), pp. 375–378. IEEE (2016)

5. Chen, C., Wang, D.: Research on association rules mining base on positive and negative items of FP-tree. In: 2016 6th International Conference on Machinery, Materials, Environment, Biotechnology and Computer. Atlantis Press, pp. 1395–1399 (2016)

6. Wang, J., et al.: Association rules mining based analysis of consequential alarm sequences in chemical processes. J. Loss Prev. Process Ind. **41**, 178–185 (2016)

7. Tarus, J.K., Niu, Z., Kalui, D.: A hybrid recommender system for e-learning based on context awareness and sequential pattern mining. Soft. Comput. **22**(8), 2449–2461 (2017). https://doi.org/10.1007/s00500-017-2720-6

8. Ge, J., Xia, Y., Wang, J., Nadungodage, C.H., Prabhakar, S.: Sequential pattern mining in databases with temporal uncertainty. Knowl. Inf. Syst. **51**(3), 821–850 (2016). https://doi.org/10.1007/s10115-016-0977-1

9. Zhai, L., et al.: Temporal association rule mining based on T-Apriori algorithm and its typical application. In: Proceedings of International Symposium on Spatio-Temporal Modeling, Spatial Reasoning, Analysis, Data Mining and Data Fusion (2005)

10. Han, J., et al.: Prefixspan: mining sequential patterns efficiently by prefix-projected pattern growth. In: Proceedings of the 17th International Conference on Data Engineering. IEEE Washington, DC, USA (2001)

Exploring Psycholinguistic Differences Between Song and Ming Emperors Bases on Literary Edicts

Shuangyu Liu[1,2] and Tingshao Zhu[1,2(✉)]

[1] Institute of Psychology, Chinese Academy of Sciences, Beijing 100101, China
tszhu@psych.ac.cn
[2] Department of Psychology, University of Chinese Academy of Sciences,
Beijing 100049, China

Abstract. Imperial edicts in ancient China were with strong political preference and practicality. It is very meaningful to interpret the psychological meaning of the emperors and necessary to analyze the sociality and individuality of emperors based on their imperial edicts. This paper analyzed the differences of psycholinguistic features in imperial edicts of Song and Ming dynasties via Classic Chinese LIWC (CC-LIWC) to obtain word frequency results for each word category in the edicts and used statistical tests to compare the differences of functional words-personal pronouns, emotional process words, cognitive insight words, physiological process words and motivational level words - in the edicts of Song and Ming dynasties. The result indicates the differences in CC-LIWC lexical frequencies between Ming and Song dynasties had an impact on the intensity of the emperors' work and their mentality in dealing with state issues.

Keywords: Psychological semantics · CC-LIWC · Psychology · Emotional motivation theory · Language comprehension · Textual semantic analysis

1 Introduction

The key to human behavior and decision making is determined by their psychological characteristics and personal ethics, so the study of the behavior of historical figures from the psychological process and personality psychology has been valued by many researchers. An emperor in feudal period of ancient China played the role of a central decision maker, his emotional process and personality tendencies were no longer an independent being but a comprehensive one of the country's even a dynasty 's fate. Therefore, the psychological analysis of ancient emperors is of great value, whether it is a study of political, economic or cultural aspects [1].

Song and Ming Dynasties were both feudal dynasties established by the Han Chinese, both were subjected to long periods of invasion by ethnic minorities and dynasties dealt with foreign barbarians by economic means through the abolition of the "peace and affection" policy. These two dynasties featured the late feudal period of ancient China and had certain similarities and inheritances in some state aspects [2].

© Springer Nature Switzerland AG 2021
Q. Zu et al. (Eds.): HCC 2020, LNCS 12634, pp. 459–468, 2021.
https://doi.org/10.1007/978-3-030-70626-5_48

Experts and scholars have studied the emperors of Song and Ming dynasties, such as Mao Zedong studied the three emperors of Song dynasty and made convincing comments on their character, psychology, policy gains and losses, especially on their political short-comings and deficiencies [3]. Mao reviewed Zhu Yuanzhang's atrocities and inferiority complex by studying on historical articles which describes the political performance in Zhu Yuangzhang's late governing, hence analyzed Zhu's character traits. As for the Mastery of Zhu Di of Ming Chengzu, Mao emphasized that during the reign duration, Zhu Di aroused all efforts to make the country prosperous and national unity by argued his views through historical events, etc. [4]. As can be seen, most of these studies focused on individual emperor of Ming or Song dynasty, rarely comparing the two dynasties. The research is biased towards the description of specific events from historical sources and there is no empirical method of data and statistics.

Text analysis is a method for quantifying language use. The LIWC method used to be the most widely recognized and used method in the field of psychological research, and many meaningful research results have been achieved: for example, the relationship between word use and mental health; the relationship between word use and personality traits; and cognitive processing. Natural language processing techniques were used to extract words from the study text, and psychological semantic analysis was performed using dictionaries based on word frequency to enable comparison of differences in psychological characteristics and to analyze the language of psychology [5].

This paper combines natural language processing techniques, psychology and history to study the psychosemantic changes of ancient emperors through the analysis of literary edicts and to establish a study of the personality of ancient emperors at the level of scientific psychology.

2 Research Methods

According to historical records, an imperial edict was a document issued by the emperor to inform the whole country. In the Zhou Dynasty of ancient China, edicts could be used by all rulers and officials. After the unification of the Six Kingdoms and the establishment of Qin monarchy, Yingzheng the First Emperor, thought he was "the head of three previous emperors and the five kings", named himself *Emperor the First* and called himself "Zhen". Han inherited the Qin system, Tang and Song abolished without using it while the imperial edict was used in Song Dynasty; Yuan Dynasty reverted the edicts into use; Ming used edicts to announce major political decrees or to admonish ministry officials.

In order to analyze the texts of the edicts, data on the edicts of the emperors of Ming and Song dynasties were first collected. The Song dynasty edicts were extracted from the *Collection of Song Great Edicts*, a book contains comprehensive records of Song edicts covering major historical events of the period. The edicts of Ming dynasty are extracted from *Ming Record*, a chronological history of Ming dynasty, recording a large amount of data about 250 years from the first emperor of Ming dynasty, Zhu Yuanzhang, to Zhu Youjiao, with important historical value and is one of the basic historical books for studying the history of Ming.

All edicts were manually screened to ensure their complete form and accuracy. The edicts of eight emperors from Taizu to Emperor Huizong of Song as contained in the collection of Song Dajie Decrees and the edicts of thirteen Emperors. The edicts of Ming spans from Emperor Taizu to Taizong in the Ming Shishu Records. There were some textual errors in the texts that it required further verified the text of the edicts by manual verification and annotated the author, year and dynasty for the edicts, finally obtained the text of the edicts in Chinese

The selected edicts of each emperor of Ming and Song dynasties were first used as a text, and the LIWCs in these texts were extracted by natural language processing techniques (CCLIWC was used in this paper). The dictionary, which is provided for psychological research for the analysis of ancient Chinese texts for psychological dictionaries, contains psychologically related word classes) of the word classes [6], compared each word classification to see if there was any differences between the two groups of words, and to perform a specific analysis on the lexical categories which were different [7].

(a) The CC-LIWC archaic dictionary performs data analysis and processing as follows (Table 1).

Table 1. Summary of edicts

Dynasty		Total number of emperors	Number of edicts
Song	Northern song	9	57
	Southern song	2	3
Ming		16	101

Classical dictionary word division: loading CC-LIWC classical word division dictionary, removing the punctuation space in the TXT text, classical word division; getting the word division results will be all the words for LIWC classification.

(b) Results of data analysis: Basic statistical indicators: TCC (total number of words in the articles), TWC (total number of words in the articles), LWC (total number of keywords involving LWC), LCR coverage (see Eq. 1).

$$LCR = \frac{LWC}{TWC} \tag{1}$$

Word classification and word frequency statistics: word frequency of LWC keywords of various categories and dimensions and their percentages.

(c) Test for differences: the dictionary uses natural language processing program comes with scipy.stats.kstest module, when the two sets of data were normal distribution, Student'sttest was used to test the difference; when the two sets of data are non-normal distribution, wilcoxtest and Mann Whitney U test were used respectively; when the sample size is less than 20, wilcoxtest test is selected; when the sample size is more than 20 or in other situations, Mann WhitneyUtest test is selected [8, 10, 12].

3 Results

We analyzed the textual data of Ming and Song dynasties' edicts through a natural language processing program, based on the ancient LIWC developed by SCLIWC2015, and saw that the lexical differences were driven by linguistic processes words (linguistic processes), other grammatical words, and the social processes, personal concern, and informal language, of which there are 19 psychologically meaningful words.

3.1 Description of Variables

See Table 2.

Table 2. Description of variables

Classification	Variable	Descriptor
Linguistic processes	I	First person singular
	We	First person singular
Social processes	Family	Social process words
	Friends	Social process words
	Posemo	Emotional process words
	Negemo	Emotional process words
	Anx	Emotional process words
	Anger	Emotional process words
	Sad	Emotional process words
	Causal	Cognitive process words
	Discrepancies	Cognitive process words
	Tentative	Cognitive process words
	Bio	Physiological process words
	Achieve	Motivation words
Personal concern	Work	Working words
	Leisure	Entertainment words
	Money	Money words
	Death	Death words

The word frequencies of CCLIWC for each edict were calculated, words with psychological significance were selected, and Song and Ming dynasties were chosen as experimental controls to compare the differences in word frequencies between the two dynasties. The word frequency of CCLIWC word classifications is different between the two dynasties, which further illustrates the differences in linguistic expressions and psychological changes of the emperors' edicts in different periods.

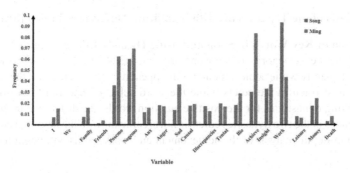

Fig. 1. Summary graph of average word frequency for words with psychological meaning

From Fig. 1, it can be found that the four most frequently used word categories in Song and Ming edicts are: positive emotion word Posemo and negative emotion word negemo; cognitive process word insight; personal concern word (Personalconcern) in the work word, the motivation word achievement word. More than half of the above parts of speech are mental process words. Then the author mainly analyzed the differences of the speech and explored the inner psychological process of the emperors in the two dynasties.

First, it used the U-test to test whether there is a difference in the frequency of these types of words in Ming and Song edicts, and Table 3 shows the p-values of the significance test from small to large ranking results. The difference is obvious as follows.

Why was there such a big difference in the frequency of the positive emotion word Posemo, the Achievement word Achieve and the Work word work used by the two emperors? Below we will go into the analysis the use of the frequency of each keyword by the emperors, from the time and space dimension to analyze whether the word frequency of these keywords are different, and analyze the reasons behind the differences in word frequency of keywords.

Table 3. Ascending results of the significance test for word frequency

Lexical category	t value	p <= 0.05
Posemo	1544	0.001**
Achieve	2023.5	0.000217
Work	2495	0.030829
Negemo	2556.5	0.04909

Note: 0.001** means much less than 0.001.

3.2   Analysis of the Top 3 Terms with Significant Differences in Significance

Comparison of Key Words in Song and Ming Dynasty Edicts. Figure 2 shows that by comparing the 3 categories of keywords in Song and Ming edicts, work words appear the most frequently while achievement words appear the least frequently; the word work appeared most often in the edicts while the word achieve appeared least often in the edicts of Ming dynasty; as for the positive emotion words, the difference in the number of occurrences in the two dynasties edicts is not significant with the keyword comparison results, an in-depth analysis of the achievement and work words will begin next.

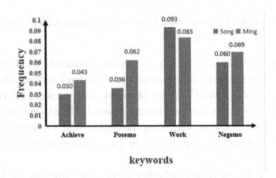

Fig. 2. Frequency statistics and ranking of distinctly different words in Song and Ming

Comparing Trends in the Frequency of Keyword Classifications in the Two Dynasties. Through the observation on Fig. 3, it can be seen that the work word, achievement word, positive emotion words and negative emotion words have different fluctuation trends, in the early Song and early Ming dynasties, these types of words change into a higher degree; in the middle period, the fluctuation is rather sharp; in the late of the two dynasties the change is not significant.

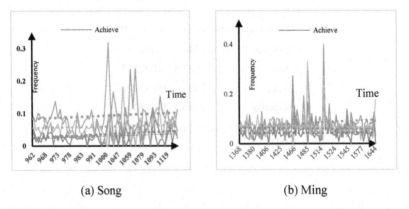

(a) Song   (b) Ming

Fig. 3. Comparison of keyword frequency trends between Song and Ming dynasties

Through Fig. 3, it can be observed that the trends of work words, achievement words, positive emotion words and negative emotion words are different in Song and Ming edicts. In Song edicts, the trend of change of these types of words is in an overall decline; work words decline significantly; negative emotion words with an upward trend; positive emotion words and achievement words both decline sharply; in Ming edicts, the overall trend of these types of words rises, the rising trend is slow.

4 Discussions

This paper uses CCLIWC word frequency features to analyze the word frequency changes of imperial edicts of the Song and Ming dynasties through psycho-semantic analysis, and discusses them from the perspectives of diligent government and ethnic policies. This paper has some shortcomings, such as this paper only selects individual lexical categories for analysis and does not consider the influence of other lexical categories on the rise and fall of dynasties, in future research, we will consider multiple lexical categories comprehensively, label the relevant indexes of social characteristics, and build a machine learning model to achieve more accurate prediction of social changes in different periods. The psycho-semantic approach proposed in this paper provides a new perspective for the study of ancient history and culture, especially for the analysis of social changes over a large time span.

History records that in the Ming Dynasty, Zhu Yuanzhang (1328–1398) and the late Ming Chongzhen Emperor (1611–1644). Zhu Yuyan (1328–1398) have been recognized as hard-working and dedicated monarchs; the other Ming emperors were not as dedicated as the first two in terms of work; in addition to the three dedicated monarchs, the other emperors were Ming Xuanzong (1399–1435) who focused on fighting crickets, Ming Wuzong (1491–1521) who liked to play, Ming Sizong (1611–1644) was addicted to alchemy to become immortal, Ming Shenzong who did not go to court for more than 30 years, and Ming Xi Zong who was also dive into carpentry. This can also be seen in Table 5(b), the word frequency of work words throughout the Ming dynasty is very low, but by Ming Si Zong there is a rapid rise, which is also consistent with history, Chongzhen emperor during his reign, very diligent, on the work is conscientious [5]. It can also be seen from Fig. 3 (b) that the frequency of achievement words is very low and the trend of change is very stable throughout the Ming dynasty. This can be analyzed from the establishment of Zhu Yuanzhang's (1328–1398) system at the beginning of the Ming Dynasty, which established the cabinet system, the East Chamber secret service organs and the system of officials. It broke the system of the previous emperor handling everything by himself. This gave the emperor more time to do other things and did not interfere with his work because he did not deal with government affairs. This caused the emperor to be less involved in his work, so the sense of accomplishment slowly diminished as he handled his work. This is also the reason for the low frequency of achievement type words in the edicts [6].

Historically, the Northern Song emperors had reached a certain height in terms of material and spiritual civilization with their artistic and literary achievements, were at the highest stage of the development of feudal society, During the Song Dynasty, the four major inventions were also the most fully utilized stage. From the Northern Song Emperor Song Taizu (927–976) to the Song Yingzong (1032–1067) period, the state was more prosperous and the emperor was more diligent and wise, to the Song Shenzong (968–1022)period the Northern Song gradually weakened, and finally due to poverty and decline in power, the Northern Song collapsed during the period of Emperor Hui Zong (1100). In the Southern Song Dynasty, the emperors were largely inactive, mostly ineffective and scraping by, but Song Xiaozong (1127–1194) did something. Later Song Li Zong (1205–1264) and Song Du Zong (1240–1274) favored sycophants, greed for beauty and pleasure, and did not set their hearts on governing the country, which exacerbated the fall of the Southern Song.

Figure 3(a) shows the overall trend of work words in edicts changed significantly during Song compared to Ming, showing that most of Song emperors took their work seriously and their outstanding contribution to cultural and educational prosperity of Northern Song period. The downward trend of achievement words in the edicts is evident. It is speculated from the following three aspects: 1. During the Five Dynasties, Zhao Kuangyin's cultural cultivation was relatively prominent. In Northern Song period, Taizu's cultural cultivation was inferior to other emperors in the middle and late emperors' effectiveness in ruling the country was inferior to that of Song Taizu. 2. Northern Song emperors' own good cultural cultivation did not contribute to effective solution to the thorny political and military problems, the excessive pursuit of culture became an important factor that prevent Northern Song emperors from doing their best to solve political and military problems. The representative of Northern Song period was Song Huizong, in sharp contrast to Song Taizu. Some emperors of the Northern Song Dynasty were deviated from the main duties of the emperors when they implemented some of their ideas into the governance of the country, leading to the emergence of the phenomenon of cultural quality deviating from the effectiveness of governance. As a result of cognitive bias, which is their failure to objectively and dialectically analyze the emergence of major problems, the subsequent decisions made there are deviations, the result will be poor, can lead to a lack of confidence and weaken the sense of achievement [8].

The positive emotion words in Fig. 3(a)(b) have flat fluctuations in the frequency of positive emotion words, while the negative emotion words have a slowly increasing trend of change. From the point of view of ethnic policy, both the Ming and Song dynasties had a long-standing concern about the external problems of ethnic minorities, and the most serious among the successive Chinese dynasties. The Song dynasty was a dynasty that favored literature over military force, with a feeble military force that made it difficult to implement a policy of peace and affection backed by force. However, the Song Dynasty had a highly prosperous economy and a rapidly developing monetary economy, which also provided a material guarantee for dealing with foreign troubles through economic means as a basis. At that time, the dependence of ethnic minorities on Chinese trade was also further strengthened. Taking such an approach had a positive effect in maintaining stability, so overall it was more optimistic. The Ming dynasty, which was next to the Song dynasty, was less effective in dealing with external barbarians in terms of Mongol participation in the force, and in solving problems by force. It also used good diplomacy with the foreign barbarians through tea and horse trade, and the Mongolian problem was quelled. Both dynasties were optimistic about solving the problem by strengthening economic and trade ties.

The last of both dynasties went from prosperity to decline, the later the state strength and military strength and external harassment, the frequency of negative words in the edicts gradually showed an upward trend. At this time, the emperor's various pressures became more and more, and there was no hope, and the mood was more negative.

5   Conclusion

In this essay, we extracted the emperor's imperial edicts from the collection of Song Dazhao Decrees and Ming Shiluan and analyzed the differences in the language expressions of the two dynasties via the psychological meaning of the vocabulary in the edicts. The number of emperors in both dynasties was less than 20, the rank-sum test was used. After comparing the frequency of work classification words, the difference in the frequency of such words in the edicts of the two dynasties' was obvious; through analysis and historical records, the difference in the duties and intensity of work of the two dynasties was the main reason for the difference. In terms of foreign policy, the two dynasties adopted the decision to achieve mutual benefit through economic and trade exchanges. Attitudes were still very positive to face problems and make positive solutions; so the frequency of positive sentiment words has been at a stable level. When the two dynasties finally fell apart, the emperor of the time would recall the prosperous scenes of the former periods, which formed a psychological gap, and some negative emotion words gradually increased.

References

1. Qi, L.: The study of feudal social history must pay attention to the analysis of the emperor's personality and psychology. Soc. Sci. **05**, 61–62 (1988)
2. Longbo, D., Peng, X.: A brief discussion of the relationship between the development of monetary economy and ethnic policies in the Song and Ming dynasties) BeiFang Literature, **2**, 86–87 (2013)
3. Xuexin, C.: Zhu Di's mastery of people in Ming Chengzu. Leader. Sci. **000**(005), 46–48 (2019)
4. Qiuqiang, Z.: A review of psychological approaches to language use analysis. Mod. SOE Res. **6**, 154–156 (2017)
5. Ding, F., Puguang, Z.: The trajectory of history: an empirical analysis of seventy years of modern and contemporary literary studies in China–centered on the statistics of the frequency of inscriptions. Stud. Lit. Art, (9) (2019)
6. Anonymous. Why the Ming Dynasty lasted 300 years with many faint rulers. Life and companionship (second half of monthly edition), **000**(010), 72–72 (2017)
7. Wu, J.: Re-understanding the history of the mid-Ming dynasty from the perspective of social changes. Ming Dynasty China from the perspective of world changes–International Symposium. 0
8. Xing, F., Zhu, T.: Large-scale Online Corpus based Classical Chinese Integrated Dictionary (2020)
9. Li, Y.: Research on the cultural quality and effectiveness of the Northern Song emperors. Hebei University (2012)

10. Zhiyuan, L., Liu Yang, Y., Cunchao, T., et al.: Quantitative observation and analysis of vocabulary semantic change and social change. Lang. Strat. Res. **000**(006), 47–54 (2016)
11. Longbo, D., Peng, X.: A brief discussion of the relationship between the development of monetary economy and ethnic policies in the song and ming dynasties. Northern Lit. Next **4**, 86–87 (2013)
12. Miaorong, F.: Critical Technology in Ancient Chinese Psychological Semantic Analysis Based on LIWC. Master Dissertation. University of Chinese Academy of Sciences, Beijing (2020)
13. Rude, S., Gortner, E.M., Pennebaker, J.: Language use of depressed and depression-vulnerable college students. Cogn. Emot. **18**(8), 1121–1133 (2014)
14. Golbeck, J., Robles, C., Edmondson, M., et al.: Predicting Personality from Twitter. In: IEEE Third International Conference on Privacy, Security, Risk and Trust. IEEE, pp. 149–156 (2011)
15. Graesser, A.C., Lu, S., Jackson, G.T., et al.: AutoTutor: a tutor with dialogue in natural language. Behav. Res. Meth. **36**(2), 180–192 (2004)
16. Xiangxue, L.: On the general policy of nationalities in the ming dynasty and its impact on border defense. J. Hubei Coll. Nationalities Philos. Soc. Sci. **022**(002), 53–57 (2004)
17. Kaoquan, H.: The distribution of power in the Ming dynasty. Chinese Out-of-School Education (Basic Education Edition), (2010)

The Influence of the Reform of Entrance Examination on University Development

Dongfang Wan[✉], Chenglin Zheng[✉], Yue Wang[✉], Wenjia Hu[✉], Xinyi Ma[✉], and Jinjiao Lin[✉]

School of Management Science and Engineering, Shandong University
of Finance and Economics, Jinan, Shandong, China
1763721028@qq.com, 2953914289@qq.com, 2088009262@qq.com,
1351621980@qq.com, 2655669128@qq.com, ljj@sdufe.edu.cn

Abstract. The university entrance examination application is a very important last step for university entrance examination students. It is not only related to the future academic career and career development of candidates, but also affects the future direction of professional construction in universities. Shandong Province has been implementing the "university + major" application filling and filing admission mode, starting from 2020, will be changed to a new mode of "major (category) + university". In order to explore the influence of the reform of university entrance examination on the development of universities, this article starts from the collection of questionnaires and analyzes the data, showing the relevant impact and reasons of the reform of university entrance examination, and puts forward constructive suggestions on the major construction and enrollment publicity of universities.

Keywords: University entrance examination · Application reform · University development

1 Introduction

The university entrance examination has always attracted much attention, and the university entrance examination application application is the last step of university entrance examination students. It is not only related to the future academic career and career development of candidates, but also affects the future direction of professional construction in universities. Therefore, it is a crucial issue for students and university. Therefore, from the reform of university entrance examination, it is a very worthy subject to explore and study to analyze its related influence.

1.1 Research Background

For a long time, Shandong Province has been implementing the "university + major" application filling and filing admission mode. However, from 2020, Shandong summer university entrance examination application filling and filing admission mode will be

© Springer Nature Switzerland AG 2021
Q. Zu et al. (Eds.): HCC 2020, LNCS 12634, pp. 469–475, 2021.
https://doi.org/10.1007/978-3-030-70626-5_49

changed from the current "university + major" to "major (category) + university" new mode.

According to the original "university + major" application filling requirements, candidates should first choose the university to apply for the examination, and then choose the major from the selected universities. In the process of admission, the university chosen by the examinee is taken as the unit of filing. As long as the examinee files are put into a application university, and the candidates fill in and submit to the professional adjustment, they will not be rejected by the university because of the low total score. However, under the new "university + major" mode, the mode of application filling and filing has changed to the major is a priority, from the university filing as a unit to a major (category) plus a university as a unit. Moreover, there will be no need to be reallocated to other major, and the vacancy plan of each major will be completed through the recruitment of applications. Besides, in order to appropriately expand the number of candidates, the new filling method sets the number of professional applications to 96 increasing 24 professional applications [3].

The adoption of the new filling method will enhance the matching degree between the accepted major and interest of candidates, and avoid the situation that a large number of candidates are not in line with their professional wishes because of being admitted by redistributed. But it is also likely for universities to pay too much attention to their minimum admission line. Under the protection of the minimum admission line, no matter how high the level of professional education, the general local universities can not compete with the "famous universities" or "key university" in everyone's mind, and the differences of majors in the university are also covered up. It may not be conducive to the professional construction and development of the University, and is not conducive to the competition of university running level and education quality within and among universities. Through the study of Jinan university entrance examination voluntary reporting, we can find many specific problems, so as to further think and solve.

1.2 Research Plan

Our team took Jinan City as an example, by means of questionnaire survey, this paper investigates the students' understanding and planning after the reform of voluntary reporting, and understands the advantages and disadvantages of the reform of application reporting in university entrance examination. Furthermore, the questionnaire data obtained are deeply interpreted and the reasons are analyzed, so as to find out the impact of college entrance examination voluntary filling reform on colleges and universities. Finally, a summary is made, linking the application reform of university entrance examination with the professional construction of universities, and combining with theoretical knowledge, providing suggestions for the construction of universities. The questionnaire used in this survey was randomly distributed, and finally 205 questionnaires were collected.

2   The Basic Situation of Research

2.1   Give Priority to the Base Case of Voluntary Reporting Before the Reform

We can see from the Fig. 1: before the reform of university entrance examination, the number of students who gave priority to university and major was about the same. Only a small number of candidates looked at the impact of rankings and geography, it can be proved that the priority of university and major is the main consideration direction of application, which is also the main object of our research.

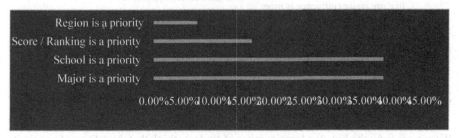

Fig. 1. The priority of voluntary reporting before the reform

2.2   Give Priority to the Base Case of Voluntary Reporting After the Reform

We can see from the Fig. 2: according to the results of the questionnaire survey, after the reform of college entrance examination voluntary reporting, the number of students who give priority to major did not increase as expected, but showed a downward trend.

Fig. 2. The priority of voluntary reporting after the reform

2.3   The Reasons for Choosing the Above Filling Form as the Preferred Method

We can see from the Fig. 3: the proportion of respondents who chose university as the priority did not change, but some respondents who chose region and other is a priority changed their choice to interest is a priority.

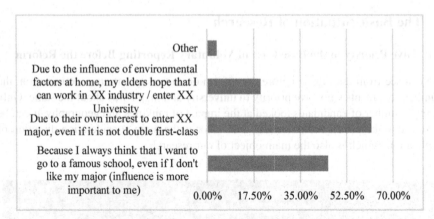

Fig. 3. The priority of voluntary reporting before the reform

2.4 The Basic Information of Whether the Reform of University Entrance Examination is Helpful or not

We can see from the Fig. 4: a large number of students think that the reform of university entrance examination is more or less helpful to their choice, while a small percentage of students think that the university entrance examination reform doesn't have an impact on them.

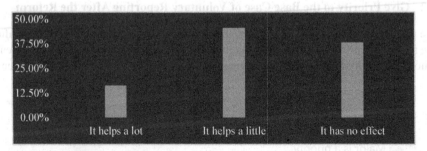

Fig. 4. Survey on whether the university entrance Examination application reporting reform has an impact on the application selection

2.5 Summary of the Questionnaire Survey

Students generally believe that the reform is helpful to further their study and increases the diversity of their choices. But there are also a few who think that there is a deficiency. Some people think that the new pattern of reporting is too complicated. It is difficult to fill in accurately without previous data as a reference.

To sum up the above arguments, no matter before or after the reform, the opinions of students' application reporting are dominated by "famous university first". It can be seen that the reform of university entrance examination application reporting has little influence on the candidates' filling concept. In addition, with less reference data and more variable conditions, it is difficult for them to fill out.

3 Analysis on the Influence of Questionnaire Survey Results

3.1 The Speed of Elimination of Inferior Majors in Universities is Accelerated

The reform in Shandong province adopts the form of "major + university" to improve the matching degree between students' majors and their interests. Therefore, the number of students who choose the inferior major in many universities will be less and less, which will lead to the cancellation of such majors. Specific to each big professional, there are 17 major majors in clothing and fashion design, 13 in educational technology, 12 in information and computing science, and 11 in product design and information management and information system [1]. There are also many cancellations in editing and publishing, fashion design and engineering, industrial design, mathematics and applied mathematics [4]. As to a few majors that are not attractive enough for students to apply will face eliminate.

The standard of the reform and adjustment is as "two bases and one reference". Universities should increase the correlation between university entrance examination and high school learning which provides more space for universities to scientifically select talents as well.

3.2 It Has Great Influence on Normal Universities

In addition, the reform has a great impact on normal universities: the teacher qualification certificate is separated from the teacher training major. In most recruitment, the teacher training major is not required, only the teacher qualification certificate is required, and the teacher qualification certificate can be registered independently. Therefore, the disadvantages of the normal major are enlarged. However, in people's mind, some normal university still account for a large proportion, because the university is of high quality. Nowadays, many people want to increase their value by taking the postgraduate entrance exam, but the postgraduate entrance exam for teachers' major is very special. For example, for teachers' major in physics, the postgraduate entrance exam is for master's degree in physics education, which is pedagogy, not physics. The teacher education major focuses on education rather than technology and science. If take pedagogy to compare, that certainly is inferior to professional study education; It's no match for a physicist who specializes in physics. Therefore, this major is very special, the number of applicants will be less and less.

3.3 Promote the Merger of Universities into Major Categories of Enrollment

The total quantity of recruit students of a university is an enormous amount, but assignment by discipline, the number of applicants for the speciality is always small. if people refer only to the number of applicants, it well reduce the number of people signing up cause no one dares to sign up for. Then a lot of universities need to be amalgamated by big categories, such a big category of professional enrollment number will be much more, it is easier to attract people to enter oneself for an examination [2]. For example, Shandong University of Finance and Economics consolidates the majors of logistics management, engineering management, e-commerce, information management and information system into management science and engineering. For a long time, we have been advocating wide caliber training in universities, but the higher the major is more and more detailed. The reform of the enrollment system causes that universities have to start from the enrollment is wide caliber, so in the cultivation of wide caliber is relatively smooth, a natural outcome. This is also a positive effect of the new university entrance examination on the running of universities.

4   Summary and Suggestions on the Impact of University Entrance Examination Application Reporting Reform

4.1   A Summary of the Impact of University Entrance Examination Application Form on Students and University

- **A summary of the impact of university Entrance Examination Application form on students.**
  After the reform of university entrance examination, it is more likely that students can choose their favorite major. For students who give priority to university, the way of university entrance examination application will not bring great help to them, but it will not have a bad influence either. Therefore, the reform of university entrance examination will enhance the matching degree between the students' admitted major and their interests and specialties, so as to avoid the situation that a large number of students do not meet their professional wishes due to the adjustment of admission. Moreover, it will be conducive to the employment of most students in the future, reducing the mismatch between work and major, so that students can better apply their professional knowledge in the society, and make more contributions to the society. But at the same time, the reform of university entrance examination may also lead to the difficult phenomena of filling out such as little data reference and great changes in the situation.

- **A summary of the impact of university entrance Examination application form on universities.**

After the reform of university entrance examination, students will pay more attention to the choice of majors, which may lead to the elimination of some inferior majors And the dominant major will be continuously strengthened and become a key to the success of university enrollment. The selection of talents of talents in universities will be more accurate. And universities will have more space to scientifically select talents.

4.2   Suggestions on the Impact of University Entrance Examination Application Reform

This year is the first year of the university entrance examination reform in Shandong Province, and the problems such as the lack of reference data put forward by the students will be gradually solved as the reform goes on. However, the relevant organizations still need to make timely statistics and update the data, so as to provide more accurate services for students and carry out more accurate resource reporting methods. The examinee then needs according to own preference, knows each university specialized construction situation more.

Universities need to pay close attention to the students' application and be ready to adjust their majors constantly. The inferior majors can be considered to eliminate. For the advantageous majors, it is necessary to continue to maintain the advantage, and strive for further development on the basis of the original.

5 Conclusion

The education of the youth is the foundation of the country. In recent years, for the sake of China's education, the Ministry of education has issued a variety of educational reform measures. The reform of university entrance examination is also a key step to help teenagers grow up. We believe that although the university entrance examination application reform will make it more difficult to fill in, it will have a positive impact on the subsequent development of candidates. And it is also conducive to universities to carry out a new adjustment of their majors, so as to achieve a new transformation, so as to promote the national education reform.

Acknowledgment. Youth Innovative on Science and Technology Project of Shandong Province (2019RWF013), Postgraduate Education Reform Research Project of Shandong University of Finance and Economics (SCJY1911), Teaching Reform Research Project of Shandong University of Finance and Economics in 2020 (jy202011), Teaching Reform Research Project of Shandong Province (M2018X169, M2020283).

References

1. Chen, M.: How universities cope with the challenges brought about by the new university entrance examination entrance examination reform. Hunan Educ. versionssa (2019).
2. Chen, Z., Liu, J.: The new university entrance examination will change with university. Guangming Daily, version 06 (2018)
3. Lu, Z., Kao, W.: Notice on printing and issuing the implementation plan of summer examination and admission work for universities in 2020 in Shandong Province, December 16 2019
4. Jiaogaohan: Notice of the Ministry of education on publishing the filing and examination and approval results of undergraduate majors in regular universities 2019, Jiaogaohan, 21 February 2020

Author Index

Printed in the United States
by Baker & Taylor Publisher Services

Printed in the United States
by Baker & Taylor Publisher Services